Springer Finance

The Springer Finance series, launched in 1998, is addressed to students, academic researchers and practitioners working on increasingly technical approaches to the analysis of financial markets. It covers mathematical and computational finance broadly, reaching into foreign exchange, term structure, risk measure and management, portfolio theory, equity derivatives, energy finance and commodities, financial economics.

All titles in this series are peer-reviewed to the usual standards of mathematics and its applications.

Fred Espen Benth • Paul Krühner

Stochastic Models for Prices Dynamics in Energy and Commodity Markets

An Infinite-Dimensional Perspective

Springer

Fred Espen Benth
Department of Mathematics
University of Oslo
Oslo, Norway

Paul Krühner
Institute for Statistics and Mathematics
Vienna University of Economics and Business
Vienna, Austria

ISSN 1616-0533 ISSN 2195-0687 (electronic)
Springer Finance
ISBN 978-3-031-40366-8 ISBN 978-3-031-40367-5 (eBook)
https://doi.org/10.1007/978-3-031-40367-5

Mathematics Subject Classification: 46N30, 60G60, 60G51, 60H15, 60J76, 62R10, 91B74, 91G30

This work was supported by Norges Forskningsråd (Grant 216096, Grant 274410)

This Springer imprint is published by the registered company Springer Nature Switzerland AG
The registered company address is: Gewerbestrasse 11, 6330 Cham, Switzerland

Paper in this product is recyclable.

Preface

The material presented in this book springs out of our research on infinite dimensional modeling of energy markets. The research project *Managing Weather Risk in Energy Markets*, funded by the Norwegian Research Council, was our starting point in 2012 to develop the theory of term-structure models in energy and commodity markets resting on stochastic calculus in Hilbert space. The theory and analysis in a series of papers are collected and further developed in this book. Power markets have been the leading point of motivation for our investigations due to their intriguing properties. But our analysis carries over to commodity markets in general as well.

We have made use of the basic theory of stochastic analysis in Hilbert spaces found in the book by Peszat and Zabczyk [165]. This is combined with some classical functional analysis, in particular semigroup theory along with mild solutions of stochastic partial differential equations in Hilbert space. Admittedly, term structure models in energy and commodity markets are closely related to fixed-income modeling. In our exposition, we have adopted a convenient state space from Filipović [99], which we further develop and analyse for our purposes.

Our aim and intention has been to make the material presented in this book self-contained; however, some pre-requisites are of course needed. The reader should know the basic theory of stochastic calculus for Lévy processes, as well as some insight from elementary functional analysis. Furthermore, we shall be rather brief on the extensive theory around stochastic partial differential equations, but give ample references for deeper insight along the way. The intended audience of this book is PhD students and researchers in mathematical finance with an interest in energy and commodity markets. We sincerely hope that practitioners also shall find interesting material for developing risk models in energy and commodity markets.

The theory presented in this monograph is by far a complete story on how to deal with risk in energy and commodity markets. Indeed, there are challenging questions around estimation and simulation of infinite dimensional stochastic processes, to mention some, that we leave mostly untouched. Furthermore, we do not pay much attention to portfolio management, which is an area with many challenges in the context of infinite dimensional stochastics in energy and commodity markets. Our hope is that the analysis in this monograph will inspire the reader for further research, both theoretically and practically, in this exciting branch of mathematical finance of undoubted societal importance.

We want to express our gratitude to Florentina Paraschiv, Marcus Eriksson at Refinitiv and Jūratė Šaltytė Benth for providing data and assisting us with the empirics presented in this book. The theory on which this book is based has been developed in collaboration with several good colleagues. We are very grateful to all of them (none mentioned, none forgotten!). The opportunity to present parts of the contents of this book at the Winter School on Mathematical Finance in Lunteren 2016 is greatly acknowledged. During the work on this book, we have also enjoyed support from the Center of Advanced Studies' *Stochastics for Environmental and Financial Economics* at the Norwegian Academy of Sciences and Letters during the academic year 2014/15. The Wolfgang Pauli Institute in Vienna and its leader professor Norbert Mauser are also thanked for the opportunity to run several thematic years on mathematical finance and high-dimensional stochastics. Financial support from UiO:Energy and the research-project *Stochastics for Time-Space Risk Models*, funded by the Research Council of Norway under grant no. 274410, are also greatly acknowledged.

Catriona Byrne, Remi Lodh and Ute McCrory at Springer are thanked for their constant support throughout the writing and preparations of this book. The constructive critics from five referees are acknowledged.

Finally, we are grateful for all the support from our beloved ones: Jūratė and Julia (FEB) and Julia and Lili (PK).

Oslo, Norway
Vienna, Austria
June 2023

Fred Espen Benth
Paul Krühner

Contents

Chapter 1
Introduction

Abstract An introduction to the specifics of energy markets and the motivation for our stochastic pricing approach are given. We also provide empirical evidence for the stochastic modelling approach taken from the markets of coal, gas and power. The material presented in this monograph is positioned in relation to other books in the area.

1.1 Energy and Commodity Markets

Commodities like metals, agricultural products and energies are traded on exchanges world-wide. The markets are typically split into spot (immediate delivery), forward and futures (future delivery) and derivatives thereon (plain vanilla call and put options), attracting producers, consumers and investment banks and insurance companies. The markets related to energy commodities is our main concern in this book, which includes oil, gas, coal, electricity, but also their adjacent markets trading in temperature, wind, transport and emissions. Major exchanges for energy trading include NYMEX[1] in the US and EEX[2] in Europe.

Energy is a key economic driver and of crucial importance in society. Climate change has led to a shift in energy markets away from fossil fuels to renewables, evidenced by increased wind and solar electricity generation substituting gas- and coal-fired power plants. Moreover, electrification takes place in various sectors of economic activity, as for example seen in transportation with an increasing share of electrical vehicles. At the same time, weather patterns are likely to be different in the future, with global warming and more extreme variation in wind and precipitation. These trends impact the markets for energy and power significantly, and justify paying a particular attention to energy markets in general

[1] Formerly the New York Mercantile Exchange, now part of the CME Group: www.cmegroup.com.

[2] European Energy Exchange: www.eex.com.

F. E. Benth, P. Krühner, *Stochastic Models for Prices Dynamics in Energy and Commodity Markets*, Springer Finance,
https://doi.org/10.1007/978-3-031-40367-5_1

and electricity in particular. Energy and power obviously interrelate closely with other commodities like agriculture and metals.

A spot market is usually organising trades with immediate physical delivery of the commodity. Often, when talking about the spot, one in reality means the closest forward and futures contract, that is, the forward and futures contract with closest delivery time in the future. There is no spot market for oil, for example, and for most commodities transport is required for physical delivery. Another example is electricity, where the spot market is a day-ahead market, in the sense that production and consumption of power is settled for each hour during a day in an auction taking place the day ahead. Power is also in principle non-storable, except in expensive batteries or indirectly in hydro reservoirs. Non-storability implies that one cannot trade power forward in time.

Forward and futures contracts can have both physical delivery or financial settlement. In the latter case, the buyer of such contracts receives the equivalent amount of money based on the spot price at delivery rather than the commodity itself. Moreover, In the EEX power market, say, the forward and futures contracts are settled financially over different delivery periods like weeks, months, quarters and years, up to 4 years ahead. In Fig. 1.1 we have plotted observed base load futures prices (in Euro/MWh) traded during the month of August 2021. The market trades many delivery periods, but we have shown the two front week contracts, the front month (September) contract as well as the two front quarters (Q4 of 2021 and Q1 of 2022). Notice that when we pass the middle of August, the second front week

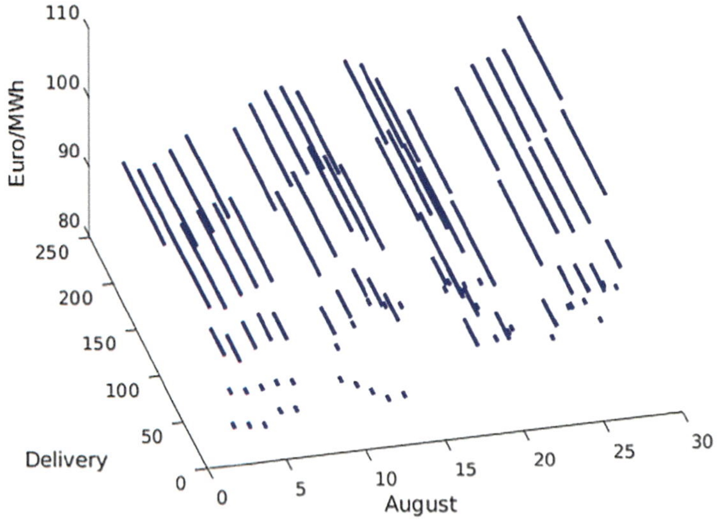

Fig. 1.1 Base load forward prices at EEX in August 2021. Two front weeks, front month (September) and two front quarters (Q4 2021 and Q1 2022) are shown. The x-axis shows the trading days in August, while the y-axis shows the start and end of delivery enumerated in number of days counting from August 1. Data are downloaded from eex.com on September 10, 2021

contract is the first week of September, while in the last trading week both front week contracts are the first and second week in September.

The weather market at CME[3] provides an interesting case where there is no spot market as such, as one cannot buy and sell the temperature in a specific geographical location, say. However, at CME, one can trade on forwards and futures which are financially settled based on indices on temperatures.

From these cases we see that delivery of the commodity may take place at a specific time in the future (e.g., oil), or it may take place over a specific time period in the future (e.g., power). In our analysis, the distinction between forward and futures contracts will not play any role, and we will simply refer to them collectively as *forwards*. Sometimes in the literature, forwards with delivery period are coined *swaps*, as they are swapping a fixed price of a commodity with a floating. We will also from time to time use the term "swap" when talking about forwards delivering over a period.

In the spot markets for power one can observe *negative* prices occasionally, and even rather frequently in certain periods. This happens when there is an oversupply of electricity, and is largely an effect resulting from of a market structuring with renewable power generation given grid priority over coal- and gas-fired plants. In shorter time periods, one may experience weather conditions leading to high generation of wind power, but as the coal and gas-fired power plants may find it costly to shut down their production, they are willing to pay for their generation. Rather surprisingly, even forward prices has turned negative, an event observed in the WTI[4] oil forwards on April 20, 2020, for delivery in May. One may also suspect that power forwards with delivery periods down to perhaps a given day may become negative, although the authors are not aware of any such incidents in the market.

In the stochastic modelling of power markets, negative prices constitute a risk that should be accounted for. But energy markets have other particularities contributing to the variability, like frequent price spikes and evident seasonalities. The problem with storability of power may lead to imbalances that give raise to large price increases (or indeed, also price drops). These are wiped out relatively fast, and appear as large spikes in the price path. Not unusually, the spikes may have values several magnitudes above normal. Seasonality is another aspect of electricity, as demand varies with season (in winter there is increased demand for heating, and in summer for cooling). With the increasing importance of renewable power generation based on wind and sun, also the supply side of the market shows seasonal variability.

In many commodity markets options on the forward and futures contracts are offered for trade. Usually, these options are plain vanilla call and put options, written on a selection of forward and futures contracts as the underlying assets. In power markets, say, the typical underlying assets are monthly forwards, with exercise time being the beginning of the delivery period. Spread options also play an important role in energy markets, providing for example a tool of hedging

[3] Chicago Mercantile Exchange: www.cmegroup.com.

[4] Western Texas Intermediate, a benchmark for US oil.

profit risk for coal and gas-fired power plants or in the valuation of such. They exist also as virtual power plant agreements (sometimes called tolling agreements) traded OTC[5]. At NYMEX one finds so-called crack spreads, which are call and put options written on the spread between different qualities of oil. To hedge risk exposure between prices and cost factors (power price vs. coal price, say), or prices and externalities (power prices and weather variables determining demand, say), is critical in market operations for producers and consumers. There exists a zoology of various tailormade contracts traded OTC in commodity and energy markets, however, we will in this book not focus much on these derivatives.

For a thorough discussion of the principles of commodity and energy markets and the basics of stochastic modeling, we refer the interested reader to the monographs by Clewlow and Strickland [72], Eydeland and Wolyniec [97] and Geman [111].

1.2 Stochastic Modelling Approaches to Price Risk

The stochastic modelling of price risk in commodity markets refers to describing the random evolution in time of spot, forward and derivatives (options) prices. The focus in this monograph is on modeling the term structure of forward prices, that is, the time dynamics of forward prices over all delivery times.

If one can trade in a liquid spot product, the so-called buy-and-hold strategy provides immediately an arbitrage-free price dynamics of the forward. More precisely, if $S(t)$ denotes the spot price at time $t \geq 0$, then the forward price at time t with delivery of the commodity at time $T \geq t$, denoted $F(t, T)$, is

$$F(t, T) = S(t)e^{r(T-t)}, \tag{1.1}$$

where r is the (constant) risk-free interest rate. If the forward price fails to satisfy this equality, a buy-and-hold strategy will give an arbitrage opportunity (see for example Geman [111]). This simple relationship between forward prices and the underlying spot dynamics does not require any stochastic model. However, if the spot dynamics follows a semimartingale stochastic process, the forward price can also be expressed as (loosely speaking, under additional appropriate conditions)

$$F(t, T) = \mathbb{E}_{\mathbb{Q}}[S(T) \,|\, \mathcal{F}_t] \tag{1.2}$$

where $(\mathcal{F}_t)_{t \geq 0}$ is a filtration modelling the information available to the market participants and $\mathbb{Q}$ is the so-called equivalent martingale measure (sometimes called the risk-neutral probability).

Trading in the commodity spot cannot in practice be done without frictions like storage and transportation. This violates the simple relationship between the

[5] OTC is the acronym for "over-the-counter".

forward and spot price argued above, and appropriate modifications have to be done (see Geman [111]). However, in energy markets, the spot may be non-existing or impossible to trade in. In the oil market say, there is no spot market, so at best one must interpret $S(t)$ as some index. This is similar to the temperature indices on which weather forward contracts are settled on, as temperature in a geographical location cannot be used for financial operations (obviously!). On the other hand, unlike oil spot, temperatures can be observed. In the power market the spot electricity is a physical product that requires production and transmission facilities and cannot be stored over time, and thus it serves as an index for the (financial) forward contracts much in the same fashion as weather forwards on temperature. To wrap up the situation we are facing in energy markets, a "spot price" can be an index which may be observable (power, temperature...) or introduced artificially for modelling purposes (oil, coal....).

Forwards are tradeable assets, and as such they must follow a (local) martingale dynamics under some risk-neutral probability to avoid arbitrage in the market (see, e.g., Duffie [88]). The risk-neutral probability $\mathbb{Q}$ must be equivalent to the market probability, but the discounted spot dynamics does *not* need to be a (local) martingale except when the spot is a liquidly tradeable asset. The spot-based approach to forward pricing in energy markets (as well as commodity markets in general), is to *define* the forward price by the relationship (1.2), where $\mathbb{Q}$ is some equivalent probability measure and S is some dynamics referring to a spot index. One refers to $\mathbb{Q}$ as the *pricing measure*, and the forward price dynamics is a (local) martingale by definition with respect to this probability. For given specifications of the stochastic dynamics of S and the pricing measure $\mathbb{Q}$, one may compute a forward dynamics from the conditional expectation. For contracts with a delivery period, a forward price dynamics can be derived as the integral of $F(t, T)$ with respect to T over the contractual delivery time. In Sect. 5.5 of Chap. 5 we are going to provide some more details regarding the connections between pricing measures $\mathbb{Q}$ and the classical concepts in commodities like convenience yield and storage.

The risk premium is defined as the difference between the forward price and the predicted spot at delivery. More precisely, it is given by

$$F(t, T) - \mathbb{E}[S(T) \,|\, \mathcal{F}_t] = \mathbb{E}_{\mathbb{Q}}[S(T) \,|\, \mathcal{F}_t] - \mathbb{E}[S(T) \,|\, \mathcal{F}_t].$$

The pricing measure $\mathbb{Q}$ models, effectively, the risk premium in the market. Choosing an appropriate pricing measure is a challenging modelling task.

Rather than deriving the forward dynamics from a spot index, one may specify the stochastic time evolution directly. This alternative modelling paradigm is frequently referred to as the Heath-Jarrow-Morton (HJM)-approach, as it is adopted from fixed-income theory and the modeling of the analogous forward *rates*. The idea is simply to specify a dynamics of $(t, T) \mapsto F(t, T)$ for $0 \leq t \leq T$ directly, as a stochastic process. Typical classical models (see, e.g., Clewlow and Strickland [72] and Benth and Koekebakker [32]) take the form

$$dF(t, T) = \alpha(t, T)dt + \sigma(t, T)dB(t) \tag{1.3}$$

for a finite dimensional Brownian motion B and parameter functions α and σ satisfying appropriate conditions. Notice that the parameters are depending on the delivery time T and may be stochastic, while the random fluctuations are modelled through a multi-dimensional Brownian motion. In fixed-income theory, one models under the risk-neutral probability $\mathbb{Q}$ directly, while in commodity markets this is not necessarily the canonical choice as the purpose may not be pricing of derivatives. However, for fixed T, the process $t \mapsto F(t,T), t \leq T$ must be a $\mathbb{Q}$-martingale (possibly local) to ensure arbitrage-freeness of the market, imposing the trivial condition that α is zero in that case. This means that the drift parameter is a model of the risk premium in the market if (1.3) is stated under the objective (market) probability, denoted $\mathbb{P}$.

What is notable in the dynamics (1.3) is the finite dimensionality of the Brownian motion. In the spot-based approach to pricing, one may assume a factor model for the dynamics of S. The classical choices are the two-factor model of Schwartz and Smith [178], where one uses a drifted Brownian motion and an Ornstein–Uhlenbeck process as factors, or the more general factor models in Benth, Šaltytė Benth and Koekebakker [25] (where even jump processes are included). Using these models, one may recover a dynamics of the forwards of the form (1.3). On the other hand, as $F(t,t) = S(t)$, an HJM-model recovers a spot dynamics which is driven by a finite number of Brownian motions. The structure of the forward curve and its evolution are encapsulated in a specification of α and σ, whereby the different Brownian motions may be viewed as noises for different segments of the term structure. For example, in a two-factor specification, we may think of two Brownian motions governing the short and long term random fluctuations, respectively.

As a final remark of this section, we want to mention that in energy and commodity markets, arithmetic models may be sensible to use for several reasons. In finance, geometric Brownian motion and generalisations of this have been preferred as they ensure positive prices. However, in energy, as we have discussed above, negative prices occur. Arithmetic models allow for this. Also, in the weather market, the underlying index is based on temperatures, which naturally can be negative as well. Additionally, arithmetic models may be preferable when studying forwards with delivery period, as the geometric case may not allow for the derivation of explicit expressions for the term structure dynamics.

1.3 The Infinite Dimensional View

The HJM-approach to forward prices already suggests that stochastic processes with an infinite dimensional state space is the natural mathematical framework. Indeed, we can view $(F(t,\cdot))_{t\geq 0}$ in (1.3) as a stochastic process which takes values in the space of functions $T \mapsto F(t,T)$ for $T \geq t$. It is preferrable to define the state space of the forward term structures independent on current time t. This can be achieved by the simple but ingenious change of time $x := T - t$ proposed by Musiela [157], leading to the Heath-Jarrow-Morton-Musiela (HJMM)-model

$f(t,x) := F(t, x+t)$. Naturally, $x \geq 0$, and signifies *time-to-delivery* rather than *time-of-delivery* which is given by T. One can then view $(f(t,\cdot))_{t\geq 0}$ as a stochastic process taking values in a suitable space of real-valued functions on the positive real line $\mathbb{R}_+$. Typically, this is chosen to be a Hilbert space. An early study of the term structure of interest rates formulated as random fields is found in Kholodnyi and Lukic [136].

Considering power markets, there is empirical evidence for high dimensionality of the noise. Koekebakker and Ollmar [141] perform a principal component analysis of the Nordic electricity forward market, and find that 10 factors are needed in an HJM-model to explain 95% of the variation of the data. Féron and Gruet [120] show, on the other hand, through an extensive statistical study that the optimal number of factors in an HJM-model for several European power markets are 5. This means that there is evidence for the need of 5 to 10 independent noise drivers (for example Brownian motions) along with possibly a definition of the volatility σ as a matrix of high dimension. Of course, 5 is not a high number, but on the other hand it leads to $5+4+3+2+1=15$ specifications to span out the variances along each noise driver as well as the covariances among these. In interest rate markets as well as commodity markets like oil, the typical number is two to three factors (see, e.g., Carmona and Tehranchi [69]), while metals may need four factors according to Geman [111, Ch. 8].

There may be several economic reasons for the higher number of factors observed in electricity markets than other markets. Different players act in different market segments, for example short term speculation vs. long term hedging, giving raise to different risks. Also, there are various sets of information along the term structure of forward prices. For example, planned new storage units or production facilities, or political decisions on market structure may affect prices on the long term, but not necessarily in the short term where other factors play a larger role. Unlike gas and oil, power is in principle non-storable, and we may face various degrees of risk factors spread along the term structure. We refer to Benth and Meyer-Brandis [39] for more on this. Recalling the study of Koekebakker and Ollmar [141], they found that as much as 25% of the variance in power data is maturity dependent, "..a feature unique to the electricity market" in their words. In relation to this study, Geman [111, p. 74] remarks that "Obviously, a model with 10 factors becomes cumbersome to implement...."

Commodity markets have experienced a *financialisation* over the last two decades, as argued in Carmona [67]. Investment banks and other speculators have become attracted to commodity forward and futures contracts as asset classes for risk diversification and profit making. The financialisation of these markets, including energy, have impacted prices. Moreover, one has also observed new trading strategies in these markets, with portfolios taking positions in different maturities along the term structure. Optimising yield by considering different maturities requires a sophisticated dynamics which is able to describe the dependencies along the term structure curve.

Instead of choosing a relatively large number of Brownian motions driving the noise in a forward term structure model, which may lead to a parameter-intensive volatility specification, we may choose to model the noise by an infinite dimensional stochastic process. The natural extension of Brownian motion would be a Wiener process taking values in some suitable Hilbert space. A Wiener process requires the definition of a covariance operator, which is mathematically encapsulating information on the spatial covariance structure of the Wiener process, where space here refers to time-to-deliveries $x \geq 0$. Hence, we have one "Brownian motion" for each time-to-delivery x, and a specification of the covariances between these "Brownian motions".

Selecting the covariance operators from a parametric family of functions, we may in fact have a low-dimensional set of parameters to be estimated from data. This is in contrast with the the situation where a high dimensional Brownian motion is used in an HJMM-model. We remark in passing that this is similar to the LIBOR-models which have been popular in interest rate theory (see, e.g., Brigo and Mercurio [64]). The covariance operator may for example be a parametric function which measures the covariance between different time-to-deliveries as a function taking into account the distance between deliveries, so that for example forwards in the long-end of the curve is only weakly correlated with products in the short end. We refer to Benth and Paraschiv [42] for an empirical analysis of forward prices from an infinite dimensional perspective. Even though our motivation mainly refers to the power markets, this approach can be used in all commodity forward markets.

Empirical evidence also strongly suggest non-Gaussian behaviour of forward prices (see Benth, Šaltytė Benth and Koekebakker [25, Ch. 8] and Frestad, Benth and Koekebakker [107]), which calls for infinite dimensional Lévy processes extending the class of Wiener processes. Certain non-Gaussian distributions have drawn particular attention in finance, like for example the hyperbolic, normal inverse Gaussian or CGMY families of distributions (see Eberlein and Keller [91], Barndorff-Nielsen [12], Carr et al. [70], Benth, Šaltytė Benth and Koekebakker [25, Ch. 8] and Frestad, Benth and Koekebakker [107]), the latter two for an energy market context. These families of distributions are relevant in energy and commodity markets.

The Samuelson effect (see Samuelson [173]) is an important characteristic of volatility in commodity markets. It says that the volatility decays with time-to-delivery, and may be explained as an effect of mean-reversion of spot prices. Additionally, it is an established empirical fact in most commodity forward markets. As in financial markets, stochastic volatility is also relevant for commodity and energy prices. In a forward market context, this calls for models explaining the random variations of volatility across the term structure curve. If the noise is infinite dimensional, we face a situation where the stochastic volatility model must be infinite dimensional as well, in fact, a random operator acting on the state space of the noise.

Generally, the dynamics of $(f(t,\cdot))_{t\geq 0}$ is modeled as a stochastic partial differential equation (SPDE). From the change from T to x in the parametrization of delivery, the no-arbitrage dynamics of $f(t,\cdot)$ will have a drift involving $\partial f(t,x)/\partial x$, the derivative of f in space, for which the shift semigroup will be an ingredient in the (mild or strong) solution of the SPDE. HJMM-models of forward prices place themselves in the theory on SPDEs with Lévy-driven noise, for which the monograph by Peszat and Zabczyk [165] is a basic reference. The Filipović space, introduced by Filipović [99], is a convenient state space for the forward curves, a separable Hilbert space of, roughly speaking, weakly differentiable real-valued functions on the positive real line, being asymptotically constant. From an economic perspective, it is reasonable from arbitrage considerations to have continuous term structure curves. Additionally, far into the future it should not matter when delivery takes place, it should in any case be the same price (less seasonality, of course). Theoretically, the Filipović space is also nice as the shift semigroup turns out to the strongly continuous and quasi-contractive on this space. Furthermore, in the HJMM-context, power forwards having a delivery period rather than delivery at a specified calendar time, can be realized through integral operators acting on solutions $f(t, \cdot)$. These integral operators turn out to be bounded linear operators on the Filipović space.

We recall that plain vanilla options in commodity markets typically are written on forward contracts. Hence, as the forward price comes from an infinite dimensional stochastic process in the HJMM-framework, the arbitrage-free derivatives price will be the discounted conditional expectation under the pricing measure of a nonlinear functional on the Hilbert space of states. These expectation values can further be analysed in terms of Fourier methods, leading to formulas for prices which can be numerically computed. Considering spread options, this entails an additional challenge as we then aim at a bivariate (or more generally, multivariate) system of forward curves. These can naturally be modelled as having state space in the product of Filipović spaces, where, for example, the driving Wiener process can be decomposed into "correlated" but "univariate" Wiener processes. Such considerations are relevant if one wants to model jointly the term structure of forward prices in gas and power, say, and next price spread options on these (or assess the real option value of a gas-fired power plant).

Let us summarise the above discussion to emphasise the benefits of our infinite dimensional approach to forward price modelling. Potentially high-dimensional, parameter-intensive (factor-)models may be substituted by random field specifications in infinite dimensional function spaces. The ingredients in these models are encoded as families of functions described by a low-dimensional set of parameters. The infinite dimensional approach also opens for more flexibility in explaining the stylized features observed in commodity forward prices in general, and power and energy in particular. Although the infinite dimensional modelling approach is mathematically abstract, it still provides an analytically tractable framework for pricing and risk management. Finally, all HJM-models define a stochastic dynamics which takes values in some space of functions, and therefore the infinite dimensional approach is the natural set-up.

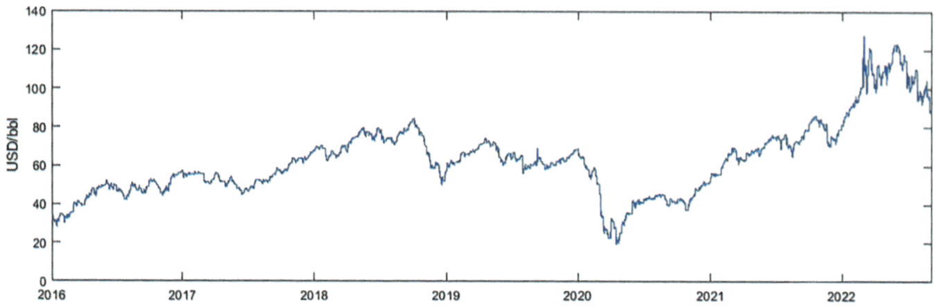

Fig. 1.2 Daily Brent Crude oil front month futures prices from January 1, 2016 to September 8, 2022

1.4 A Review of Stylized Features of Commodity and Energy Prices

This section reviews some stylized features of commodity and energy forward prices. It serves as a motivational background for our infinite dimensional approach to forward price modelling.

In Fig. 1.2 we show a time series plot of daily front month futures prices for Brent Crude oil traded at the Intercontinental Exchange (ICE).[6] The prices are denominated in US$/barrel, and ranges from January 1, 2016 until September 8, 2022. As is usual in finance, we consider the logarithmic returns (logreturns, for short), defined over the interval $[t, t + \Delta t]$ by

$$X(t) = \ln\left(\frac{S(t + \Delta t)}{S(t)}\right).$$

The logreturns of the Brent Crude oil futures prices are depicted in Fig. 1.3. As is apparent from the time series, there are clusterings of more volatile periods, for example in the first half of 2020. Volatility clustering is a frequently observed phenomenon in financial price series, and usually taken as empirical evidence of stochastic volatility. The empirical distribution of the logreturn data (in red stars) along with the fitted normal distribution (blue line) are shown in Fig. 1.4. The small daily variations in prices, reflected in the center of the empirical distribution, have higher probability than predicted by the fitted normal. On the other hand, the "medium-range" variations of size around ± 0.025 are far less probable than explained by the normal distribution. Not very evident in this plot are the extreme logreturns, that is, the tails of the empirical distribution. In Fig. 1.5, where the empirical and fitted normal distribution are plotted together on a logarithmic

[6] We are grateful to Marcus Eriksson at Refinitiv for providing us with these data, as well as the API2 coal and TTF gas futures prices considered below.

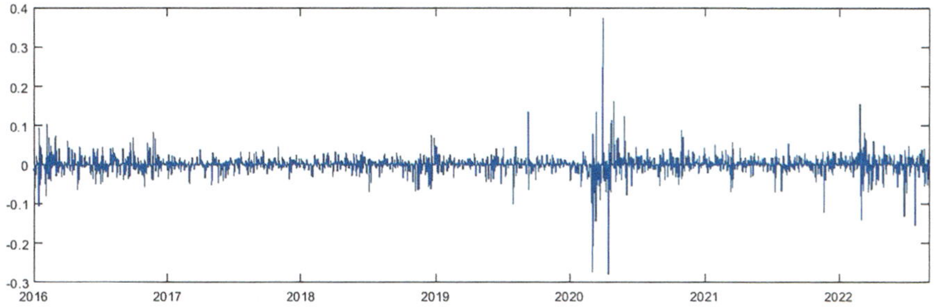

Fig. 1.3 Logreturns of daily Brent Crude oil front month futures prices from January 2, 2016 to September 8, 2022

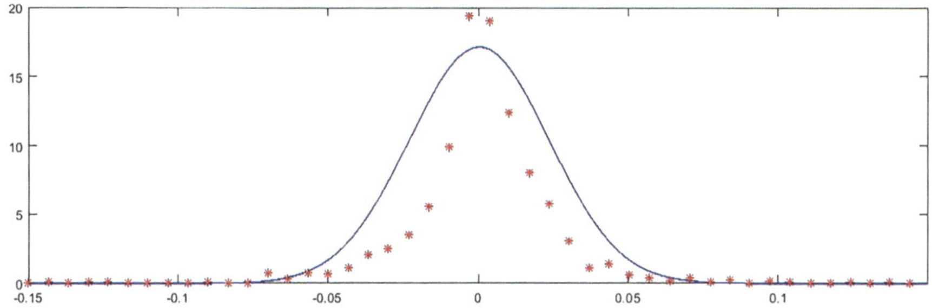

Fig. 1.4 Empirical distribution (in red stars) of logreturns of daily Brent Crude oil front month futures prices from January 2, 2016 to September 8, 2022. Blue line is the fitted normal distribution

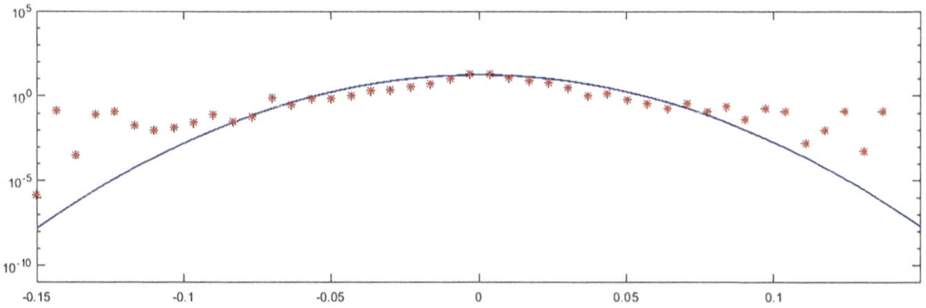

Fig. 1.5 Empirical distribution (in red stars) of logreturns of Daily Brent Crude oil front month futures prices from January 2, 2016 to September 8, 2022. Blue line is the fitted normal distribution, and the frequency axis is on a logarithmic scale

frequency axis, we see that the tails of the empirical distribution are heavier than the normal.

From this first simple statistical analysis of front month Brent Crude oil futures we find statistical evidence of a non-Gaussian dynamics, with indications of stochastic volatility.

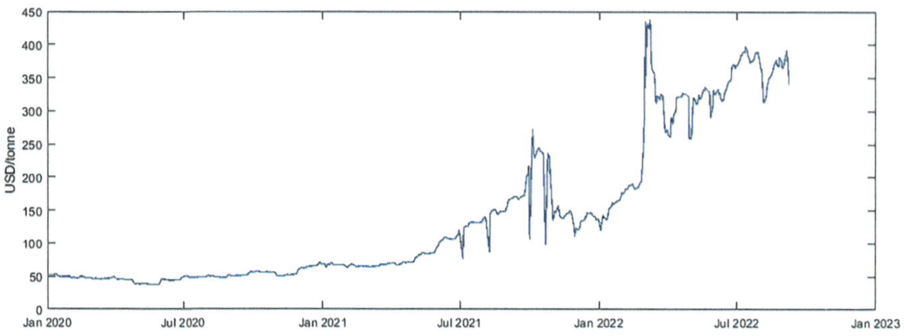

Fig. 1.6 Daily coal API2 front month futures prices from January 2, 2020 to September 8, 2022

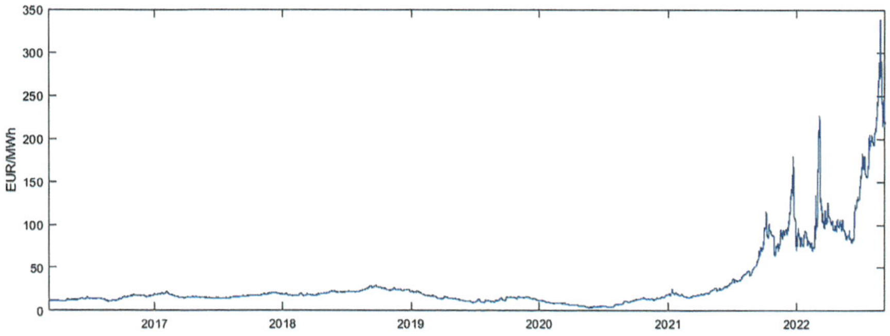

Fig. 1.7 Daily TTF gas front month futures prices from March 1, 2016 to September 8, 2022

Similar evidence is found for gas and coal prices, as we now briefly present. In Figs. 1.6 and 1.7 we show, respectively, the daily API2 coal (top) and TTF gas (bottom) front month futures prices collected from the ICE exchange. The API2 coal prices ranges from January 2, 2020 until September 8, 2022, while the TTF gas data set starts at March 1, 2016 with the same end date as API2 coal. The API2 coal futures prices are denominated in US$/tonne and the TTF gas prices in EURO/MWh. There is a dramatic increase of coal and gas prices from about 2021 due to the energy crisis in Europe. In the logreturn time series of the two commodities depicted in Figs. 1.8 and 1.9 we clearly see volatility clustering. Moreover, there also seems to be an increase in the volatility over time in the TTF gas prices. Again the empirical density plots of the logreturns show a peak in the center compared with the normal, see Figs. 1.10 and 1.11. Evidently, the coal prices are much more peaky than the gas prices.

To explain the logreturn distribution one must use different models than the normal. One possibility is offered by assuming the logreturns to be normal inverse Gaussian distributed (NIG). The NIG distribution was introduced by Barndorff-Nielsen and has later become popular in financial applications (see Barndorff-Nielsen [12]). We will return to this class of distributions in Chap. 2, but here we provide a plot (see Fig. 1.12) of the fit of this four-parameter distribution to the TTF

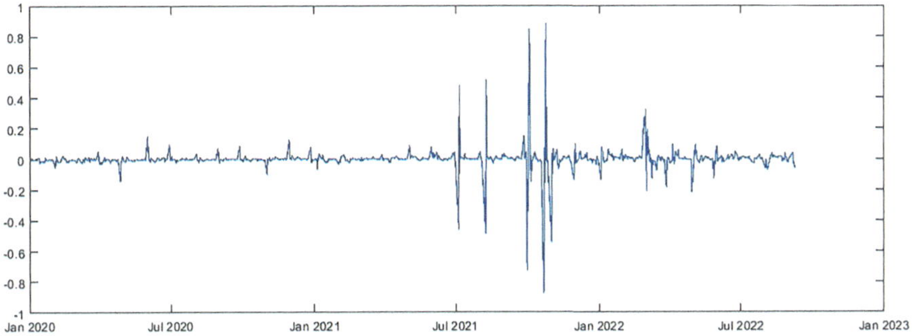

Fig. 1.8 Logreturns of API2 coal front month futures prices from January 2, 2020 to September 8, 2022

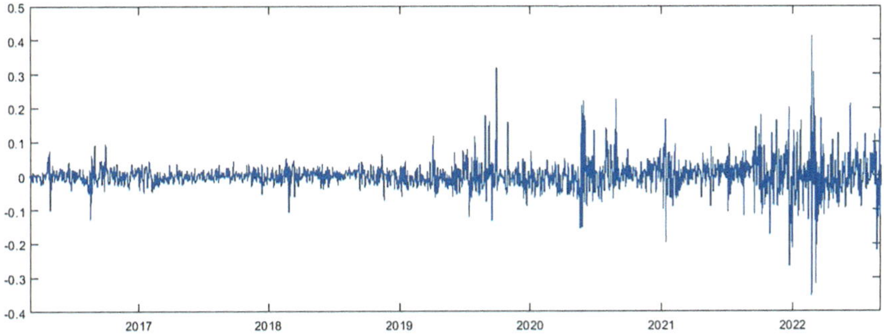

Fig. 1.9 Logreturns of daily TTF gas front month futures prices from March 1, 2016 to September 8, 2022

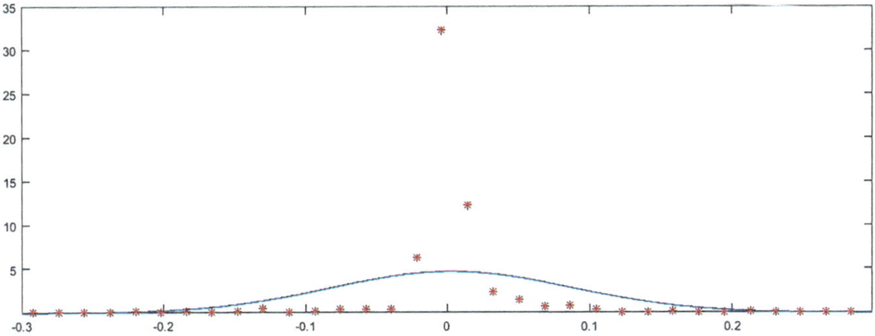

Fig. 1.10 Empirical distribution (in red stars) of logreturns of daily API2 coal front month futures prices from January 2, 2020 to September 8, 2022. Blue line is the fitted normal distribution

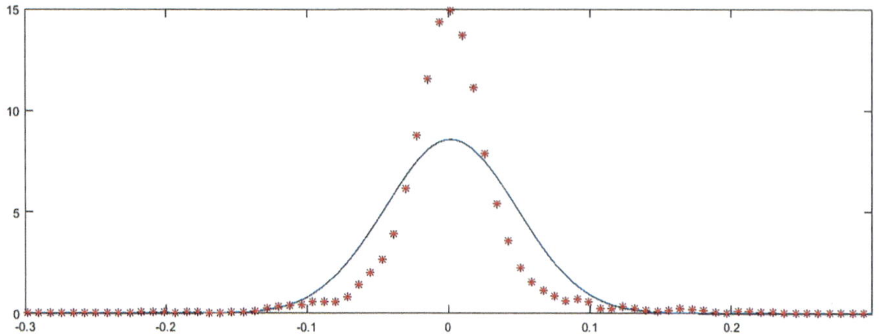

Fig. 1.11 Empirical distribution (in red stars) of logreturns of daily TTF gas front month futures prices from March 1, 2016 to September 8, 2022. Blue line is the fitted normal distribution

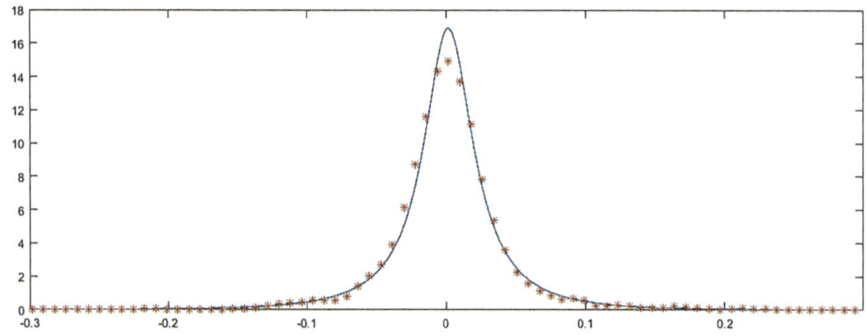

Fig. 1.12 Empirical distribution (in red stars) of logreturns of daily TTF gas front month futures prices from March 1, 2016 to September 8, 2022. Blue line is the fitted normal inverse Gaussian distribution

gas logreturn data. We observe a much better fit than the normal, clearly indicating the need to introduce non-Gaussian models for the price dynamics of futures. Lévy processes is the natural class of stochastic processes for this purpose, and we give a theoretical introduction of these processes in Chap. 2.

In Fig. 1.13 we plot the volatility of the logreturns of daily futures prices for various contracts on API2 coal. Using available data series on the front quarter and year contract, we have estimated the volatility for the different contracts. One can observe a decay in the volatility with time to delivery, which is a typical feature of the term structure of volatility in commodity futures markets. As said earlier, this feature is often referred to as the *Samuelson effect* after [173] (see Geman [111] for further discussion about this effect in commodity markets).

Estimation of the correlation between the logreturn price data, reveals that the month contract is close to zero correlated with the quarter and year contracts (resp. 0.06 and -0.08). On the other hand, the quarterly and yearly logreturn price series are correlated by 0.84. Thus, far out on the term structure curve there is a strong

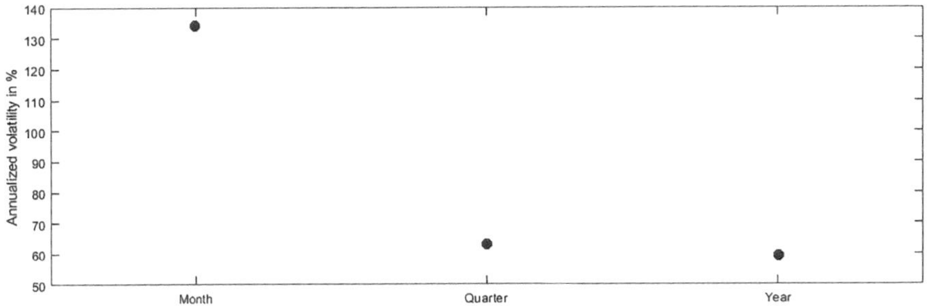

Fig. 1.13 The volatility of front month, quarter and year daily futures logreturns on API2 coal. The volatility is annualised (and in %)

correlation among prices, but still not perfectly correlated. On the other hand, there seems to be negligible correlation between the short end (month) and long end (quarters and farther). Without performing any formal principal component analysis, these estimates for the correlation indicate that there is at least a need for three factors to explain the price movements along the term structure curve.

Let us consider the class of electricity forwards. In Benth and Paraschiv [42], an infinite dimensional modelling approach to power forward prices traded on the EEX is proposed and studied empirically, and we report here some of the findings.

We have available hourly forward price curves from the EEX each day ranging from January 1, 2009 to July 15, 2015, i.e., constituting 2386 hourly forward curves with times to delivery covering two years into the future. These term structure curves are generated by the Institute of Operations Research and Computational Finance, University of St. Gallen in Switzerland, using a smoothing approach to transform the observed swap prices into hourly-delivery. The data has been further cleaned from seasonality effects. The approach is described in detail in Benth and Paraschiv [42, Sect. 3]. For the discussion to come, each hourly forward curve is sampled at daily delivery times, where we choose the noon as sampling hour. Hence, we have available a data set of 2386 de-seasonalized daily forward price term structures $f(t,x)$, where the daily time to deliveries x ranges over two years (i.e., $2\times 365 = 730$ daily delivery times). In the further analysis, we focus on the Samuelson effect, the risk premium and the correlation structure across the maturities observed from the data.

As we have previously indicated, the continuous-time dynamics of the forward price curve can be modelled as a solution to a stochastic partial differential equation. In Benth and Paraschiv [42], the underlying model is

$$df(t,x) = \left(\frac{\partial}{\partial x} f(t,x) + \theta(x) f(t,x)\right) dt + d\omega(t,x)$$

where $\omega(t,x)$ is some time-space random field describing the noise. Furthermore, the term $\partial f(t,x)/\partial x$ is the maturity-correction coming from the Musiela parametrisation using *time to delivery* x rather than *time of delivery* T, while $\theta(x)f(t,x)$ is a modelling of the risk premium in the market. At the current stage of this monograph, we interpret the above dynamics at an informal level, where the notation $df(t,x)$, dt and $d\omega(t,x)$ means a change over a small time increment dt. In fact, most of this book is devoted to build up a rigorous theory for the dynamics of f. In discrete-time and space, we have

$$f(t+\Delta t,x)-f(t,x)=\frac{\Delta t}{\Delta x}\left(f(t,x+\Delta x)-f(t,x)\right)+\theta(x)f(t,x)\Delta t+\widetilde{\epsilon}(t,x)$$

for given Δt, $\Delta x>0$, where $\widetilde{\epsilon}(t,x):=\omega(t+\Delta t,x)-\omega(t,x)$.

In our empirical analysis, we choose $\Delta t=\Delta x=1$, and we estimate the risk premium proportionality constant $\theta(x)$ by a linear regression as follows: first define

$$z(t,x):=f(t+\Delta t,x)-f(t,x)-\frac{\Delta t}{\Delta x}\left(f(t,x+\Delta x)-f(t,x)\right)$$

and observe that

$$z(t,x)=\theta(x)f(t,x)\Delta t+\widetilde{\epsilon}(t,x)\,.$$

Next, for fixed x, consider the function

$$G(\theta(x)):=\sum_{t=1}^{N}\left(z(t,x)-\theta(x)f(t,x)\Delta t\right)^2$$

where in the present case $N=2386$ and $\Delta t=1$. Minimizing $G(\theta)$ with respect to θ, yields

$$\widehat{\theta}(x)=\frac{\sum_{t=1}^{N}z(t,x)f(t,x)}{\Delta t\sum_{t=1}^{N}f(t,x)^2}\,.$$

In Fig. 1.14 we have plotted the estimated proportionality factor θ of the risk premium across maturities up to one year. We observe that the premium is overall slightly negative for longer delivery times (above approximately 60 days or so), which is in accordance with the standard theory of producers inducing a hedging pressure. However, in the short end of the term structure the risk premium changes sign and shows more irregularity. Geman and Vašíček [114] argue that a positive risk premium in power markets can occur due to lack of storability of the commodity and retailers wishing to hedge price spike risk. We refer to Bessembinder and Lemon [57] and Benth, Cartea and Kiesel [26] for further theoretical and empirical evidence of a positive risk premium in power markets.

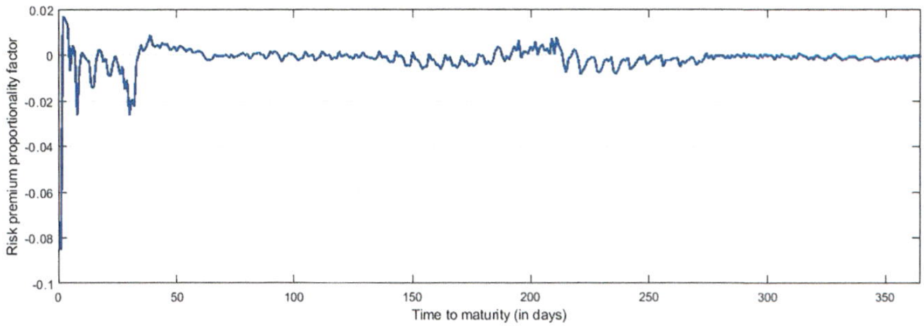

Fig. 1.14 The estimated risk premium proportionality factor $\widehat{\theta}(x)$ for EEX power forwards

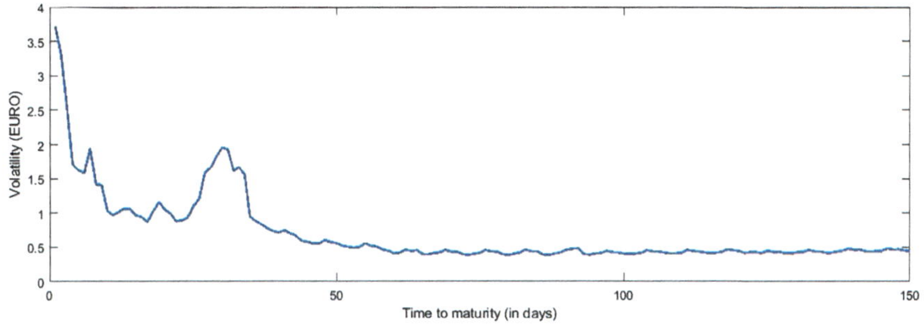

Fig. 1.15 The estimated volatility term structure for EEX power forwards

One may expect that there is a Samuelson effect also in power markets. To investigate this, we suppose that

$$\widetilde{\epsilon}(t, x) := \sigma(x)\epsilon(t, x)$$

where $\sigma(x)$ is the volatility at time to delivery x. Estimating the standard deviation of the time series $(\widetilde{\epsilon}(t, x))_{t=1}^{N}$ for each x, gives us an estimate of $\sigma(x)$ which is depicted in Fig. 1.15. We clearly observe a decay of the volatility with increasing time to delivery. But the shape of the volatility term structure is not monotone, with a clear hump somewhere around 30 days to delivery. As argued in Benth and Paraschiv [42], such a hump in the volatility can be attributed to the transition from weekly to monthly forward contracts for trading beyond four weeks, and the differences in trading volumes of these contracts. Indeed, there is further evidence of volatility clustering, as demonstrated in Fig. 1.16 which shows the time series of the normalised noise at seven days (one week) to maturity $(\epsilon(t, 7))_{t=1}^{N}$. We re-emphasise that volatility clustering is a sign of non-Gaussianity and stochastic volatility. In Benth and Paraschiv [42], the normal inverse Gaussian distribution is shown to be a good model for the residual noise time series for the different

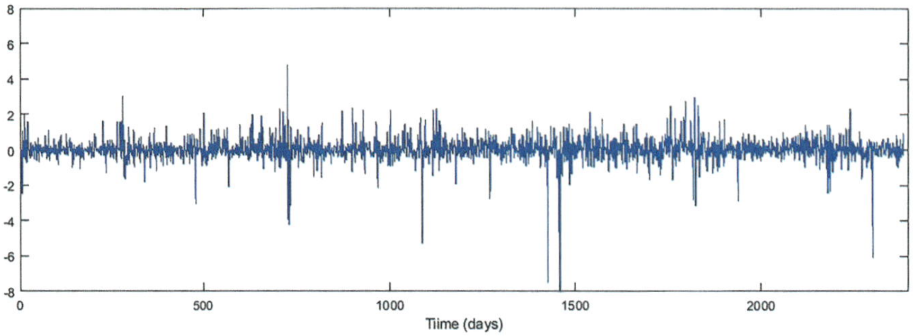

Fig. 1.16 Time series of the normalised noise for time to maturity in 7 days

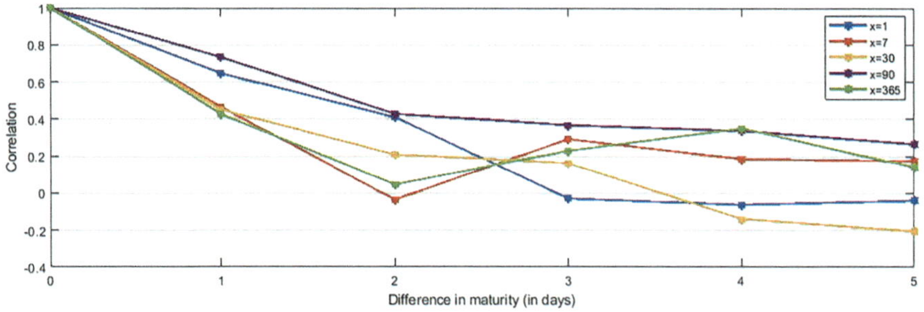

Fig. 1.17 The correlation structure for various fixed maturities x with closest 5 maturity days

maturities. The parameters of the fitted normal inverse Gaussian distribution vary with time to maturity x.

Let us end this case study of EEX forward prices by investigating the dependency structure of the residual noise across times to delivery. In Fig. 1.17 we have depicted the correlation of the noise at a given maturity with the 5 consecutive maturities. In the figure, the maturities $x = 1, 7, 30, 90$ and $x = 365$ are chosen, where we plot the correlation between the time series $(\epsilon(t, x))_{t=1}^N$ with $(\epsilon(t, x + y))_{t=1}^N$ for $y = 1, \ldots, 5$. Overall, we see that residual noise has a decaying dependency with its neighboring delivery times. With y equal to one day, the correlations range between 0.4 to 0.8, while for $y = 2$ the range is 0 to 0.4. Thereafter, we find positive correlations between 0.2 and 0.4, with $x = 1$ and $x = 30$ being exceptions. The maturity $x = 1$, which is close to the spot, has seemingly a zero correlation structure for contracts delivering 3 days or later. For the contract delivering in one month $x = 30$, we observe negative correlations of around -0.2. This is apparently a specific feature happening for maturities $x = 30$ or slightly above, indicating that trading volumes and the market design transitioning from week to month contracts for these maturities may play a specific role. We refer to Benth and Paraschiv [42] for a complete plot of the correlation structure of all various choices of x and y. Apparent is the stronger persistence in positive correlation for bigger values of y the farther out we are on the term structure x, which we can also see signs of in

Fig. 1.17 for $x = 90$ and $x = 365$. However, there is also a persistent bump for $y = 2$ in most maturities x. We remark that there are still weekend-effects appearing in the correlation structure, which further complicates the picture.

As we may conclude from these considerations, modelling the term structure dynamics of forward prices is a delicate problem which involves the term structure of the risk premium and volatility, as well as non-Gaussianity across different maturities. Furthermore, we may encounter complex dependency structures for the noise in the dynamics of the forward prices. One may in a stylized manner conveniently express such forward price dynamics using infinite dimensional random field models. This monograph builds up the theory for such models.

1.5 Outline and Positioning of the Book

This book is separated into two parts, where the first part introduces some necessary mathematical tools while in the second part we present theory for modelling the stochastic forward price dynamics.

The first part, named *Mathematical Tools* consists of three chapters. Lévy processes, which are the drivers of randomness, are necessary to build up models for the stochastic dynamics of forward prices. Chapter 2 introduces and analyses Lévy processes in Hilbert space with a focus on subordinated Wiener processes. Particular cases of Lévy processes, like the normal inverse Gaussian (NIG), are defined in this way. We shall also discuss Wiener processes taking values in product Hilbert spaces, which are useful in cross-commodity term structure models.

The so-called Filipović space is defined in Chap. 3. The Filipović space is a convenient state space for forward prices as functions of time-to-maturity. It is a separable Hilbert space where the evaluation functional as well as the shift semigroups possess required properties. Linear operators closely linked to forward curve modelling, including covariance operators, are investigated. Some additional basic material on Filipović spaces is collected in an Appendix for easy reference.

Stochastic integration theory in Hilbert space is surveyed in Chap. 4, along with a brief collection of results and notation for stochastic partial differential equations. The forward price evolution in the Musiela parametrisation is naturally a solution to a stochastic partial differential equation. We also include results on the application of linear functionals on infinite stochastic integrals and their representation in terms of stochastic integration with respect to finite dimensional Wiener or Lévy processes.

The second part of the book, entitled *Modelling the Forward Price Dynamics and Derivatives Pricing* consists of three chapters as well. In Chap. 5 we analyse the classical approach of deriving the forward price from a spot model. Factor models of various kinds are presented, including continuous-time autoregressive moving average processes, and we calculate the forward price by taking the conditional expectation of the spot at delivery, under some pricing measure modeling the risk premium of the market. As we show, these models give rise to a term structure which is related to a stochastic partial differential equation.

The stochastic partial differential equation from Chap. 5 provides the motivation for a Heath-Jarrow-Morton approach to term structure modeling. In Chap. 6 we define simple arithmetic and geometric forward price dynamics as a random field in *time-to-maturity*, solving a stochastic partial differential equation. Stochastic volatility is a central modeling aspect in financial markets, and we bring this into infinite dimensional term structure models. We also extend to Markovian models, and have a strong focus on drift conditions ensuring arbitrage-free models. From a practical viewpoint, factor models may be attractive. We treat such in some detail as well, along with questions on modeling forwards with delivery period (swaps) and seasonality and "$\mathbb{P}$-" vs. "$\mathbb{Q}$-models".

In the final Chap. 7 we look at the price of various energy-related options on forwards and their Greeks. Depending on the model, one can derive Black-76 formulas, or one can resort to Fourier-based pricing methodology in the case of non-Gaussian models. The delta of the options is particularly interesting, as this can be represented as a sensitivity of the price towards changes in the whole term structure.

At the end of each chapter we have included some notes on the relevant literature for, and around, our analysis. Admittedly, the discussion in these sections is limited by our knowledge of existing work. We have tried to make a fair account of the papers which have both inspired and directly contributed to our theory, but needless to say, lots of great scientific contributions are not mentioned. On the other hand, we believe that examining the papers we highlight, the interested reader will also come across many of the contributions we do not mention. Our bibliographical notes also aim at touching on further lines of research that we do not treat in this book, again without being at all exhaustive on these matters. For an excellent survey of the stochastic modelling of electricity markets, we refer to Deschatre, Féron and Gruet [82]. Here one also finds links to commodity market models.

The material in all the chapters builds on the previous chapters, so the book is naturally read in the sequence of the appearance of chapters. Basic knowledge of functional analysis, in particular Hilbert spaces and operators, as well as stochastic calculus for Lévy processes in infinite dimensions are to a large degree presumed. However, we have aimed at presenting the core material as self-contained as possible.

The present book works as a continuation of the monographs by Aïd [1], Benth, Šaltytė Benth and Koekebakker [25] and Benth and Šaltytė Benth [49]. In [25], a general class of stochastic jump processes, so-called independent increment processes, is used as a modelling class for energy and commodity price models. Independent increment processes can be thought of as time inhomogeneous Lévy processes. The authors in [1] and [25] consider both spot price and Heath-Jarrow-Morton-based forward price models, distinguishing between the arithmetic and geometric classes. The independent increment processes are the noise drivers in spot factor models, where the factors are defined as Ornstein-Uhlenbeck dynamics. Introducing pricing measures by Girsanov and Esscher transforms, forward prices for contracts with fixed-delivery and periodic delivery are derived, being extensively discussed in [25]. The Heath-Jarrow-Morton approach is also analysed in [1] and [25], using a finite number of noise drivers, again based on jump processes. However, the proper mathematical framework of function spaces for the forward prices in the Heath-Jarrow-Morton approach is not touched, as there is no analysis of forward prices as stochastic processes taking values in function spaces. In the present book we pose these dynamical models in their right mathematical context, and thereby also open up for new and more versatile stochastic models. For example, as discussed in the previous section, we can allow for random fields in the noise driving the forward prices. On the other hand, this also asks for new developments of mathematical theory, as we provide in the current book. Worth noticing is the explicit mentioning of the need for infinite dimensional models, in particular Lévy processes, by Aïd [1, p. 35]. In fact, therein also ambit fields are mentioned, which is a class of random fields that has been used in Heath-Jarrow-Morton models for electricity by Barndorff-Nielsen et al. [14, 15]. Ambit fields may also be naturally cast in an infinite dimensional mathematical framework as presented in the current book.

There is a lot of empirical analysis and examples of energy market prices in Aïd [1] and Benth et al. [25], most notably, electricity prices. Their findings clearly point towards the need for non-Gaussian models, for example. Thus, infinite dimensional Lévy processes are called for in addition to operator-valued stochastic volatility processes. Another market with a "non-storable spot" is the temperature market. This is also analysed in Benth et al. [25] (and briefly mentioned in Aïd [1]), but further extended to other weather markets for wind and precipitation in Benth and Šaltytė Benth [49]. In that monograph, so-called continuous-time autoregressive moving average processes play a central role in modelling weather dynamics, and many empirical studies are provided. In the current book we provide a general class of models which encapsulates the most popular commodity spot price dynamics, including the continuous-time autoregressive moving average processes. Moreover, we show how to price forwards based on this general class and how they provide examples of infinite dimensional dynamics with finite-factor noise drivers.

The three monographs Aïd [1], Benth et al. [25] and Benth and Šaltytė Benth [49] provide an empirical foundation together with a rich number of finite dimensional stochastic price models motivating the studies of the current book. As it is natural to use a certain infinite dimensional Hilbert space as state space, mathematical questions regarding operators and representations appear which have clear impact on practice. We also extend the pricing of derivatives in these markets using the methodology of fast Fourier transform, say, paving the way for a practical use of the abstract mathematical models.

In Aïd [1] there is also an analysis of different electricity derivatives involving a control, like tolling contracts and other swing derivatives. In the current book we confine our discussions to different types of spread and basket options, in addition to more plain-vanilla call and put options traded on exchanges. Pirrong [167] applies stochastic control in terms of dynamic programming to propose and analyse structural models for storage of general commodities, and load for the particular case of electricity. Structural models are also discussed in Aïd [1], and take into account important fundamental variables in the commodity spot price determination. Pirrong [167] links pricing of forwards and other commodity derivatives to the solution of partial differential equations via the Feynman-Kac formula. He performs several empirical case studies using finite difference numerical methods for different one- and two-factor Gaussian-based models. In the current book, we do not pay any particular attention to structural models but focus on what is called *reduced-form models*, which aim at explaining stylized *statistical and probabilistic* features of commodity market prices.

Of the above-mentioned three monographs, in particular Aïd [1] presents a detailed introduction to the mechanisms of electricity markets. More general overviews of power markets are provided by Geman [111] and Eydeland and Wolyniec [97], alongside with a presentation of many of the classical pricing models and empirics for these. These two monographs also give a detailed analysis of the more financial and economic sides of power markets. We note that Geman [111] is not restricting herself to power, but also includes a rich analysis of the commodity markets for agriculture and metals.

The pricing paradigms in the treatments mentioned above were first introduced in a rigorous way by Clewlow and Strickland [72]. Clewlow and Strickland [72] brought over ideas from interest rate theory and formulated the spot-based and Heath-Jarrow-Morton-based forward pricing approaches in a commodity setting. Already here jump processes are introduced as a necessary modeling device in power markets, along with various stochastic volatility models giving motivation to further applied and theoretical studies leading up to our infinite dimensional framework.

Indeed, we have also been influenced by the infinite dimensional interest rate theory of Filipović [100]. Although our book does not cover Heath-Jarrow-Morton models for forward rates, we make use of the same state space as developed in Filipović [100] in our analysis of forward prices in commodity markets. We introduce this space and its properties, but also develop a theory on linear operators, in particular Hilbert-Schmidt operators, on this space which are applicable to defining covariance and volatility operators. Moreover, linear operators also appear when studying forward contracts with delivery of the underlying commodity over a specified time period, the typical contract in electricity markets, say. Filipović [100] focuses on finite dimensional noise drivers, whereas a general infinite dimensional approach is developed by Carmona and Tehranchi [69]. In the latter monograph, a general term structure infinite dimensional model with cylindrical Gaussian noise is proposed. The main focus is on consistency and finite dimensional realizations of the model, as well as bond portfolio theory, mostly involving models with finite dimensional Gaussian noise.

The main difference between modeling the dynamics of forward prices on commodities and forward rates in fixed income theory is the no-arbitrage drift condition. For commodities, this condition takes a much simpler form than in the interest rate case, where the volatility enters the drift in a non-trivial way. We believe, however, that the material presented in this book may be useful also in further studies in interest rate theory beyond Filipović [100] and Carmona and Tehranchi [69]. Apart from the analysis of linear operators, the theory on infinite dimensional Lévy processes creates a basis for new forward rate models, to mention one aspect. We refer to Eberlein et al. [90] for Lévy processes in interest rate theory.

As a final note, we would like to mention the recent transfer away from LIBOR rates to new risk-free rates like SOFR and SONIA in fixed income markets. The CME offers trade in forward contracts on the (arithmetic and geometric) average of these overnight rates, which builds a bridge from interest rate markets over to the theory presented in this book. We refer to Skov and Skovmand [180] for a recent contribution.

1.6 Some Frequently Used Notation

Let $\mathbb{F}$ be $\mathbb{R}$ or $\mathbb{C}$. For a measure space $(A, \mathcal{A}, \mu)$, we denote by $L^p(A, \mathcal{A}, \mu)$ the space of measurable functions $f : A \to \mathbb{F}$ such that $\int_A |f(x)|^p \mu(dx) < \infty$, $p \in \mathbb{N}$. Frequently we write simply $L^p(A)$ when the σ-algebra $\mathcal{A}$ and the measure μ is clear from the context. In particular, we write $L^2(\mathbb{R})$ for the space $L^2(\mathbb{R}, \mathcal{B}_{\mathbb{R}}, Leb)$, where Leb is the Lebesgue measure and $\mathcal{B}_{\mathbb{R}}$ is the Borel σ-algebra. Usually it is clear from the context whether we choose $\mathbb{F}$ to be the complex or real numbers, and we will not precise this further.

For a separable Hilbert space H, we will stick to the generic notation $|\cdot|$ for its norm, and let the inner product be $\langle\cdot,\cdot\rangle$. We write ONB for an *orthonormal basis* in the Hilbert space. Frequently, we a add sub-index to the norm or inner product to signify a specific Hilbert space. For bounded linear operators $Q \in L(H, K)$ between two separable Hilbert spaces H and K, we denote by $\|Q\|_{\text{op}}$ its operator norm. If $H = K$, we simply write $L(H)$ instead of $L(H, H)$. We denote by $\|Q\|_{\text{HS}}$ the Hilbert-Schmidt norm, with $\langle\cdot,\cdot\rangle_{\text{HS}}$ being the inner product. The space of all Hilbert-Schmidt operators is denoted $L_{\text{HS}}(H, K)$, with $L_{\text{HS}}(H) = L_{\text{HS}}(H, H)$.

We will throughout the book assume that we have given a filtered probability space $(\Omega, \mathcal{F}, (\mathcal{F}_t)_{t\geq 0}, \mathbb{P})$, satisfying the *usual conditions*. A *random variable* X with state space H is a measurable map from Ω into H, where H is (here usually) a separable Hilbert space equipped with its Borel σ-algebra. X is said to be *integrable* if $\mathbb{E}[|X|] < \infty$, where $\mathbb{E}[\cdot]$ is the expectation operator with respect to the probability $\mathbb{P}$. By square integrable, we mean that $|X|^2$ has finite expectation. We use the abbreviation IID for a sequence $(X_i)_{i\in\mathbb{N}}$ of *independent and identically distributed* random variables. A stochastic process $(X(t))_{t\geq 0}$ is said to be RCLL if it has paths which are right-continuous with left limits.

By ∂_x we mean $\partial/\partial x$, the partial derivative with respect to the variable x. If $\mathbf{x} \in \mathbb{R}^n$, we intepret $\mathbf{x}$ as a column vector and its transpose $\mathbf{x}^\top$ as a row vector.

Part I
Mathematical Tools

Chapter 2
Lévy processes on Hilbert Spaces

Abstract We introduce and study Lévy process in Hilbert space. These processes are the basic noise drivers in the forward price dynamics. Explicit constructions based on subordination of Wiener process to define normal inverse Gaussian, stable and variance-gamma Lévy processes with values in Hilbert space are provided.

When modelling the evolution of forward prices in energy markets, we will use Hilbert space valued Lévy processes to describe the noise in the dynamics. In this chapter we build up the theory for defining such processes using the method of subordination, which paves the way for explicit construction of many interesting special classes of Lévy processes as well as an analytic treatment of their probabilistic features.

Hilbert space valued Lévy processes can be applied to modelling of the spatio-temporal dynamics of weather variables like wind and temperature, and the evolution of forward rates in fixed income markets. Other areas of application of Hilbert space valued Lévy processes include quantum physics and turbulence (see for instance Barndorff-Nielsen and Schmiegel [17] for an approach to turbulence modelling based on ambit fields). A general approach to forward price modelling in energy markets using ambit fields can be found in Barndorff-Nielsen, Benth and Veraart [14], where empirical studies argue for spatio-temporal Lévy processes as the modelling device for the stochastic evolution. Empirical studies of the spatio-temporal dynamics of temperatures and wind speeds also show signs of a Lévy random field structure of normal inverse Gaussian type (see Benth and Šaltytė Benth [48, 49]). Carmona and Teranchi [69] and Peszat and Zabczyk [165] discuss infinite dimensional modelling of forward rates in fixed income markets, where in particular the latter work applies Hilbert space valued Lévy processes.

The theory presented in this chapter is based on Benth and Krühner [35], and follows the presentation there closely. We discuss the construction of *multivariate* Lévy processes, in the sense that we consider Lévy processes in a product Hilbert space. The reason for using a product Hilbert space as the state space is to accomodate for cross-commodity models, where the dynamics of the forward curves in different markets are specified in a joint curve model. This is conveniently

F. E. Benth, P. Krühner, *Stochastic Models for Prices Dynamics in Energy and Commodity Markets*, Springer Finance,
https://doi.org/10.1007/978-3-031-40367-5_2

formulated in a product space. At the end of the Chapter, we include some analysis of the covariance operator of bivariate Wiener processes in Hilbert space.

2.1 Subordinated Hilbert Space Valued Lévy Processes

In this section we define subordinated Lévy processes with values in a Hilbert space. Let us first start to introduce some notation that will be convenient. We let H_j be a separable Hilbert space with inner product $\langle\cdot,\cdot\rangle_j$ and associated norm $|\cdot|_j$. Assume U_j to be an H_j-valued Lévy process with RCLL (or càdlàg[1]) paths for $j = 1, \dots, d$, $d \in \mathbb{N}$, such that $U_1, \dots, U_d$ are independent, cf. Peszat and Zabczyk [165, Section 4]. Further, (b_j, Q_j, ν_j) denotes the characteristic triplet of U_j, with $b_j \in H_j$, Q_j a symmetric positive-definite trace class operator on H_j and ν_j being a Lévy measure on H_j, e.g., a non-negative measure concentrated on $H_j\backslash\{0\}$ such that

$$\int_{H_j} \{|u|_j^2 \wedge 1\}\, \nu_j(du) < \infty\,.$$

According to Parthasarathy [162, Theorem VI.4.10], such a triplet exists and is unique. We recall from Peszat and Zabcyzk [165, Section 4.6] the *Lévy exponent* $\varphi_j : H_j \to \mathbb{C}$ of U_j to be

$$\mathbb{E}\left[e^{i\langle U_j(t),u\rangle_j}\right] = \exp(t\varphi_j(u))\,, \tag{2.1}$$

for any $t \in \mathbb{R}_+$, $u \in H_j$, for all $j = 1, \dots, d$. The left-hand side in (2.1) is referred to as the *characteristic functional* of the process U_j. Like in the finite dimensional case there is a connection between the characteristic triplet of a Lévy process and its Lévy exponent. It follows from Peszat and Zabczyk [165, Theorem 4.27] that

$$\varphi_j(u) = i\langle u, b_j\rangle_j - \frac{1}{2}\langle Q_j u, u\rangle_j + \int_{H_j} \left\{e^{i\langle u,x\rangle_j} - 1 - i\langle u, \chi(x)\rangle_j\right\} \nu_j(dx)\,,$$

for any $u \in H_j$ and every $j = 1, \dots, d$. Here and throughout we work with the truncation function $\chi(x) := x1_{\{|x|\le 1\}}$, where $|\cdot|$ generically refers to the norm induced by the inner product $\langle\cdot,\cdot\rangle$. The intended inner product will be clear from the context.

Define $\mathbf{U} := (U_1, \dots, U_d)$, $H := H_1 \times \cdots \times H_d$ and $\langle \mathbf{u}, \mathbf{v}\rangle := \sum_{j=1}^d \langle u_j, v_j\rangle_j$ for $\mathbf{u}, \mathbf{v} \in H$. For $\mathbf{a} \in \mathbb{R}^d$ and $\mathbf{u} \in H$ we define $\mathbf{au} := (a_1u_1, \dots, a_du_d) \in H$. For bounded linear operators $\mathcal{T}_1, \dots, \mathcal{T}_d$ on $H_1, \dots, H_d$ we define $\mathcal{T}_1 \times \cdots \times \mathcal{T}_d : H \to H, \mathbf{u} \mapsto (\mathcal{T}_1u_1, \dots, \mathcal{T}_du_d)$ and for $\mathbf{a} \in \mathbb{R}_+^d$ and $\mathcal{T} := \mathcal{T}_1 \times \cdots \times \mathcal{T}_d$ we

[1] the French acronym *càdlàg* means continuous from the right, limits from the left.

also define $\mathbf{a}\mathcal{T} := a_1\mathcal{T}_1 \times \cdots \times a_d\mathcal{T}_d$. Let $(\mathbf{b}, Q, \nu)$ be the characteristic triplet of $\mathbf{U}$, which can be expressed in the characteristic triplets of $U_1, \ldots, U_d$: It holds that $\mathbf{b} = (b_1, \ldots, b_d), Q = Q_1 \times \cdots \times Q_d$ and

$$\nu(A) = \sum_{j=1}^{d} \nu_j^{\eta_j}(A)$$

for any $A \in \mathcal{B}(H)$ where η_j is the natural embedding from H_j into H, e.g., $\eta_1 : H_1 \to H$ is such that $\eta_1(u) = (u, 0, \ldots, 0)$. For $\boldsymbol{\theta} \in \mathbb{R}_+^d$ we define $\mathbf{U}(\boldsymbol{\theta}) := (U_1(\theta_1), \ldots, U_d(\theta_d))$.

Remark 2.1 Let $(\mathcal{F}_t)_{t\geq 0}$ be a filtration such that $\mathbf{U}(t) - \mathbf{U}(s)$ is independent of $\mathcal{F}_s$, $t \geq s$. Then the following statements are equivalent.

- $\mathbf{U}$ is a $(\mathcal{F}_t)_{t\geq 0}$-martingale.
- $\mathbf{U}$ is a martingale with respect to its own (right-continuous) filtration.
- $\mathbf{U}$ is mean zero, i.e. $\mathbf{U}$ has finite expectation and $\mathbb{E}[\mathbf{U}(t)] = 0$ for any $t \geq 0$.
- $\int_{\{|\mathbf{x}|>1\}} |\mathbf{x}|\, \nu(d\mathbf{x}) < \infty$ and $0 = \mathbf{b} + \int_{\{|\mathbf{x}|>1\}} \mathbf{x}\, \nu(d\mathbf{x})$.

The following projection result does not rely on the specific construction of H:

Proposition 2.1 *Let $\mathcal{L} \in L(H, \mathbb{R}^n)$ for $n \in \mathbb{N}$, i.e., $\mathcal{L}$ is a bounded linear operator from H to $\mathbb{R}^n$. If $\mathbf{U}$ is an H-valued Lévy process, then $X := \mathcal{L}(\mathbf{U})$ is an $\mathbb{R}^n$-valued Lévy process with RCLL (càdlàg) paths.*

Proof Obviously, by linearity of $\mathcal{L}$, $X(0) = \mathcal{L}(\mathbf{U}(0)) = \mathcal{L}(0) = 0$. Furthermore, as $\mathcal{L}$ is continuous, the càdlàg paths of $\mathbf{U}$ is inherited by X.

Since the increments of $\mathbf{U}(t) - \mathbf{U}(s)$ for $t \geq s$ is independent of $\mathcal{F}_s$. We find for $z \in \mathbb{R}^n$ and $t \geq s \geq 0$,

$$\begin{aligned}
\mathbb{E}[e^{iz^\top (X(t)-X(s))} \mid \mathcal{F}_s] &= \mathrm{e}[e^{iz^\top \mathcal{L}(\mathbf{U}(t)- \mathcal{L}(\mathbf{U}(s))} \mid \mathcal{F}_s] \\
&= \mathbb{E}[e^{i\langle \mathbf{U}(t)-\mathbf{U}(s), \mathcal{L}^* z\rangle} \mid \mathcal{F}_s] \\
&= \mathbb{E}[e^{i\langle \mathbf{U}(t)-\mathbf{U}(s), \mathcal{L}^* z\rangle}] \\
&= \mathbb{E}[e^{i\langle \mathbf{U}(t-s), \mathcal{L}^* z\rangle}] \\
&= \mathbb{E}[e^{iz^\top \mathcal{L}(\mathbf{U}(t-s))}] \\
&= \mathbb{E}[e^{iz^\top X(t-s)}].
\end{aligned}$$

Here, $\mathcal{L}^* \in L(\mathbb{R}^n, H)$ is the dual operator of $\mathcal{L}$. In the third and fourth equality we used the independent and stationary increment property of $\mathbf{U}$. We conclude that X has independent and stationary increments.

It holds that $|X(t) - X(s)| \leq \|\mathcal{L}\|_{\text{op}}|\mathbf{U}(t) - \mathbf{U}(s)|$. Hence, if $|X(t) - X(s)| > \epsilon$, then $|\mathbf{U}(t) - \mathbf{U}(s)| > \epsilon/\|\mathcal{L}\|_{\text{op}}$, and

$$\lim_{t \to s} \mathbb{P}(|X(t) - X(s)| > \epsilon) \leq \lim_{t \to s} \mathbb{P}(|\mathbf{U}(t) - \mathbf{U}(s)| > \epsilon/\|\mathcal{L}\|_{\text{op}}) = 0,$$

by stochastic continuity of $\mathbf{U}$. Stochastic continuity of X follows, and thus X is an $\mathbb{R}^n$-valued Lévy process. □

Notice that for any $u \in H$, $z^\top \mathcal{L}(u) = \langle u, \mathcal{L}^*(z) \rangle$, and thus there exist $f_1, \dots, f_n \in H$ such that $\mathcal{L}(u) = (\langle u, f_1 \rangle, \dots, \langle u, f_n \rangle)$, namely, $f_i := \mathcal{L}^*(e_i)$ where e_i is the ith canonical basis vector in $\mathbb{R}^n$.

Let $\boldsymbol{\Theta}$ be a Lévy process independent of $\mathbf{U}$, with càdlàg paths taking values in $\mathbb{R}^d$ such that Θ_j is a subordinator for all $j = 1, \dots, d$ (cf. Sato [175, Definition 21.4] or Skorokhod [179]). We recall that a subordinator is a Lévy process with increasing paths, which means a pure-jump Lévy process with only positive jumps and a non-negative drift. Since the paths of $\boldsymbol{\Theta}$ are of bounded variation, the characteristic triplet of $\boldsymbol{\Theta}$ is $(\mathbf{a}, 0, \kappa)$ with $\int_{\mathbb{R}^d_+} \chi(\boldsymbol{\theta})\,\kappa(d\boldsymbol{\theta}) < \infty$ (see Sato [175, Theorem 21.9]). Define

$$\mathbf{a}_0 := \mathbf{a} - \int_{\mathbb{R}^d_+} \chi(\boldsymbol{\theta})\,\kappa(d\boldsymbol{\theta}), \tag{2.2}$$

and the function $\psi : (\mathbb{R}_- + i\mathbb{R})^d \to \mathbb{C}$

$$\psi(\mathbf{z}) = \mathbf{a}_0\mathbf{z} + \int_{\mathbb{R}^d_+} \{e^{\mathbf{z}\boldsymbol{\theta}} - 1\}\,\kappa(d\boldsymbol{\theta}). \tag{2.3}$$

From Sato [175, Theorem 8.1] it can be seen that

$$\mathbb{E}[e^{\mathbf{z}\boldsymbol{\Theta}(1)}] = \exp(\psi(\mathbf{z})),$$

for any $\mathbf{z} \in (\mathbb{R}_- + i\mathbb{R})^d$ and Sato [175, Theorem 21.5] yields $\mathbf{a}_0 \in (\mathbb{R}_+)^d$ and κ is concentrated on $(\mathbb{R}_+)^d$. We refer to ψ in (2.3) as the *Laplace exponent* of the subordinator $\boldsymbol{\Theta}$.

The *multivariate subordinated Lévy process* is defined to be

$$\mathbf{L}(t) := (U_1(\Theta_1(t)), \dots, U_d(\Theta_d(t))) = \mathbf{U}(\boldsymbol{\Theta}(t)), \tag{2.4}$$

for any $t \geq 0$. Since $\Theta_1, \dots, \Theta_d$ and $U_1, \dots, U_d$ all have càdlàg paths, the process $\mathbf{L}$ has càdlàg paths. We recall that multivariate subordination of $\mathbb{R}^d$-valued Lévy processes has been analysed in Barndorff-Nielsen, Pedersen and Sato [16], while the extension to Hilbert space valued processes (and even to infinite variate subordination) can be found in Pérez-Abreu and Rocha-Arteaga [164]. In the remaining of this section, we recall some statements from Pérez-Abreu and Rocha-Arteaga [164] which are useful to our setting and further analysis.

The Lévy exponent of the subordinated Lévy process $\mathbf{L}$ can be expressed in the Lévy exponent of the original Lévy processes $U_1, \ldots, U_d$ and the Laplace exponent of the subordinator $\boldsymbol{\Theta}$. The precise result is stated in the following theorem:

Theorem 2.1 *The process* $\mathbf{L}$ *is a Lévy process and its Lévy exponent* $\rho : H \to \mathbb{C}$ *is given by*

$$\rho(\mathbf{u}) = \psi\left((\varphi_1(u_1), \ldots, \varphi_d(u_d))\right),$$

where ψ *is the Laplace exponent of* $\boldsymbol{\Theta}$ *in* (2.3) *and* φ_j *is the Lévy exponent of* U_j, $j = 1, \ldots, d$.

Proof The result follows from a straightforward adaption of the corresponding finite dimensional statement in Barndorff-Nielsen, Pedersen and Sato [16, Theorem 3.3]. □

Moreover, the characteristic triplet of the subordinated Lévy process can be expressed in terms of the triplets of the original Lévy processes, the subordinators and the distribution of the original Lévy processes:

Theorem 2.2 *The characteristic triplet of* $\mathbf{L}$ *is* $(\boldsymbol{\gamma}, \Gamma, \ell)$, *where*

$$\begin{aligned}
\boldsymbol{\gamma} &= \mathbf{a}_0 \mathbf{b} + \int_{\mathbb{R}_+^d} \mathbb{E}[\chi(\mathbf{U}(\boldsymbol{\theta}))]\, \kappa(d\boldsymbol{\theta}), \\
\Gamma &= \mathbf{a}_0 Q, \\
\ell(A) &= \sum_{j=1}^{d} a_{0,j} \nu_j^{\eta_j}(A) + \int_{\mathbb{R}_+^d} \mathbb{P}^{\mathbf{U}(\boldsymbol{\theta})}(A)\, \kappa(d\boldsymbol{\theta})
\end{aligned}$$

for any Borel sets $A \subseteq H$, *with* $\mathbb{P}^{\mathbf{U}(\boldsymbol{\theta})}$ *being the probability distribution of* $\mathbf{U}(\boldsymbol{\theta})$, $(\mathbf{b}, Q, \nu)$ *and* $(\mathbf{a}, 0, \kappa)$ *the characteristic triplets of* $\mathbf{U}$ *and* $\boldsymbol{\Theta}$, *respectively, and* $\mathbf{a}_0$ *defined in* (2.2). *Moreover, it holds that* $\int_{\mathbb{R}_+^d} |\mathbb{E}[\chi(\mathbf{U}(\boldsymbol{\theta}))]|\, \kappa(d\boldsymbol{\theta}) < \infty$.

Proof The conclusion of the theorem is a special case of the statement in Pérez-Abreu and Rocha-Arteaga [164, Theorem 5.1]. We note that the restrictive statement in [164, Equation (38)] forces all Lévy measures appearing there to be finite. However, by careful inspection, neither the proof nor any other statement in that paper needs this restrictive condition on the Lévy measures. Indeed, it is sufficient to require it only for those measurable sets which have a positive distance to the origin. We can thus apply [164, Theorem 5.1] to show our result. □

2.2 Growth Behaviour

In this section we analyse the expected multi-time growth behaviour of integrable Hilbert space valued Lévy processes. The growth properties derived here are interesting in their own right, but will also be used in the next section.

First, let us define the *growth function* of $\mathbf{U}$:

Definition 2.1 The *growth function* $g_{\mathbf{U}} : \mathbb{R}_+^d \to \mathbb{R}_+ \cup \{\infty\}$ of the process $\mathbf{U}$ is the function

$$g_{\mathbf{U}}(\boldsymbol{\theta}) = \mathbb{E}[|\mathbf{U}(\boldsymbol{\theta})|] .$$

If $\mathbf{U}$ is integrable, then obviously $g_{\mathbf{U}}(\boldsymbol{\theta}) < \infty$ for any $\boldsymbol{\theta} \in \mathbb{R}_+^d$. The growth function $g_{\mathbf{U}}$ satisfies some growth and monotonicity properties, as the next Lemma shows:

Lemma 2.1 *Assume* $\mathbf{U}$ *is integrable and* $\boldsymbol{\theta}_1, \boldsymbol{\theta}_2 \in \mathbb{R}_+^d$. *Then*

$$g_{\mathbf{U}}(\boldsymbol{\theta}_1 + \boldsymbol{\theta}_2) \le g_{\mathbf{U}}(\boldsymbol{\theta}_1) + g_{\mathbf{U}}(\boldsymbol{\theta}_2) .$$

If moreover $\mathbf{U}$ *has mean zero, then*

$$g_{\mathbf{U}}(\boldsymbol{\theta}_1) \le g_{\mathbf{U}}(\boldsymbol{\theta}_1 + \boldsymbol{\theta}_2) .$$

Proof First, note by the triangle inequality that

$$|\mathbf{U}(\boldsymbol{\theta}_1 + \boldsymbol{\theta}_2)| \le |\mathbf{U}(\boldsymbol{\theta}_1 + \boldsymbol{\theta}_2) - \mathbf{U}(\boldsymbol{\theta}_1)| + |\mathbf{U}(\boldsymbol{\theta}_1)| .$$

Moreover, but stationarity of increments,

$$U_j(\theta_{1,j} + \theta_{2,j}) - U_j(\theta_{1,j}) \stackrel{d}{=} U_j(\theta_{2,j})$$

with $\stackrel{d}{=}$ meaning equality in distribution. Hence,

$$|\mathbf{U}(\boldsymbol{\theta}_1 + \boldsymbol{\theta}_2) - \mathbf{U}(\boldsymbol{\theta}_1)| \stackrel{d}{=} |\mathbf{U}(\boldsymbol{\theta}_2)| ,$$

and the first claim follows.

As for the second claim, it follows by Peszat and Zabczyk [165, Prop. 3.15 (vi)] that

$$|\mathbb{E}\left[\mathbf{U}(\boldsymbol{\theta}_1 + \boldsymbol{\theta}_2) \,\middle|\, \mathcal{F}_{\boldsymbol{\theta}_1}\right]| \le \mathbb{E}\left[|\mathbf{U}(\boldsymbol{\theta}_1 + \boldsymbol{\theta}_2)| \,\middle|\, \mathcal{F}_{\boldsymbol{\theta}_1}\right] ,$$

where $\mathcal{F}_{\boldsymbol{\theta}_1}$ is the σ-algebra generated by $\mathbb{R}_+^d \ni \boldsymbol{\theta} \mapsto \mathbf{U}(\boldsymbol{\theta})$ for $\boldsymbol{\theta} \leq \boldsymbol{\theta}_1$.[2] Hence,

$$\begin{aligned}\mathbb{E}\left[\mathbf{U}(\boldsymbol{\theta}_1+\boldsymbol{\theta}_2)\,\middle|\,\mathcal{F}_{\boldsymbol{\theta}_1}\right] &= \mathbb{E}\left[\mathbf{U}(\boldsymbol{\theta}_1+\boldsymbol{\theta}_2)-\mathbf{U}(\boldsymbol{\theta}_1)+\mathbf{U}(\boldsymbol{\theta}_2)\,\middle|\,\mathcal{F}_{\boldsymbol{\theta}_1}\right]\\ &= \mathbb{E}\left[\mathbf{U}(\boldsymbol{\theta}_1+\boldsymbol{\theta}_2)-\mathbf{U}(\boldsymbol{\theta}_1)\right]+\mathbf{U}(\boldsymbol{\theta}_1)\,.\end{aligned}$$

The last equality follows from the independent increment property of $\mathbf{U}$ and $\mathcal{F}_{\boldsymbol{\theta}_1}$-measurablity of $\mathbf{U}(\boldsymbol{\theta}_1)$. From the assumption of zero mean and the tower property of conditional expectation, if follows

$$\mathbb{E}\left[|\mathbf{U}(\boldsymbol{\theta}_1)|\right] \leq \mathbb{E}\left[|\mathbf{U}(\boldsymbol{\theta}_1+\boldsymbol{\theta}_2)|\right]\,,$$

and we conclude the second assertion. □

We notice that If g_{U_j} is the growth function of U_j, then we have $g_{\mathbf{U}}(\boldsymbol{\theta}) \leq \sum_{j=1}^d g_{U_j}(\theta_j) \leq \sqrt{d}\, g_{\mathbf{U}}(\boldsymbol{\theta})$ for any $\boldsymbol{\theta} \in \mathbb{R}_+^d$.

Lemma 2.2 *Let $g_{\mathbf{U}}$ be the growth function of $\mathbf{U}$ and assume $g_{\mathbf{U}}(\boldsymbol{\Theta}(1))$ is integrable. Then $\mathbf{L}$ is integrable.*

Proof By the tower property of conditional expectation, we have

$$\begin{aligned}\mathbb{E}[|\mathbf{L}(1)|] &= \mathbb{E}[|\mathbf{U}(\boldsymbol{\Theta}(1))|]\\ &= \mathbb{E}\left[\mathbb{E}\left[|\mathbf{U}(\boldsymbol{\Theta}(1))|\,\middle|\,\boldsymbol{\Theta}(1)\right]\right]\\ &= \mathbb{E}[g_{\mathbf{U}}(\boldsymbol{\Theta}(1))]\\ &< \infty\,.\end{aligned}$$

The lemma follows. □

If $\mathbf{U}$ is a mean zero square integrable Lévy process, then it can be shown that its growth function $g_{\mathbf{U}}$ increases like a root function.

Proposition 2.2 *Let $\mathbf{U}$ be a zero mean square integrable Lévy process and $\sigma_j^2 := \mathbb{E}\left[|U_j(1)|^2\right]$ for $j = 1, \ldots, d$. Then for the growth function $g_{\mathbf{U}}$ of $\mathbf{U}$ there exist $\boldsymbol{v} \in \mathbb{R}_+^d$ and $c \in (0, 1]$ such that*

$$c\sqrt{\boldsymbol{\theta}^\top \boldsymbol{\sigma}^2} \leq g_{\mathbf{U}}(\boldsymbol{\theta}) \leq \sqrt{\boldsymbol{\theta}^\top \boldsymbol{\sigma}^2}\,,$$

for any $\boldsymbol{\theta} \in \mathbb{R}_+^d$ with $\theta_j \geq v_j$, $j = 1, \ldots, d$.

Proof By the Cauchy-Schwarz inequality, it holds that

$$\mathbb{E}[|\mathbf{U}(\boldsymbol{\theta})|] \leq \mathbb{E}[|\mathbf{U}(\boldsymbol{\theta})|^2]^{1/2} = \left(\sum_{j=1}^d \mathbb{E}[|U_j(\theta_j)|_j^2]\right)^{1/2}.$$

[2] By $\boldsymbol{\theta} \leq \boldsymbol{\theta}_1$ we mean that $\theta_j \leq \theta_{1,j}$ for all $j = 1, \ldots, d$.

But from Peszat and Zabczyk [165, Thm. 4.44], $\mathbb{E}[|U_j(\theta_j)|_j^2] = \theta_j\sigma_j^2$ since $\mathbf{U}$ is square integrable. Hence, the upper bound on $g_{\mathbf{U}}$ holds.

Let us next show the lower bound on the growth function. Consider the case $d = 1$. Since the claim is trivial if $\sigma^2 = 0$ we may assume that $\sigma^2 > 0$. Let $e \in H$ be normed and define $M(\theta) := \langle U(\theta), e\rangle$, $\theta \geq 0$, which is a real-valued mean zero square integrable Lévy process. Let $s^2 := \mathbb{E}\left[|M(1)|^2\right]$ and $c_1 := \frac{s^2}{\sigma^2}$. Since e is normed, we find that $s^2 \leq \sigma^2$, and thus $c_1 < 1$. For any bounded function $g : \mathbb{R} \to \mathbb{R}_+$ such that $g(x) \leq |x|$, we find

$$\mathbb{E}\left[|M(\theta)|\right] = \sqrt{\theta s^2}\mathbb{E}\left[\left|\frac{M(\theta)}{\sqrt{\theta s^2}}\right|\right] \geq \sqrt{\theta s^2}\mathbb{E}\left[g\left(\frac{M(\theta)}{\sqrt{\theta s^2}}\right)\right]$$

for any $\theta \geq 0$. Furthermore, by the central limit theorem,

$$\lim_{\theta\to\infty} \mathbb{E}\left[g\left(\frac{M(\theta)}{\sqrt{\theta s^2}}\right)\right] = \mathbb{E}\left[g(Y)\right] ,$$

where Y is a standard normal random variable. Now, choose $g(x) := |x|1_{\{|x|\leq K\}}$ where $K > 0$ is such that for given $\epsilon > 0$ it holds that

$$\int_{-K}^{K} |y|\frac{1}{\sqrt{2\pi}} \exp(-y^2/2)dy > \epsilon .$$

Define $c_2 := \int_{-K}^{K} |y|\frac{1}{\sqrt{2\pi}} \exp(-y^2/2)dy - \epsilon > 0$. Consequently, we can find $\upsilon \geq 0$ such that

$$\mathbb{E}\left[g\left(\frac{M(\theta)}{\sqrt{\theta s^2}}\right)\right] \geq \mathbb{E}\left[g(Y)\right] - \epsilon ,$$

for any $\theta \geq \upsilon$. Hence,

$$\begin{aligned}\mathbb{E}\left[|U(\theta)|\right] &\geq \mathbb{E}\left[|M(\theta)|\right] \\ &\geq \sqrt{\theta s^2}(\mathbb{E}\left[g(Y)\right] - \epsilon) \\ &= c_2\sqrt{\theta s^2} \\ &= c_2\sqrt{c_1}\sqrt{\theta\sigma^2}\end{aligned}$$

for any $\theta \in \mathbb{R}_+$ with $\theta \geq \upsilon$. By tuning K and ϵ, we can obtain c_2 to be less than 1. Hence, the lower bound follows in the case $d = 1$. The general case $d > 1$ follows by a simple extension of the above arguments. □

We can now state a bound for the growth function of a square-integrable Lévy process which shows that it grows slower than a linear function after a long time and slower than a square-root function in short time.

Proposition 2.3 *Assume* $\mathbf{U}$ *is integrable. Then there are constants* $c_1, c_2 \in \mathbb{R}_+$ *such that the growth function* $g_{\mathbf{U}}$ *of* $\mathbf{U}$ *admits the inequality*

$$g_{\mathbf{U}}(\boldsymbol{\theta}) \leq c_1|\boldsymbol{\theta}| + c_2|\boldsymbol{\theta}|^{1/2},$$

for any $\boldsymbol{\theta} \in \mathbb{R}_+^d$.

Proof From the Lévy-Kintchine decomposition (see Peszat and Zabczyk [165, Theorem 4.23]), the Lévy process $\mathbf{U}$ can be written as a sum of a deterministic Lévy process (the drift) $\mathbf{A}(\boldsymbol{\theta}) := \mathbf{b}\boldsymbol{\theta}$ for some $\mathbf{b} \in H$, a square integrable martingale $\mathbf{M}$ which is a Lévy process, and an integrable compound Poisson process $\mathbf{C}$. The triangle inequality for norms allows us to treat each summand separately. The drift $\mathbf{A}$ obviously gives rise to a linear growth as $|\mathbf{A}(\boldsymbol{\theta})| \leq |\mathbf{b}||\boldsymbol{\theta}|$. From Prop. 2.2 we know that the square integrable martingale part $\mathbf{M}$ has a square root growth behaviour. Each coordinate in the compound Poisson process $\mathbf{C}(\boldsymbol{\theta})$ can be represented as

$$C_j(\theta_j) = \sum_{i=1}^{N_j(\theta_j)} Z_{i,j}$$

where $(Z_{i,j})_{i\in\mathbb{N}}$ is a sequence of IID random variables with values in H_j and N_j is a (real-valued) Poisson process. Hence, with Z_j being a generic random variable which is equal to $Z_{i,j}$ in distribution for all $i \in \mathbb{N}$,

$$\mathbb{E}\left[|C_j(\theta_j)|_j\right] \leq \mathbb{E}\left[\left[\sum_{i=1}^{N_j(\theta_j)} |Z_{i,j}|_j \;\middle|\; N_j(\theta_j)\right]\right] = \mathbb{E}\left[N_j(\theta_j)\right]\mathbb{E}\left[|Z_j|_j\right].$$

This shows the linear scaling induced by the compound Poisson process, and the result follows. □

We note that for a one-dimensional standard Brownian motion B, then $\mathbb{E}[|B(\theta)|] = \sqrt{\theta}\mathbb{E}[|B(1)|]$ for any $\theta \in \mathbb{R}_+$. If $U(\theta) = \theta$ for any $\theta \in \mathbb{R}_+$, then trivially $\mathbb{E}[|U(\theta)|] = \theta$. These two examples show that the powers appearing for the long, respectively short term growth bound for the growth function are sharp.

We end this section by collecting two results on integrability and subadditive or multiplicative functions which will be very useful for proving existence of the first and second moment of $\mathbf{L}$. First, a definition (extending Sato [175, Definition 25.2] for submultiplicative functions):

Definition 2.2 A function $f : H \to \mathbb{R}_+$ is called *submultiplicative* if $f(\mathbf{x}+\mathbf{y}) \leq af(\mathbf{x})f(\mathbf{y})$ for some constant $a > 0$. A function $f : H \to \mathbb{R}_+$ is called *subadditive* if $f(\mathbf{x}+\mathbf{y}) \leq f(\mathbf{x}) + f(\mathbf{y})$.

The next theorem generalizes Sato [175, Theorem 25.3] to Hilbert space valued Lévy processes.

Theorem 2.3 *Let f be a measurable, submultiplicative function which is bounded on a neighbourhood of zero in H. Then $\mathbb{E}[f(\mathbf{U}(t))] < \infty$ for some (and hence all) $t > 0$ if and only if*

$$\int_{\{|\mathbf{x}|>1\}} f(\mathbf{x})\,\nu(d\mathbf{x}) < \infty.$$

Proof This is a straightforward adaption of the finite dimensional case, cf. Sato [175, Theorem 25.3]. □

Remark 2.2 Sato [175] assumes in Theorem 25.3 that the submultiplicative function is *locally bounded*, meaning that it is bounded on every compact. A careful inspection of the proof reveals that the submultiplicative function only needs to be bounded on the unit ball (see Lemma 25.5 in [175]).

As in the finite dimensional case the next Proposition follows straightforwardly from Theorem 2.3, and we skip the proof.

Proposition 2.4 *Let $f : H \to \mathbb{R}_+$ be a measurable, subadditive function which is bounded in a neighbourhood of zero of H. Then $f(\mathbf{U}(t))$ is integrable for some (and hence all) $t > 0$ if and only if*

$$\int_{\{|\mathbf{x}|>1\}} f(\mathbf{x})\,\nu(d\mathbf{x}) < \infty\,.$$

In particular, $\mathbb{E}[|\mathbf{U}(1)|] < \infty$ if and only if

$$\int_{\{|\mathbf{x}|>1\}} |\mathbf{x}|\,\nu(d\mathbf{x}) < \infty\,.$$

We remark that both the above results will be used to characterise integrability of $\mathbf{L}$. In particular, we will apply Proposition 2.4 and Theorem 2.3 for $\mathbf{U}$ (as formulated), $\mathbf{L}$ (with the obvious change of Lévy measure) and $\boldsymbol{\Theta}$ (letting $H = \mathbb{R}^d_+$). For the latter, we recall from Lemma 2.1 that the growth function $g_{\mathbf{U}}$ is subadditive on $\mathbb{R}^d_+$.

2.3 Integrability, Mean and Covariance of Subordinated Lévy Processes

In this section we investigate the probabilistic features of the subordinated Lévy process $\mathbf{L}(t) = \mathbf{U}(\boldsymbol{\Theta}(t))$, i.e. we give necessary and sufficient conditions for $\mathbf{L}$ to have finite first or second moment and provide formulas for those moments.

In particular, in the latter case, this provides us with explicit knowledge of the covariance operator of $\mathbf{L}$.

Following the analysis in Peszat and Zabczyk [165, Section 4.9] one can characterise finiteness of the second moment of $\mathbf{L}$ completely in terms of the moments of $\mathbf{U}$ and $\boldsymbol{\Theta}$. If $\mathbf{U}$ is a square-integrable martingale, then integrability of $\boldsymbol{\Theta}$ is sufficient and essentially necessary to ensure that $\mathbf{L}$ is square integrable, see Theorem 2.4 below. In the case $\mathbf{U}$ is not a martingale, square integrability of $\mathbf{U}$ and $\boldsymbol{\Theta}$ are sufficient and essentially necessary for square integrability of $\mathbf{L}$. We also show that $\mathbf{L}$ is integrable if $\mathbf{U}$ and $\boldsymbol{\Theta}$ are integrable, see Theorem 2.7 below. If $\mathbf{U}$ is a (non-trivial) square-integrable martingale, then it is sufficient and necessary that $\sqrt{|\boldsymbol{\Theta}(1)|}$ is integrable, see Theorem 2.6 below. If $\mathbf{U}$ is a martingale but not square-integrable, then it is possible that integrability of $\sqrt{|\boldsymbol{\Theta}(1)|}$ is not sufficient to ensure integrability of $\mathbf{L}$, as we will show in Proposition 2.10 at the end of Section 2.4.2.

The notion of covariance operator for a random variable in H with finite second moment is central in our analysis in this Section (and in fact also in other parts of the book).

Definition 2.3 Let $\mathbf{Y}$ be any H-valued random variable with finite second moment. Then the *covariance operator* $\mathrm{Cov}(\mathbf{Y})$ of $\mathbf{Y}$ is defined by the equation

$$\langle \mathrm{Cov}(\mathbf{Y})\mathbf{x}, \mathbf{y}\rangle = \mathbb{E}\left[\langle \mathbf{Y} - \mathbb{E}\left[\mathbf{Y}\right], \mathbf{x}\rangle\langle \mathbf{Y} - \mathbb{E}\left[\mathbf{Y}\right], \mathbf{y}\rangle\right] ,$$

for any $\mathbf{x}, \mathbf{y} \in H$.

We remark that the covariance operator of a square integrable H-valued random variable $\mathbf{Y}$ can be expressed in the covariance operators of $Y_1, \dots, Y_d$ if $Y_1, \dots, Y_d$ are independent. Indeed, if this is the case we have

$$\mathrm{Cov}(\mathbf{Y}) = \mathrm{Cov}(Y_1) \times \cdots \times \mathrm{Cov}(Y_d) .$$

Recall that $U_1, \dots, U_d$ are assumed independent, which yields that $\mathrm{Cov}(\mathbf{U}(1)) = \mathrm{Cov}(U_1(1)) \times \cdots \times \mathrm{Cov}(U_d(1))$. In the next Proposition we recall some properties of square integrable Lévy processes formulated in our context.

Proposition 2.5 *The Lévy process* $\mathbf{U}$ *is square integrable if and only if* $\int_H |\mathbf{x}|^2 \, \nu(d\mathbf{x}) < \infty$. *If* $\mathbf{U}$ *is square integrable, then*

(i) $\mathbb{E}[\mathbf{U}(t)] = t\left(\mathbf{b} + \int_{|\mathbf{x}|>1} \mathbf{x}\, \nu(d\mathbf{x})\right)$,
(ii) $\mathbb{E}[|\mathbf{U}(t) - \mathbb{E}[\mathbf{U}(t)]|^2] = t\left(\mathrm{Tr}(Q) + \int_H |\mathbf{x}|^2 \, \nu(d\mathbf{x})\right)$,
(iii) $\mathbf{M}(t) := \mathbf{U}(t) - t\mathbb{E}[\mathbf{U}(1)]$ *is a mean zero and square integrable Lévy process with the characteristic triplet* $(\mathbf{b} - \mathbb{E}[\mathbf{U}(1)], Q, \nu)$, *and*
(iv) $\langle \mathrm{Cov}(\mathbf{U}(1))\mathbf{x}, \mathbf{y}\rangle = \langle Q\mathbf{x}, \mathbf{y}\rangle + \int_H \langle \mathbf{x}, \mathbf{z}\rangle\langle \mathbf{y}, \mathbf{z}\rangle \, \nu(d\mathbf{z})$ *for any* $\mathbf{x}, \mathbf{y} \in H$.

Proof The claims in the proposition are collected from Peszat and Zabczyk [165, Theorem 4.47 and Theorem 4.49]. □

Our first concern is to characterise the square integrability of the subordinated process $\mathbf{L}$. The main result is formulated in Theorem 2.4 below, while the proof is devided into three parts where the next two lemmas each contain a part. It is essentially necessary that $\mathbf{U}$ is square integrable to have a finite second moment of $\mathbf{L}$. However, square integrability of the multivariate subordinator $\boldsymbol{\Theta}$ is only needed if $\mathbf{U}$ is not a martingale (i.e, the mean is different than zero). If $\mathbf{U}$ is a square integrable martingale, then integrability for $\boldsymbol{\Theta}$ is sufficient to ensure that $\mathbf{L}$ has a finite second moment. This is the statement of the next Lemma.

Lemma 2.3 *Assume* $\mathbf{U}$ *is square integrable and has mean equal to zero. If* $\boldsymbol{\Theta}$ *is integrable, then* $\mathbf{L}$ *is mean zero, square integrable and*

$$\mathrm{Cov}(\mathbf{L}(1)) = \mathbb{E}\left[\boldsymbol{\Theta}(1)\right]\mathrm{Cov}(\mathbf{U}(1))\,.$$

Proof Let $g(\boldsymbol{\theta}) := \mathbb{E}[|\mathbf{U}(\boldsymbol{\theta})|^2]$ for any $\boldsymbol{\theta} \in \mathbb{R}_+^d$. Proposition 2.5(ii) applied componentwise yields,

$$g(\boldsymbol{\theta}) = \sum_{j=1}^{d} \theta_j \mathbb{E}[|U_j(1)|_j^2]$$

for any $\boldsymbol{\theta} \in \mathbb{R}_+^d$. By conditioning on $\boldsymbol{\Theta}$ we get

$$\begin{aligned}
\mathbb{E}[|\mathbf{L}(t)|^2] &= \mathbb{E}[g(\boldsymbol{\Theta}(t)] \\
&= \sum_{j=1}^{d} \mathbb{E}[\Theta_j(t)]\mathbb{E}[|U_j(1)|_j^2] \\
&< \infty\,,
\end{aligned}$$

for any $t \geq 0$. Thus $\mathbf{L}$ is square integrable. Conditioning on $\boldsymbol{\Theta}$ yields $\mathbb{E}[\mathbf{L}(t)] = 0$ for any $t \geq 0$ and

$$\begin{aligned}
\langle \mathrm{Cov}(\mathbf{L}(1))\mathbf{x}, \mathbf{y}\rangle &= \mathbb{E}\left[\langle \mathbf{L}(1)), \mathbf{x}\rangle\langle \mathbf{L}(1), \mathbf{y}\rangle\right] \\
&= \sum_{j=1}^{d} \mathbb{E}[\Theta_j(1)]\langle \mathrm{Cov}(U_j(1))x_j, y_j\rangle_j \\
&= \left\langle \mathbb{E}[\boldsymbol{\Theta}(1)]\mathrm{Cov}(\mathbf{U}(1))\mathbf{x}, \mathbf{y}\right\rangle,
\end{aligned}$$

for any $\mathbf{x}, \mathbf{y} \in H$. The lemma follows. □

The next lemma shows that if $\mathbf{U}$ is square integrable but with non-zero mean, the finiteness of the second moment of $\mathbf{L}$ is ensured if $\boldsymbol{\Theta}$ is square integrable. Also in this case we have an expression for the covariance functional of $\mathbf{L}(1)$, however, to state this we need to introduce some notation. For an element $\mathbf{c} \in H$, define the operator $\mathbf{c} \otimes \mathbf{c} : H \to H$ by

$$(\mathbf{c} \otimes \mathbf{c})\mathbf{x} := \langle \mathbf{c}, \mathbf{x} \rangle \mathbf{c} . \tag{2.5}$$

This is of course the standard tensor product between two elements in the Hilbert space H. Since,

$$\langle (\mathbf{c} \otimes \mathbf{c})\mathbf{x}, \mathbf{y} \rangle = \langle \mathbf{c}, \mathbf{x} \rangle \langle \mathbf{c}, \mathbf{y} \rangle = \sum_{i,j=1}^{d} \langle c_i, x_i \rangle_i \langle c_j, x_j \rangle_j = \sum_{i,j=1}^{d} \langle c_i \otimes c_j x_i, y_j \rangle_j ,$$

where $c_i \otimes c_j : H_i \to H_j, x_i \mapsto \langle c_i, x_i \rangle_i c_j$, we can view $\mathbf{c} \otimes \mathbf{c}$ as a $d \times d$-matrix where elements are the operators $c_i \otimes c_j$. For example, for $d = 2$, we find

$$\begin{bmatrix} c_1 \otimes c_1 & c_2 \otimes c_1 \\ c_1 \otimes c_2 & c_2 \otimes c_2 \end{bmatrix} \begin{bmatrix} x_1 \\ x_2 \end{bmatrix} = \langle \mathbf{c}, \mathbf{x} \rangle \mathbf{c} = (\mathbf{c} \otimes \mathbf{c})\mathbf{x} ,$$

for $\mathbf{x} = (x_1, x_2) \in H_1 \times H_2$. Further, if A is a $d \times d$-matrix, we intepret $A(\mathbf{c} \otimes \mathbf{c})$ as the Hadamard product between the two matrices A and $\mathbf{c} \otimes \mathbf{c}$, i.e., the $d \times d$-matrix of operators $A_{i,j}(c_i \otimes c_j)$. We are now ready to formulate our result:

Lemma 2.4 *Assume* $\mathbf{U}$ *and* $\boldsymbol{\Theta}$ *are square integrable. Then* $\mathbf{L}$ *is square integrable,*

$$\begin{aligned} \mathbb{E}[\mathbf{L}(1)] &= \mathbb{E}[\boldsymbol{\Theta}(1)]\mathbb{E}[\mathbf{U}(1)] \quad \textit{and} \\ \mathrm{Cov}(\mathbf{L}(1)) &= \mathbb{E}[\boldsymbol{\Theta}(1)]\mathrm{Cov}(\mathbf{U}(1)) + \mathrm{Cov}(\boldsymbol{\Theta}(1))(\mathbb{E}[\mathbf{U}(1)] \otimes \mathbb{E}[\mathbf{U}(1)]) . \end{aligned}$$

Proof Note that

$$\begin{aligned} |\mathbf{L}(1)|^2 &= |\mathbf{U}(\boldsymbol{\Theta}(1))|^2 \\ &= |\mathbf{U}(\boldsymbol{\Theta}(1)) - \boldsymbol{\Theta}(1)\mathbb{E}[\mathbf{U}(1)] + \boldsymbol{\Theta}(1)\mathbb{E}[\mathbf{U}(1)]|^2 \\ &= |\widetilde{\mathbf{U}}(\boldsymbol{\Theta}(1))|^2 + 2\langle \widetilde{\mathbf{U}}(\boldsymbol{\Theta}(1)), \boldsymbol{\Theta}(1)\mathbb{E}[\mathbf{U}(1)] \rangle + |\boldsymbol{\Theta}(1)\mathbb{E}[\mathbf{U}(1)]|^2 . \end{aligned}$$

Here, $\widetilde{\mathbf{U}}(t) := \mathbf{U}(t) - t\mathbb{E}[\mathbf{U}(1)]$ is a mean-zero Lévy process which is square integrable since $\mathbf{U}$ is square integrable, see Proposition 2.5(iii). From the proof of Lemma 2.3,

$$\mathbb{E}[|\widetilde{\mathbf{U}}(\boldsymbol{\Theta}(1))|^2] = \sum_{j=1}^{d} \mathbb{E}[\Theta_j(1)]\mathbb{E}[|\widetilde{U}_j(1)|_j^2] < \infty ,$$

since $\boldsymbol{\Theta}$ and $\mathbf{U}$ are square integrable. By conditioning on $\boldsymbol{\Theta}(1)$, we find

$$\mathbb{E}\left[\langle \widetilde{\mathbf{U}}(\boldsymbol{\Theta}(1)), \boldsymbol{\Theta}(1)\mathbb{E}[\mathbf{U}(1)] \rangle\right] = \mathbb{E}\left[\mathbb{E}\left[\langle \widetilde{\mathbf{U}}(\boldsymbol{\Theta}(1)), \boldsymbol{\Theta}(1)\mathbb{E}[\mathbf{U}(1)] \rangle \,\middle|\, \boldsymbol{\Theta}(1)\right]\right] = 0$$

since $\widetilde{\mathbf{U}}(1)$ has mean zero. Finally,

$$\mathbb{E}[|\boldsymbol{\Theta}(1)\mathbb{E}[\mathbf{U}(1)]|^2] = \sum_{j=1}^{d} \mathbb{E}[\Theta_j^2(1)]|\mathbb{E}[U_j(1)]|_j^2 < \infty\,,$$

since $\boldsymbol{\Theta}(1)$ is square integrable by assumption. Hence, $\mathbf{L}(1)$ is square integrable.
For the mean value of $\mathbf{L}(1)$, we find

$$\mathbb{E}[L_j(1)] = \mathbb{E}\left[\mathbb{E}[U_j(\Theta_j(1)) \mid \Theta_j(1)]\right] = \mathbb{E}[\Theta_j(1)]\mathbb{E}[U_j(1)]\,,$$

and thus, $\mathbb{E}[\mathbf{L}(1)] = \mathbb{E}[\boldsymbol{\Theta}(1)]\mathbb{E}[\mathbf{U}(1)]$.
Next, we derive the covariance operator for $\mathbf{L}(1)$. For $\boldsymbol{\theta} \in \mathbb{R}_+^d$, it holds

$$\begin{aligned}
&\mathbb{E}\left[\langle \mathbf{U}(\boldsymbol{\theta}), \mathbf{x}\rangle \langle \mathbf{U}(\boldsymbol{\theta}), \mathbf{y}\rangle\right] \\
&\quad = \langle \mathrm{Cov}(\mathbf{U}(\boldsymbol{\theta}))\mathbf{x}, \mathbf{y}\rangle + \langle \mathbb{E}[\mathbf{U}(\boldsymbol{\theta})], \mathbf{x}\rangle \langle \mathbb{E}[\mathbf{U}(\boldsymbol{\theta})], \mathbf{y}\rangle \\
&\quad = \sum_{j=1}^{d} \theta_j \langle \mathrm{Cov}(U_j(1))x_j, y_j\rangle_j + \sum_{i,j=1}^{d} \theta_i\theta_j \langle \mathbb{E}[U_i(1)], x_i\rangle_i \langle \mathbb{E}[U_j(1)], y_j\rangle_j \\
&\quad = \sum_{j=1}^{d} \theta_j \langle \mathrm{Cov}(U_j(1))x_j, y_j\rangle_j \\
&\qquad + \sum_{i,j=1}^{d} \theta_i\theta_j \langle (\mathbb{E}[U_i(1)] \otimes \mathbb{E}[U_j(1)])x_i, y_j\rangle_j\,.
\end{aligned}$$

Since, similarly

$$\langle \mathbb{E}[\mathbf{L}(1)], \mathbf{x}\rangle \langle \mathbb{E}[\mathbf{L}(1)], \mathbf{y}\rangle = \sum_{i,j=1}^{d} \mathbb{E}[\Theta_i(1)]\mathbb{E}[\Theta_j(1)]\langle (\mathbb{E}[U_i(1)] \otimes \mathbb{E}[U_j(1)])x_i, y_j\rangle_j\,,$$

the result follows by conditioning of $\boldsymbol{\Theta}(1)$. □

We fully characterise the square integrability of $\mathbf{L}$.

Theorem 2.4 $\mathbf{L}$ *is square integrable if and only if* L_j *is square integrable for all* $j = 1, \ldots, d$. *Let* $j \in \{1, \ldots, d\}$. *Then* L_j *is square integrable if and only if any of the following statements hold.*

(i) U_j *and* Θ_j *are square integrable.*
(ii) U_j *is mean zero and square integrable and* Θ_j *is integrable.*
(iii) $\Theta_j = 0$, *a.s.*
(iv) $U_j = 0$ *a.s.*

Moreover, L_j is mean zero and square integrable if and only if (ii), (iii) or (iv) holds. If (i) holds for any $j = 1, \dots, d$, then

$$\mathbb{E}[\mathbf{L}(1)] = \mathbb{E}[\boldsymbol{\Theta}(1)]\mathbb{E}[\mathbf{U}(1)] \quad \textit{and}$$
$$\operatorname{Cov}(\mathbf{L}(1)) = \mathbb{E}[\boldsymbol{\Theta}(1)]\operatorname{Cov}(\mathbf{U}(1)) + \operatorname{Cov}(\boldsymbol{\Theta}(1))(\mathbb{E}[\mathbf{U}(1)] \otimes \mathbb{E}[\mathbf{U}(1)]) .$$

If (ii) holds for any $j = 1, \dots, d$, then

$$\operatorname{Cov}(\mathbf{L}(1)) = \mathbb{E}[\boldsymbol{\Theta}(1)]\operatorname{Cov}(\mathbf{U}(1)) .$$

Proof The first claim follows directly from the equation $|\mathbf{L}(1)|^2 = \sum_{j=1}^d |L_j(1)|_j^2$. The formulas for the expectation and covariance operator of $\mathbf{L}$ at the end of the Theorem follow from Lemmas 2.3 and 2.4.

For the characterisation of square integrability of L_j we can assume without loss of generality that $d = 1$ where $H = H_1$. The *if* part follows from Lemmas 2.3 and 2.4. It remains to argue for the *only if* part: Assume that L is square integrable. Let (γ, Γ, ℓ) be the characteristic triplet of L as given in Theorem 2.2. Proposition 2.5 yields

$$\begin{aligned} \infty &> \int_H |x|^2 \, \ell(dx) \\ &= \int_H |x|^2 \, (a_0 \nu)(dx) + \int_0^\infty \mathbb{E}[|L(\theta)|^2] \, \kappa(d\theta) . \end{aligned}$$

Thus, $\int_0^\infty \mathbb{E}[|U(\theta)|^2] \, \kappa(d\theta) < \infty$ and $\int_H |x|^2 \, (a_0\nu)(dx) < \infty$. We now distinguish between two cases, where the second case has three sub-cases:

Case 1: Suppose U is not square integrable. Then Proposition 2.5 implies that $\int_H |x|^2 \, \nu(dx) = \infty$. Thus $\kappa = 0$ and $a_0 = 0$, and therefore $\Theta = 0$, $a.s.$ We conclude with Statement (iii) in the Theorem.

Case 2: Suppose U is square integrable. Denote by $v := \mathbb{E}[|U(1) - \mathbb{E}U(1)|^2]$ and $m := \mathbb{E}[U(1)]$ and note that $\mathbb{E}[|U(\theta)|^2] = \theta v + \theta^2 |m|^2$ for any $\theta \geq 0$. Hence, we have

$$\int_0^\infty \theta v \, \kappa(d\theta) < \infty \quad \text{and} \int_0^\infty \theta^2 |m|^2 \, \kappa(d\theta) < \infty .$$

We have three sub-cases to treat:

Case 2.1: If $m \neq 0$, then $\int_0^\infty \theta^2 \, \kappa(d\theta) < \infty$. Hence Sato [175, Corollary 25.8] yields that Θ is square integrable which is Statement (i).

Case 2.2: If $m = 0$ but $v \neq 0$, then U is mean zero and $\int_0^\infty \theta \, \kappa(d\theta) < \infty$. Hence Sato [175, Corollary 25.8] yields that Θ is integrable. Thus we have Statement (ii).

Case 2.3: Suppose $m = 0$ and $v = 0$. Since $0 = v = \mathbb{E}[|U(1)|^2]$ we have $U = 0$, $a.s.$, which is Statement (iv).

Lemma 2.3 yields that if (ii), (iii) or (iv) holds, then $\mathbf{L}$ is mean zero and square integrable. If $\mathbf{L}$ is mean zero and square integrable and (i) holds, then we have

$$0 = \mathbb{E}[\mathbf{L}(1)] = \mathbb{E}[\boldsymbol{\Theta}(1)]\mathbb{E}[\mathbf{U}(1)].$$

Thus $\mathbb{E}[\boldsymbol{\Theta}(1)] = 0$ which yields Statement (iii) or $\mathbb{E}[\mathbf{U}(1)] = 0$ which implies Statement (ii). □

Theorem 2.4 above is a complete characterisation of the second order structure of the Lévy process $\mathbf{L}$. However, there are Lévy processes without finite second moment. In that case the first order structure and the martingale property are still interesting. We next aim at developing necessary and sufficient conditions for the existence of a first moment, and give a condition that suffices to show that $\mathbf{L}$ is a martingale. We pay special attention to the case when $\mathbf{U}$ is a Wiener process.

Theorem 2.5 *Let* $\mathbf{U}$ *and* $\boldsymbol{\Theta}$ *be integrable. Then* $\mathbf{L}$ *is integrable and*

$$\mathbb{E}[\mathbf{L}(1)] = \mathbb{E}[\boldsymbol{\Theta}(1)]\mathbb{E}[\mathbf{U}(1)]\,.$$

Proof If $\mathbf{L}$ is integrable, then conditioning on $\boldsymbol{\Theta}$ yields the formula for the expectation of $\mathbf{L}(1)$. To show integrability of $\mathbf{L}$, let $g_{\mathbf{U}}$ be the growth function of $\mathbf{U}$, cf. Definition 2.1. Then, Proposition 2.3 yields that there is $C > 0$ and $g_{\mathbf{U}}(\boldsymbol{\theta}) \leq 1+C|\boldsymbol{\theta}|$ for any $\boldsymbol{\theta} \in \mathbb{R}^d_+$. Lemma 2.2 yields that $\mathbf{L}$ is integrable if $g_{\mathbf{U}}(\boldsymbol{\Theta}(1))$ is integrable. However, by the assumption of integrability of $\boldsymbol{\Theta}$,

$$\mathbb{E}[g_{\mathbf{U}}(\boldsymbol{\Theta}(1))] \leq 1 + C\mathbb{E}[|\boldsymbol{\Theta}(1)|] < \infty\,,$$

and the Theorem follows. □

We have seen that the martingale property of $\mathbf{U}$ allows us to put weaker assumptions on $\boldsymbol{\Theta}$ to ensure that $\mathbf{L}$ is square integrable. The same is true to ensure integrability of $\mathbf{L}$, as we show in the next Theorem.

Theorem 2.6 *Let* $\mathbf{U}$ *be a square integrable martingale such that* $U_1, \ldots, U_d$ *are nontrivial. Then* $\mathbf{L}$ *is a martingale if and only if* $\sqrt{|\boldsymbol{\Theta}(1)|}$ *is integrable, or, equivalently,* $\int_{|\boldsymbol{\theta}|>1} \sqrt{|\boldsymbol{\theta}|}\,\kappa(d\boldsymbol{\theta}) < \infty$.

Proof Proposition 2.4 yields that $\mathbb{E}[\sqrt{|\boldsymbol{\Theta}(1)|}] < \infty$ if and only if

$$\int_{|\boldsymbol{\theta}|>1} \sqrt{|\boldsymbol{\theta}|}\,\kappa(d\boldsymbol{\theta}) < \infty\,.$$

Let $g_{\mathbf{U}}$ be the growth function of $\mathbf{U}$ in the sense of Defintion 2.1. Then Proposition 2.2 yields that there is a constant $c > 0$ such that $c\sqrt{|\boldsymbol{\theta}|} \leq g_{\mathbf{U}}(|\boldsymbol{\theta}|) \leq \frac{1}{c}\sqrt{|\boldsymbol{\theta}|}$, $\boldsymbol{\theta} \in \mathbb{R}^d_+$. Lemma 2.2 yields that $\mathbf{L}$ is integrable if and only if $\sqrt{|\boldsymbol{\Theta}(1)|}$ is integrable. Finally, $\mathbb{E}[\mathbf{L}(1) \mid \boldsymbol{\Theta}(1)] = \boldsymbol{\Theta}(1)\mathbb{E}[\mathbf{U}(1)] = 0$. Hence, $\mathbf{L}$ is mean zero and therefore a martingale. □

If $\mathbf{U}$ is only integrable, we can recover integrability of $\mathbf{L}$ through a condition on the Lévy measure of $\boldsymbol{\Theta}$:

Theorem 2.7 *Let $\boldsymbol{\Theta}$ be non trivial, i.e. $\mathbb{P}(\boldsymbol{\Theta} \neq 0) > 0$. Then $\mathbf{L}$ is integrable if and only if $\mathbf{U}$ is integrable and $\int_{|\boldsymbol{\theta}|>1} \mathbb{E}[|\mathbf{U}(\boldsymbol{\theta})|]\,\kappa(d\boldsymbol{\theta}) < \infty$. If $\mathbf{U}$ and $\boldsymbol{\Theta}$ (and hence $\mathbf{L}$) are integrable, then*

$$\mathbb{E}[\mathbf{L}(1)] = \mathbb{E}[\boldsymbol{\Theta}(1)]\mathbb{E}[\mathbf{U}(1)]\,.$$

If L_j is integrable but Θ_j is not integrable, then L_j is mean zero where $j \in \{1, \dots, d\}$.

Proof We first conclude from Theorem 2.5 the formula for the expectation of $\mathbf{L}(1)$. Let $(\boldsymbol{\gamma}, \Gamma, \ell)$ be the characteristic triplet of $\mathbf{L}$ as given in Theorem 2.2 and let $g_{\mathbf{U}}$ be the growth function of $\mathbf{U}$, cf. Definition 2.1. It holds that

$$\int_{\{|\mathbf{x}|>1\}} |\mathbf{x}|\,\ell(d\mathbf{x}) = \int_{\{|\mathbf{x}|>1\}} |\mathbf{a}_0\mathbf{x}|\,\nu(d\mathbf{x}) + \int_{\mathbb{R}_+^d} \mathbb{E}[|\mathbf{U}(\boldsymbol{\theta})|1_{|\mathbf{U}(\boldsymbol{\theta})|>1})\,\kappa(d\boldsymbol{\theta}).$$

If $\mathbf{L}$ is integrable, it follows from Proposition 2.4 that $\int_{\{|\mathbf{x}|>1\}} |\mathbf{x}|\,\ell(d\mathbf{x}) < \infty$, which, again by appealing to Proposition 2.4 implies that $\mathbf{U}$ is integrable. On the other hand, if $\mathbf{U}$ is integrable and $\int_{|\boldsymbol{\theta}|>1} \mathbb{E}[|\mathbf{U}(\boldsymbol{\theta})|]\,\kappa(d\boldsymbol{\theta}) < \infty$, then

$$\int_{|\boldsymbol{\theta}|>1} g_{\mathbf{U}}(\boldsymbol{\theta})\,\kappa(d\boldsymbol{\theta}) = \int_{|\boldsymbol{\theta}|>1} \mathbb{E}[|\mathbf{U}(\boldsymbol{\theta})|]\,\kappa(d\boldsymbol{\theta}) < \infty\,.$$

Proposition 2.4 yields that $\mathbb{E}[g_{\mathbf{U}}(\boldsymbol{\Theta}(1))] < \infty$, and from Lemma 2.2 we conclude that $\mathbf{L}$ is integrable.

Now let $j \in \{1, \dots, d\}$ and assume that L_j is integrable but Θ_j is not. We have already shown that this implies that U_j is integrable. Let $g(\theta) := \mathbb{E}[U_j(\theta)] = \theta\mathbb{E}[U_j(1)]$ for any $\theta \in \mathbb{R}_+$. Thus, by conditioning,

$$\mathbb{E}[L_j(1)] = \mathbb{E}\left[\mathbb{E}\left[U_j(\Theta_j(1)) \mid \boldsymbol{\Theta}(1)\right]\right] = \mathbb{E}[g(\Theta_j(1))].$$

We see from this that $\Theta_j(1)\mathbb{E}[U_j(1)]$ is integrable. Since $\Theta_j(1)$ is not integrable we conclude that $\mathbb{E}[U_j(1)] = 0$ and hence $\mathbb{E}[L_j(1)] = 0$. □

To construct interesting Lévy processes $\mathbf{L}$ by subordination, the case when $\mathbf{U}$ is a Wiener process is of particular importance. We find:

Corollary 2.1 *Let $\mathbf{U}$ be Gaussian with mean zero and assume that* $\mathrm{Tr}(Q_j) \neq 0$ *for all $j \in \{1, \dots, d\}$. Then $\mathbf{L}$ is integrable if and only if $\sqrt{|\boldsymbol{\Theta}(1)|}$ is integrable. In that case $\mathbf{L}$ is mean zero.*

Proof This is a special case of Theorem 2.6. □

In the next Section we will see examples of Lévy processes based on $\mathbf{U}$ being a Wiener process.

2.4 Examples of Subordinated Wiener Processes

We apply the approach of subordination to extend some (univariate) Lévy processes to infinite dimensions. Moreover, with the application of energy markets in mind, we focus on the popular normal inverse Gaussian, α-stable and variance Gamma Lévy processes

To begin with, we prove a general result on the relationship between infinite dimensional Lévy processes and their one dimensional projections.

Proposition 2.6 *Assume* $\mathbf{Y}$ *is an* H*-valued stochastic process with independent increments such that* $\langle \mathbf{u}, \mathbf{Y} \rangle$ *is a Lévy processes for every* $\mathbf{u} \in H$. *Then* $\mathbf{Y}$ *is a Lévy process in* H *and its law is determined by the laws of the real-valued processes* $\langle \mathbf{u}, \mathbf{Y} \rangle$, $\mathbf{u} \in H$.

Proof Define $f : \mathbb{R}_+ \times H \to \mathbb{C}, (t, u) \mapsto \mathbb{E}[\exp(\mathrm{i}\langle \mathbf{u}, \mathbf{Y}(t) \rangle)]$, for which we observe that $f(t, \cdot)$ is the characteristic functional of $\mathbf{Y}(t)$ for any $t \geq 0$. Moreover, $f(t, \mathbf{u}) = \exp(t\rho(\mathbf{u}))$ for some function $\rho : H \to \mathbb{C}$ because $\langle \mathbf{u}, \mathbf{Y} \rangle$ is a Lévy process and thus $\rho(\mathbf{u})$ is its Lévy exponent. For $t, h \in \mathbb{R}_+$,

$$\mathbb{E}\big[\exp(\mathrm{i}\langle \mathbf{u}, \mathbf{Y}(t+h) - \mathbf{Y}(h) \rangle)\big] = \mathbb{E}\big[\exp(\mathrm{i}(\langle \mathbf{u}, \mathbf{Y} \rangle(t+h) - \langle \mathbf{u}, \mathbf{Y} \rangle(h)))\big] = f(t, \mathbf{u}),$$

where we again have used that $\langle \mathbf{u}, \mathbf{Y} \rangle$ is a Lévy process. Thus, $\mathbf{Y}$ has stationary increments and $\mathbf{Y}$ is a Lévy process in H.

Since the distribution of a Lévy process is determined by its distribution at time 1 and the distribution of $\mathbf{Y}(1)$ is determined by its characteristic functional, we conclude that the distribution of $\mathbf{Y}$ is determined by the laws of the random variables $\langle \mathbf{u}, \mathbf{Y}(1) \rangle$, $\mathbf{u} \in H$. □

In the remaining of this Chapter, we focus on the case $d = 1$, so that $H = H_1$. The particular Lévy processes we are going to consider are constructed by subordinating a H-valued Wiener process by certain subordinators, that are well-known in the real-valued case. We can extend to the multivariate case, i.e., $d > 1$, by choosing independent copies of these subordinators. We will comment briefly on this.

2.4.1 Hilbert Space Valued Normal Inverse Gaussian Lévy Process

Multivariate normal inverse Gaussian distributions (MNIG-distributions) have been first introduced by Rydberg [172], extending the univariate normal inverse Gaussian distribution defined in Barndorff-Nielsen [11]. These distributions can also be generated from a multivariate Brownian motion and an inverse Gaussian Lévy process by subordination, that is, the subordinated Brownian motion becomes a Lévy process where the marginal distributions are MNIG. We generalise this approach to construct Hilbert space-valued normal inverse Gaussian (HNIG) Lévy processes.

Definition 2.4 An H-valued Lévy process Y is an *HNIG-process* if there are $s, c \in \mathbb{R}_+$, $b \in H$ and a positive semi-definite trace class operator Q on H such that its Lévy exponent $\rho : H \to \mathbb{C}$ is given by

$$\rho(u) = s\left(c - \sqrt{c^2 + \langle Qu, u\rangle - i2\langle u, b\rangle}\right),$$

where $\sqrt{\cdot}$ denotes the main branch of the root function. Here, (s, c, b, Q) are the *parameters of the HNIG-process* Y. A *degenerate HNIG-process* is an HNIG-process where its second parameter is 0, i.e. there are $s \in \mathbb{R}_+$, $b \in H$ and a positive semi-definite trace class operator Q on H such that $(s, 0, b, Q)$ are the parameters of Y.

Knowing the parameters (s, c, b, Q) of an HNIG-process Y defines it uniquely in the sense that its characteristic triplet, and therefore its law, is unique. Indeed, if X is another HNIG-process with the same parameters as Y, then X and Y have the same Lévy exponent, or, in other words, the same characteristic triplet. Moreover, since the set of functions

$$\{f_u : H \to \mathbb{C}, x \mapsto e^{i\langle u,x\rangle} : u \in H\}$$

is a monotone class, Ethier and Kurtz [96, Corollary A.4.4] yields that the laws of two H-valued random variables Z_1 and Z_2 coincide if and only if

$$\mathbb{E}[e^{i\langle u,Z_1\rangle}] = \mathbb{E}[e^{i\langle u,Z_2\rangle}],$$

for any $u \in H$. Hence, $X(1)$ has the same law as $Y(1)$.

An H-valued random variable Z is *symmetric* if Z and $-Z$ has the same law. We find the following Proposition based on the discussion above:

Proposition 2.7 *An HNIG-process Y with parameters $(s, c, 0, Q)$ where $c > 0$ is symmetric.*

Proof Since

$$\mathbb{E}[e^{i\langle u,-Y\rangle}] = \mathbb{E}[e^{i\langle -u,Y\rangle}] = \exp(\rho(-u)),$$

for any $u \in H$, the Lévy exponent of $-Y$ is $\rho(-u)$. But as $b = 0$, we find

$$\rho(-u) = s\left(c - \sqrt{c^2 - \langle Q(-u), (-u)\rangle}\right) = \rho(u)$$

and therefore $-Y$ has the same Lévy exponent as Y. It follows that the law of $Y(t)$ is equal to the law of $-Y(t)$. □

We first show how we can construct a non-degenerate HNIG process by subordination with an inverse Gaussian Lévy process, and discuss some of its properties.

Theorem 2.8 *Let $s, c \in \mathbb{R}_+$, $c > 0$, $b \in H$ and Q a positive semi-definite trace class operator on H. Then there exists a square integrable HNIG-process Y with parameters (s, c, b, Q). The characteristic triplet (γ, Γ, ℓ) of Y is given by*

$$\gamma = \frac{s}{c}b - \int_{|x|>1} x\,\ell(dx),$$

$$\Gamma = 0 \quad \textit{and}$$

$$\ell(A) = \int_0^\infty \Phi_\theta(A)\frac{s}{\sqrt{2\pi\theta^3}}e^{-c^2\theta/2}d\theta$$

for any Borel set $A \subseteq H$ where Φ_θ denotes the Gaussian measure on H with mean θb and covariance operator θQ, $\theta \in \mathbb{R}_+$. Moreover,

$$\mathbb{E}[Y(1)] = \frac{s}{c}b\,,$$

$$\operatorname{Cov}(Y(1)) = \frac{s}{c^3}(b \otimes b) + \frac{s}{c}Q\,.$$

Proof Let U be an H-valued drifted Wiener process

$$U(t) = bt + W(t)\,,$$

where $b \in H$ and W is an H-valued Wiener process with covariance operator Q. Its Lévy exponent $\varphi : H \to \mathbb{C}$ is

$$\varphi(u) = i\langle u, b\rangle - \frac{1}{2}\langle Qu, u\rangle\,.$$

Define Θ to be an inverse Gaussian Lévy process on $\mathbb{R}$ with parameters s, c, that is, a pure-jump subordinator with Lévy measure given by

$$\kappa(d\theta) = \frac{s}{\sqrt{2\pi\theta^3}}e^{-c^2\theta/2}1_{\{\theta>0\}}\,d\theta\,,$$

cf. Rydberg [172]. Then its Laplace exponent $\psi : \mathbb{R}_- + i\mathbb{R} \to \mathbb{C}$ is given by

$$\psi(z) = s\left(c - \sqrt{c^2 - 2z}\right)\,,$$

where $\sqrt{\cdot}$ denotes the main branch of the root function. Theorem 2.1 yields that the Lévy exponent $\rho : H \to \mathbb{C}$ of the Lévy process $Y(t) := U(\Theta(t))$ is

$$\rho(u) = \psi(\varphi(u)) = s\left(c - \sqrt{c^2 + \langle Qu, u\rangle - i2\langle u, b\rangle}\right).$$

Hence, Y is an HNIG-process with parameters (s, c, b, Q).

Observe that $\psi'(0) = s/c$. On the other hand, from (2.2) we find after using the definition of $\kappa(d\theta)$ that $\psi'(0) = a_0 + s/c$, and thus $a_0 = 0$. Theorem 2.2 then yields the desired characteristic triplet (γ, Γ, μ).

Theorem 2.4 implies that Y is square integrable since both U and Θ are square integrable. Moreover, $\mathbb{E}[\Theta(1)] = s/c$, $\mathbb{E}[U(1)] = b$, $\mathrm{Cov}(U(1)) = Q$ and $\mathrm{Var}(\Theta(1)) = \psi''(0) = s/c^3$, and the given expectation and covariance operator follow. □

From Proposition 2.1 we have that $\mathcal{L}(Y)$ is a Lévy process on $\mathbb{R}^n$ for $\mathcal{L} \in L(H, \mathbb{R}^n)$, $n \in \mathbb{N}$. We further find that $\mathcal{L}(Y)(t) = (\mathcal{L} \circ U)(\Theta(t))$ and $\mathcal{L}(U)(t) = (\mathcal{L}b)t + \mathcal{L}W(t)$ is a Gaussian process on $\mathbb{R}^n$. Indeed, the drift of $\mathcal{L}(U)(t)$ is $\mathcal{L}b$ and the covariance operator becomes the $n \times n$-matrix $\mathcal{L}Q\mathcal{L}^*$. We conclude that the distribution of $\mathcal{L}(Y)(1)$ is MNIG, cf. Øigard and Hanssen [160, page 2], being a conditional Gaussian distribution with inverse Gaussian mean and variance.

If $(e_i)_{i=1}^n$ are the canonical basis vectors in $\mathbb{R}^n$, we find that

$$\mathcal{L}(Y)(t) = (\langle Y(t), f_1\rangle, \ldots, \langle Y(t), f_n\rangle)$$

where $f_i = \mathcal{L}^*(e_i) \in H, i = 1, \ldots, n$. The Lévy exponent of $\mathcal{L}(Y)$ becomes

$$\rho_n(z) = s\left(c - \sqrt{c^2 + z^\top \Sigma z - i2z^\top \beta}\right)$$

for $z \in \mathbb{R}^n$. Here $\beta \in \mathbb{R}^n$ is given by

$$\beta^\top = (\langle b, f_1\rangle, \ldots, \langle b, f_n\rangle),$$

and $\Sigma \in \mathbb{R}^{n\times n}$ has elements $(\langle Qf_i, f_j\rangle)_{i,j=1}^n$. In particular, for $n = 1$, we identify the NIG distribution parameters $\mu = 0$, $\delta = s\langle Qf, f\rangle^{1/2}$, $\beta = \langle f, b\rangle/\langle Qf, f\rangle$ and

$$\alpha^2 = \frac{c^2}{\langle Qf, f\rangle} + \beta^2.$$

Here we have identified the parameter quadruple $(\mu, \beta, \alpha, \delta)$ associated with the univariate NIG distribution. We observe that whenever $b = 0$, we find $\beta = 0$, which is the symmetry parameter of the univariate NIG distribution. Thus, we confirm the symmetry of $\mathcal{L}(Y)$ in this univariate case, recall Proposition 2.7.

In order to construct degenerate HNIG-processes we use a different subordinator, namely 0.5-stable subordinator. Subordination of Brownian motion with an α-stable subordinator will be investigated in Section 2.4.2 in more detail.

Proposition 2.8 *Let $s \in \mathbb{R}_+$, $b \in H$ and Q a positive semi-definite trace class operator on H. Then there is an HNIG-process Y with parameters $(s, 0, b, Q)$. Moreover, Y is not integrable.*

Proof Let U be a Wiener process with drift b and covariance operator Q. Let $s \geq 0$ and Θ be the 0.5-stable subordinator with Lévy measure

$$\kappa(d\theta) = s\frac{\theta^{-1.5}}{\Gamma(-0.5)}d\theta\,.$$

Then its Laplace exponent is given by

$$\psi : \mathbb{R}_- + i\mathbb{R} \to \mathbb{C},\, v \mapsto \begin{cases} s\exp(-0.5\mathrm{Log}(-v)) & v \neq 0, \\ 0 & v = 0 \end{cases}$$

where Log denotes the main branch of the logarithm. Theorem 2.1 yields that the Lévy exponent of the Lévy process $Y(t) := U(\Theta(t))$ is given by

$$\rho(u) = \psi(\varphi(u)) = s\sqrt{-\varphi(u)}$$

as desired. Observe that $\sqrt{\Theta(1)}$ is not integrable. Hence Corollary 2.1 yields the claim if $b = 0$ and Theorem 2.7 yields the claim if $b \neq 0$. □

Proposition 2.9 *Let Y be a process on H. Then Y is an HNIG-process if and only if $\mathcal{L}Y$ is an MNIG-process for every finite dimensional linear operator $\mathcal{L}$ on H.*

Proof This can be simply read off from the characteristic function. □

For example, Frestad, Benth and Koekebakker [107] show that logreturns of forward prices in power markets can be modelled very well by the normal inverse Gaussian distribution, where estimated parameters vary with time to maturity and length of delivery. We also recall the findings in the empirical studies of Section 1.4 in Chapter 1. These studies point towards the need for normal inverse Gaussian random fields in time and space as the model class for the stochastic drivers of forward prices in power markets.

2.4.2 α-Stable Hilbert Space Valued Lévy Processes

In this section we will investigate some properties of symmetric stable Lévy processes constructed by subordination of infinite dimensional Brownian motion. An explicit construction of stable Hilbert space valued Lévy processes is found in Peszat and Zabczyk [165, Example 4.38]. Like the finite dimensional case integrability properties of symmetric stable Lévy processes are related to the index of the process. Many other properties can be derived as in the finite dimensional case, cf. Sato [175, Chapter 3]. We remark that in a study by Benth *et al.* [31], (finite dimensional) α-stable processes have been applied for modelling the evolution of power spot and forward prices.

Let us first recall the definition of strictly α-stable processes.

Definition 2.5 Let $\alpha \in \mathbb{R}_+$. A stochastic process Y is a *strictly α-stable process* if $Y(t^\alpha)$ and $tY(1)$ have the same distribution for any $t \in \mathbb{R}_+$.

We find the following result for α-stable Lévy processes on Hilbert space.

Theorem 2.9 *For each $\alpha \in (0, 2]$ and each positive semi-definite trace class operator $Q \neq 0$ on H there is a symmetric H-valued strictly α-stable Lévy process Y with Lévy exponent*

$$\rho(u) = -\langle Qu, u\rangle^{\alpha/2}.$$

Such a strictly α-stable process is square integrable if and only if $\alpha = 2$ and it is integrable and mean zero if and only if $\alpha > 1$.

Let Y be a symmetric strictly α-stable Lévy process which is non-trivial, i.e. $\mathbb{P}(Y \neq 0) > 0$. Then $\alpha \in (0, 2]$. Moreover, there is a symmetric continuous function $f : S_H \to \mathbb{R}_+$ such that the Lévy exponent of Y is given by

$$\rho(u) = -|u|^\alpha f\left(\frac{u}{|u|}\right),$$

where $S_H := \{x \in H : |x| = 1\}$ denotes the sphere in H.

Proof Let Q be a positive definite trace class operator on H and U be a mean zero Gaussian Lévy process with covariance operator $2Q$. For $\alpha \in (0, 2)$, introduce an $\alpha/2$-stable subordinator Θ. According to Bertoin [56, page 73], its Laplace exponent is given by

$$\psi(z) = \begin{cases} -\exp(\alpha/2\mathrm{Log}(-z)) & \text{if } z \neq 0, \\ 0 & \text{otherwise} \end{cases},$$

where Log denotes the main branch of the logarithm. Hence Theorem 2.1 yields that the characteristic function of $X(t) := U(\Theta(t))$ is given by

$$\rho(u) = -\langle Qu, u\rangle^{\alpha/2}.$$

We have that

$$\begin{aligned}\mathbb{E}\left[e^{i\langle X(t^\alpha),u\rangle}\right] &= \mathbb{E}\left[e^{i\langle U(\Theta(t^\alpha)),u\rangle}\right]\\ &= \exp\left(-\langle Qu,u\rangle^{\alpha/2}t^\alpha\right)\\ &= \exp\left(-\langle Q(tu),tu\rangle^{\alpha/2}\right)\\ &= \mathbb{E}\left[e^{i\langle tU(\Theta(1)),u\rangle}\right].\end{aligned}$$

Thus, $X(t^\alpha)$ is equal in distribution to $tX(1)$ for any $t \in \mathbb{R}_+$, and therefore X is a symmetric H-valued strictly α-stable Lévy process. If $\alpha = 2$, then $X = U$ and hence it is square integrable. Theorem 2.4 implies that X is not square-integrable if $\alpha \neq 2$ since then Θ is not integrable. Corollary 2.1 yields that X is integrable if and only if $\sqrt{\Theta(1)}$ is integrable, which by Proposition 2.4 is true if and only if $\int_1^\infty \theta^{1/2}\kappa(d\theta) < \infty$. But the $\alpha/2$-stable subordinator Θ has Lévy measure $\kappa(d\theta) = c\theta^{-1-\alpha/2}d\theta$ for some constant $c > 0$ (see [175, page 80]), and therefore X is integrable if and only if $\alpha > 1$.

Now let $\alpha \in \mathbb{R}_+$ be arbitrary and Y be a strictly α-stable non-trivial Lévy process. Then $\langle Y,u\rangle$ is a Lévy process which is strictly α-stable for any $u \in H$. Applying Sato [175, Theorem 13.15] to the process $\langle Y,u\rangle$ yields that $\alpha \in (0,2]$. Let ρ be the Lévy exponent of Y and define $f := -\rho|_{S_H}$. Let $u \in H\backslash\{0\}$ and define $t := |u|$ and $v := u/t \in S_H$. Then

$$\exp(\rho(u)) = \mathbb{E}\left[e^{i\langle Y(1),u\rangle}\right] = \mathbb{E}\left[e^{i\langle Y(1),tv\rangle}\right] = \mathbb{E}\left[e^{i\langle Y(t^\alpha),v\rangle}\right] = \exp(-t^\alpha f(v)).$$

Thus $\rho(u) = -t^\alpha f(v)$. Since Y is symmetric, ρ is real valued and so is f. $\mathrm{Re}(\rho)$ is bounded by 0 because the characteristic function of Y is bounded by 1. Hence $f(v) \in \mathbb{R}_+$ for any $v \in S_H$. As Y is symmetric, it follows that ρ is symmetric, and therefore also f is symmetric. □

Let W be an H-valued mean zero Wiener process with covariance operator Q, and introduce the ONB $(e_n)_{n\in\mathbb{N}}$ as the eigenvectors of Q with corresponding eigenvalues $(\lambda_n)_{n\in\mathbb{N}}$ where $\lambda_n = 1/n^2$, cf. Peszat and Zabczyk [165, Section 4.4]. Then, for $t \geq 0$,

$$W(t) = \sum_{n\in\mathbb{N}} n^{-1}B_n(t)e_n,$$

where $(B_n)_{n\in\mathbb{N}}$ are independent standard Brownian motions. Introduce now $X(t) := W(\Theta(t))$, where Θ is an $\alpha/2$-stable subordinator for $\alpha \in (0,2)$. Then, from Theorem 2.9, X is an α-stable Lévy process. Moreover, we have the series representation $X(t) = \sum_{n\in\mathbb{N}} X_n(t)e_n$ with $X_n(t) = \langle X(t), e_n\rangle$. If $\alpha > 1$, we find, by conditioning,

$$\begin{aligned}\mathbb{E}[|X_n(t)|] &= \frac{1}{n}\mathbb{E}\left[\mathbb{E}[|B_n(\Theta(t))| \,\big|\, \Theta(t)]\right] \\ &= \frac{1}{n}\sqrt{\frac{2}{\pi}}\mathbb{E}\left[\sqrt{\Theta(t)}\right] \\ &= \frac{t^{1/\alpha}}{n}\sqrt{\frac{2}{\pi}}\mathbb{E}[\sqrt{\Theta(1)}],\end{aligned}$$

after appealing to the $\alpha/2$-stability of Θ. First of all, we here see explicitly how integrability of X_n is equivalent to integrability of $\sqrt{\Theta(1)}$. On the other hand, we *cannot* conclude from integrability of each X_n that X is integrable, since $\sum_{n\in\mathbb{N}} \mathbb{E}|X_n(t)| = \infty$. By Theorem 2.9 we know, however, that X is integrable. This demonstrates that one cannot always resort to the finite dimensional approximation to conclude properties on the infinite dimensional object.

We end this subsection with a result on integrability of subordinated Lévy processes, showing that integrability of $\sqrt{|\Theta(t)|}$ is not always sufficient.

Proposition 2.10 *Let $\alpha \in (1, 2)$ and U be an integrable strictly α-stable Lévy process such that U is non-trivial, i.e. $\mathbb{P}(U \neq 0) > 0$. Then $L := U(\Theta)$ is integrable if and only if $|\Theta(1)|^{1/\alpha}$ is integrable.*

Proof Let f be the growth function of U. Then

$$f(\theta) = \mathbb{E}|U(\theta)| = \sum_{j=1}^{d} \theta_j^{1/\alpha}\mathbb{E}|U_j(1)|.$$

Thus, theorem 2.7 yields that L is integrable if and only if $\int_{\mathbb{R}_+^d} |\theta|^{1/\alpha}\kappa(d\theta) < \infty$. Proposition 2.4 implies the claim. □

2.4.3 Hilbert Space Valued Variance Gamma Process

Univariate and multivariate variance gamma processes have been introduced by Madan and Seneta [152]. Univariate variance gamma processes can be constructed as a difference of two independent gamma processes or by subordinating a Brownian motion with a gamma process. The latter approach can be easily generalised to Hilbert space valued Lévy processes which we do in this section.

First, a definition of what we mean by a Hilbert space valued variance gamma process:

Definition 2.6 A Lévy process Y is a *Hilbert space valued variance gamma process* or *HVG-process*, if there are $a \in \mathbb{R}_+$, $b \in H$ and a positive semi-definite trace class operator Q on H such that its Lévy exponent is given by

$$\rho(u) = a\mathrm{Log}\,(1 + 1/2\langle Qu, u\rangle - i\,\langle b, u\rangle)\ ,$$

where Log denotes the main branch of the logarithm. (a, b, Q) are the *parameters of the HVG-process Y*.

We construct an HVG process Y in the following theorem.

Theorem 2.10 *Let $a \in \mathbb{R}_+$, $b \in H$ and Q a positive semi-definite trace class operator on H. Then there is an HVG-process Y with parameters (a, b, Q). The characteristic triplet (γ, Γ, ℓ) of Y is given by*

$$\gamma = ab - \int_{|x|>1} x\,\ell(dx),$$
$$\Gamma = 0 \quad \textit{and}$$
$$\ell(A) = \int_0^\infty \Phi_t(A) a t^{-1} e^{-t} dt$$

for any Borel set $A \subseteq H$ where Φ_t denotes the Gaussian measure on H with mean tb and covariance operator tQ. Moreover, $\mathbb{E}[Y(1)] = ab$ and $\mathrm{Cov}(Y(1)) = a(b \otimes b) + aQ$. $\langle Y, u\rangle$ is a variance gamma process for any $u \in H$.

Proof Let U be an H-valued drifted Wiener process

$$U(t) = bt + W(t)\,,$$

where $b \in H$ and W is an H-valued Wiener process with covariance operator Q. Define Θ to be a gamma process with parameters $(a, 1)$, i.e. a pure-jump subordinator with Lévy measure given by

$$\kappa(d\theta) = a\theta^{-1}e^{-\theta}1_{\{\theta>0\}}d\theta\ ,$$

cf. Bertoin [56, page 73]. Then its Laplace exponent is given by

$$\psi(z) = a\mathrm{Log}(1 - z)\,,$$

where Log denotes the main branch of the logarithm. Theorem 2.1 yields that the Lévy exponent of the Lévy process $L(t) := U(\Theta(t))$ is

$$\rho(u) = a\mathrm{Log}(1 + 1/2\langle Qu, u\rangle - i\,\langle b, u\rangle)\,.$$

Theorem 2.2 yields the specific form of the characteristics of L. Theorem 2.4 yields that L is square integrable and that its expectation and its covariance operator are given as in the theorem.

Let $u \in H$. Then $\langle L(t), u\rangle = \langle U, u\rangle(\Theta(t))$, and since $B := \langle U, u\rangle$ is a Gaussian Lévy process on $\mathbb{R}$ with drift $\langle b, u\rangle$ and covariance $\langle Qu, u\rangle$ it follows that $\langle L(t), u\rangle = B(\Theta(t))$ and consequently it is a variance gamma process.

Let Y be any HVG-process with parameters (a, b, Q). Then L and Y have the same characteristic functional and hence they have the same moments and therefore also the same characteristic triplet. Since L and Y have the same characteristic functional they have the same characteristics. □

We note that if Y is an HVG-process with parameters $(a, 0, Q)$, then its Lévy exponent satisfies $\rho(u) = \rho(-u)$ since $b = 0$, and thus Y is symmetric.

Since their introduction, variance gamma processes have been used extensively in financial modelling (see, e.g., Madan, Carr and Chang [151]).

2.5 The Bivariate Case with Dependent Variables

Recall the assumption of independence of the H_j-valued Lévy processes U_j, $j = 1, \dots, d$. The multivariate subordinator $\boldsymbol{\Theta}$ introduces a dependency among the components L_j of the Lévy process $\mathbf{L}$ unless Θ_j, $j = 1, \dots, d$ are independent, as can be seen from Theorem 2.1. In particular, if $\boldsymbol{\Theta}$ is square integrable, $\mathbf{U}(t) = \mathbf{b}t + \mathbf{W}(t)$ for some $\mathbf{b} \in H$ and square integrable Wiener process $\mathbf{W} = (W_1, \dots, W_d)$ in H with covariance operator $Q = Q_1 \times \dots \times Q_d$, it follows from Lemma 2.4 that

$$\mathrm{Cov}(\langle L_j(1), u\rangle_j, \langle L_k(1), v\rangle_k) = \mathrm{Cov}(\Theta_j(1), \Theta_k(1))\langle b_j, u\rangle_j\langle b_k, v\rangle_k\,,$$

for $j, k = 1, \dots, d$ and $(u, v) \in H_j \times H_k$. Many practical applications require Lévy processes being functions in space, that is, H_j is a space of functions on some Borel measurable subset D_j of $\mathbb{R}^n$, $n \in \mathbb{N}$. If the evaluation map $\delta_x : H_j \mapsto \mathbb{R}; u \mapsto u(x)$ for $x \in D_j$ is a continuous linear operator, then there exists a $u_x \in H_j$ such that $\delta_x := \langle \cdot, u_x\rangle_j$. Therefore, we find that the covariance between $L_j(1, x)$ and $L_k(1, y)$ for $x \in D_j$ and $y \in D_k$ is given by

$$\mathrm{Cov}(L_j(1, x), L_k(1, y)) = \mathrm{Cov}(\Theta_j(1), \Theta_k(1))b_j(x)b_k(y)\,.$$

Indeed, the correlation between $L_j(1, x)$ and $L_k(1, y)$ is

$$\mathrm{Corr}(L_j(1, x), L_k(1, y)) = \mathrm{sgn}(b_j(x)b_k(y))\mathrm{Corr}(\Theta_j(1), \Theta_k(1))$$

which shows that there is not very much variability across spatial points x and y in the correlation structure.

In this Section we consider the case when $\mathbf{U}$ is square integrable and where the components U_j, $j = 1, \ldots, d$ are *not* assumed to be independent. This will provide us with a rather flexible structure for the covariance operator, where we can achieve more realistic correlations over space than is the case in the subordinated example above. To lessen the technicalities, we assume $d = 2$. Our main concern will be to analyse the covariance operator of two *dependent* square integrable random variables in $H = H_1 \times H_2$. The overall goal is to represent two dependent Wiener processes in terms of an independent Wiener process scaled by the "correlation", much in parallel to the real-valued case. Our analysis here is taken from Benth and Krühner [37].

First, consider two square integrable random variables $X_j \in H_j$ defining a square-integrable random variable $(X_1, X_2) \in H_1 \times H_2$. Without loss of generality, we suppose that the variables have zero mean (i.e., they are centered). We denote the covariance operator of (X_1, X_2) by $\mathcal{Q}$, and argue next that it can be represented as a 2×2 matrix-valued operator:

To this end, introduce the projection operators $\Pi_i : H_1 \times H_2 \to H_i, i = 1, 2$ by

$$\Pi_1(u, v) = u\,, \qquad \Pi_2(u, v) = v\,. \tag{2.6}$$

We note that the dual operator $\Pi_i^* : H_j \to H_1 \times H_2$ is the natural embedding, e.g., $\Pi_1^* u = (u, 0)$ for $u \in H_1$, and similarly for Π_2^*. Our first lemma identifies the covariance operators of X_1 and X_2 from $\mathcal{Q}$ and the matrix-representation:

Lemma 2.5 *If $\mathcal{Q}$ is the covariance operator of the square-integrable $H_1 \times H_2$-valued random variable (X_1, X_2), then $\mathcal{Q}_i := \Pi_i \mathcal{Q} \Pi_i^*$ is the covariance operator of X_i for $i = 1, 2$. Define the linear operator $\mathcal{Q}_{12} \in L(H_1, H_2)$ by $\mathcal{Q}_{12} := \Pi_2 \mathcal{Q} \Pi_1^*$. Then it follows that*

$$\mathcal{Q} = \begin{bmatrix} \mathcal{Q}_1 & \mathcal{Q}_{12}^* \\ \mathcal{Q}_{12} & \mathcal{Q}_2 \end{bmatrix}, \tag{2.7}$$

is a representation of the covariance operator $\mathcal{Q}$.

Proof Let $i = 1, 2$. Since for $u \in H_i$, $\langle X_i, u\rangle_i = \langle (X_1, X_2), \Pi_i^*(u)\rangle$, it follows that

$$\mathbb{E}[\langle X_i, u\rangle_i \langle X_i, v\rangle_i] = \langle \mathcal{Q}\Pi_i^*(u), \Pi_i^*(v)\rangle = \langle \Pi_i \mathcal{Q} \Pi_i^* u, v\rangle_i$$

for all $u, v \in H_i$. Obviously, $\mathcal{Q}_i = \Pi_i \mathcal{Q} \Pi_i^*$ is symmetric and positive semidefinite operator on H_i. Moreover,

$$\mathrm{Tr}(\mathcal{Q}_i) = \|\mathcal{Q}^{1/2}\Pi_i^*\|_{\mathrm{HS}}^2 = \|\Pi_i \mathcal{Q}^{1/2}\|_{\mathrm{HS}}^2 \le \|\Pi_i\|_{\mathrm{op}}^2 \mathrm{Tr}(\mathcal{Q}) < \infty,$$

and hence $\mathcal{Q}_i$ is trace class. Here, $\|\cdot\|_{\mathrm{HS}}$ denotes the Hilbert-Schmidt norm on operators on H_i.

For the second part, we notice that $(u, v) = \Pi_1^*(u) + \Pi_2^*(v)$ for any $(u, v) \in H_1 \times H_2$. Hence, if $(u, v), (x, y) \in H_1 \times H_2$, we find

$$\begin{aligned}\langle Q(u, v), (x, y)\rangle &= \langle \Pi_1 Q(u, v), x\rangle_1 + \langle \Pi_2 Q(u, v), y\rangle_2 \\ &= \langle \Pi_1 Q(\Pi_1^*(u) + \Pi_2^*(v)), x\rangle_1 + \langle \Pi_2 Q(\Pi_1^*(u) + \Pi_2^*(v)), y\rangle_2 \\ &= \langle Q_1 u, x\rangle_1 + \langle Q_{12}^* v, x\rangle_1 + \langle Q_{12} u, y\rangle_2 + \langle Q_2 v, y\rangle_2 .\end{aligned}$$

The representation follows. □

As we will see, one can bound Q_{12} by the square roots of Q_1 and Q_2, as well as characterizing its range by the same covariances.

Proposition 2.11 *For $i = 1, 2$, let X_i be a square integrable random variable on H_i with covariance operator Q_i. Then the operator Q_{12} in* (2.7) *satisfies,*

(i) $|\langle Q_{12} u, v\rangle_2| \leq |\mathcal{R}_1 u|_1 |\mathcal{R}_2 v|_2$ *for any $u \in H_1$ and $v \in H_2$,*
(ii) $ran(Q_{12}) \subset \overline{ran(Q_2)}$ *and* $ran(Q_{12}^*) \subset \overline{ran(Q_1)}$,

with $\mathcal{R}_i$ being the unique positive definite square root of Q_i.

Proof First, by Pedersen [163, Prop. 3.2.11], there exists a unique positive definite operator $\mathcal{R}_i$ to the positive definite operator Q_i, $i = 1, 2$. Recall the covariance operator Q of (X_1, X_2) in (2.7). For $u \in H_1$, $v \in H_2$ and $\beta \in \mathbb{R}$, we find that

$$\begin{aligned}0 &\leq \langle Q(\beta u, v), (\beta u, v)\rangle \\ &= \beta^2 \langle Q_1 u, u\rangle_1 + \langle Q_2 v, v\rangle_2 + 2\beta \langle Q_{12} u, v\rangle_2 \\ &= \beta^2 |\mathcal{R}_1 u|_1^2 + |\mathcal{R}_2 v|_2^2 + 2\beta \langle Q_{12} u, v\rangle_2 ,\end{aligned}$$

or,

$$-\beta \langle Q_{12} u, v\rangle_2 \leq \frac{1}{2}\left(\beta^2 |\mathcal{R}_1 u|_1^2 + |\mathcal{R}_2 v|_2^2\right) .$$

Choosing β such that it has the same sign as $-\langle Q_{12} u, v\rangle_2$ implies that

$$|\beta| |\langle Q_{12} u, v\rangle_2| \leq \frac{1}{2}\left(\beta^2 |\mathcal{R}_1 u|_1^2 + |\mathcal{R}_2 v|_2^2\right) .$$

If $|\mathcal{R}_1 u|_1^2 \neq 0$, set $|\beta| = |\mathcal{R}_2 v|_2 / |\mathcal{R}_1 u|_1$ to obtain the desired inequality. In case $|\mathcal{R}_1 u|_1 = 0$, we find that $|\langle Q_{12} u, v\rangle_2| = 0$, otherwise we reach a contradiction after letting $|\beta| \to \infty$. Thus, (i) follows.

Let $v \in \ker(Q_2)$. Then $|\mathcal{R}_2 v|_2^2 = \langle Q_2 v, v\rangle_2 = 0$, and therefore we get from (i) that $|\langle Q_{12} u, v\rangle_2| = 0$ for any $u \in H_1$. Hence, $\ker(Q_{12})$ is orthogonal to $\ker(Q_2)$. As Q_2 is symmetric, its kernel and the closure of the range are closed orthogonal spaces. Correspondingly we find that $\ker(Q_{12}^*)$ is orthogonal to $\ker(Q_1)$. Thus, claim (ii) follows. □

We have a reverse result under some restrictions on the off-diagonal operator Q_{12}, found in the next theorem:

Theorem 2.11 *For $i = 1, 2$, let Q_i be positive semidefinite operators with $\mathcal{R}_i = \sqrt{Q_i}$. Let $Q_{12} \in L(H_1, H_2)$ be such that*

$$|\langle Q_{12}u, v\rangle| \leq |\mathcal{R}_1 u|_1 |\mathcal{R}_2 v|_2$$

for any $u \in H_1, v \in H_2$. Then,

$$Q := \begin{bmatrix} Q_1 & Q_{12}^* \\ Q_{12} & Q_2 \end{bmatrix},$$

defines a positive semidefinite operator on $H_1 \times H_2$. Moreover, Q is positive definite if and only if Q_1 and Q_2 are positive definite and

$$|\langle Q_{12}u, v\rangle| < |\mathcal{R}_1 u|_1 |\mathcal{R}_2 v|_2$$

for any $u \in H_1\backslash\{0\}$, $v \in H_2\backslash\{0\}$.

Proof Obviously, $Q = Q^*$. For $u \in H_1$ and $v \in H_2$, we find from the bound on Q_{12}

$$\begin{aligned} \langle Q(u, v), (u, v)\rangle &= \langle Q_1 u, u\rangle_1 + 2\langle Q_{12}u, v\rangle_2 + \langle Q_2 v, v\rangle_2 \\ &\geq |\mathcal{R}_1 u|_1^2 + |\mathcal{R}_2 v|_2^2 - 2|\mathcal{R}_1 u|_1 |\mathcal{R}_2 v|_2 \\ &= (|\mathcal{R}_1 u|_1 - |\mathcal{R}_2 v|_2)^2 \\ &\geq 0 . \end{aligned}$$

The first inequality is strict under the additional strict bound on Q_{12}. This proves the theorem. □

Note that Q in the Theorem above is not necessarily trace class even when Q_1 and Q_2 are trace class. Indeed, if we select an ONB $(e_i)_{i\in\mathbb{N}}$ being the eigenvectors of Q_1, with the associated eigenvalues $(\lambda_i)_{i\in\mathbb{N}}$, and correspondingly $(f_j)_{j\in\mathbb{N}}$ the eigenvectors of Q_2 with eigenvalues $(\eta_j)_{j\in\mathbb{N}}$, we find that $((e_i, f_j))_{i,j\in\mathbb{N}}$ is an ONB for $H_1 \times H_2$ and

$$\mathrm{Tr}(Q) = \sum_{i,j\in\mathbb{N}} \langle Q(e_i, f_j), (e_i, f_j)\rangle = \mathrm{Tr}(Q_1) + 2 \sum_{i,j\in\mathbb{N}} \langle Q_{12}e_i, f_j\rangle_2 + \mathrm{Tr}(Q_2) .$$

From the assumption on Q_{12} in the Theorem above, we find that

$$\sum_{i,j\in\mathbb{N}} |\langle Q_{12}e_i, f_j\rangle_2| \leq \sum_{i,j\in\mathbb{N}} |\mathcal{R}_1 e_i|_1 |\mathcal{R}_2 f_j|_2 = \sum_{i,j\in\mathbb{N}} \sqrt{\lambda_i \eta_j}\,.$$

If we, for example, have $\lambda_i = 1/i^2$ and $\eta_j = 1/j^2$, the sum diverges although the traces of Q_1 and Q_2, being the sums of λ_i and η_j, resp., are finite. Thus, imposing the conditions that Q_1 and Q_2 are finite trace class operators, will not be sufficient to ensure that Q becomes a finite trace class operator. We can conclude the following, though:

Corollary 2.2 *Let Q_1 and Q_2 be positive definite finite trace class operators and $Q_{12} \in L(H_1, H_2)$ such that $|\langle Q_{12}u, v\rangle| \leq |\mathcal{R}_1 u|_1 |\mathcal{R}_2 v|_2$ where $\mathcal{R}_1$ and $\mathcal{R}_2$ are the square roots of Q_1 and Q_2, resp. If $\mathcal{R}_1$ and $\mathcal{R}_2$ are finite trace class operators, then*

$$Q := \begin{bmatrix} Q_1 & Q_{12}^* \\ Q_{12} & Q_2 \end{bmatrix},$$

is a positive semidefinite finite trace class operator on $H_1 \times H_2$.

The Corollary states sufficient conditions to ensure that Q defines a covariance operator on $H_1 \times H_2$.

From elementary probability theory we know that for a bivariate Gaussian random variable (X, Y) there exists a $c \in \mathbb{R}$ and a univariate Gaussian random variable Z which is independent of X such that $Y = cX + Z$. We now extend this result to our infinite dimensional context.

First, recall $(X_1, X_2) \in H_1 \times H_2$ being square integrable with mean zero and covariance operator denoted Q. Consider a $C \in L(H_1, H_2)$ and define

$$Z := X_2 - CX_1. \tag{2.8}$$

Then Z is an H_2-valued random variable which is square integrable and having mean zero. In the next lemma we derive its covariance operator:

Lemma 2.6 *The covariance operator of Z in* (2.8) *is*

$$Q_Z = Q_2 - CQ_{12}^* - Q_{12}C^* + CQ_1C^*$$

Proof For $u, v \in H_2$, we find that

$$\mathbb{E}[\langle Z, u\rangle_2 \langle Z, v\rangle_2] = \mathbb{E}[\langle X_2, u\rangle_2 \langle X_2, v\rangle_2] - \mathbb{E}[\langle X_2, u\rangle_2 \langle X_1, C^* v\rangle_1] \\ - \mathbb{E}[\langle X_1, C^* u\rangle_1 \langle X_2, v\rangle_2] + \mathbb{E}[\langle X_1, C^* u\rangle_1 \langle X_1, C^* v\rangle_1]$$

Appealing to the representation of Q in (2.7) yields the result. □

Obviously,

$$X_2 = CX_1 + Z.$$

We ask the question if we can identify an operator C such that Z is uncorrelated from X_1, that is, $\mathbb{E}[\langle X_1, u\rangle_1 \langle Z, v\rangle_2] = 0$ for all $(u, v) \in H_1 \times H_2$.

Lemma 2.7 *Z defined in (2.8) is uncorrelated from X_1 if and only if $C \in L(H_1, H_2)$ is such that $Q_{12} = CQ_1$.*

Proof For $(u, v) \in H_1 \times H_2$, we have

$$\begin{aligned}\mathbb{E}[\langle X_1, u\rangle_1 \langle Z, v\rangle_2] &= \mathbb{E}[\langle X_1, u\rangle_1 \langle X_2, v\rangle_2] - \mathbb{E}[\langle X_1, u\rangle_1 \langle X_1, C^* v\rangle_1] \\ &= \langle (Q_{12} - CQ_1)u, v\rangle_2.\end{aligned}$$

The result follows. □

Note that if Q_1 would be invertible, then we could define $C = Q_{12}Q_1^{-1}$. However, as Q_1 is a covariance operator and thus being a compact operator, it does not permit any inverse in $L(H_1)$. We idenfity C via the *pseudo-inverse* instead.

For this purpose, we introduce the *pseudo-inverse* of an operator following the definition of Albert [4], which is based on the spectral theorem for normal compact operators on Hilbert spaces (see, e.g., Conway [73, Statement 7.6]):

Proposition 2.12 *Let H be a separable Hilbert space and C be a symmetric compact operator. Then there is an orthonormal basis $(e_i)_{i\in\mathbb{N}}$ of H and a family of real numbers $(\lambda_i)_{i\in\mathbb{N}}$ such that*

$$Cu = \sum_{i=1}^{\infty} \lambda_i \langle u, e_i\rangle e_i,$$

for any $u \in H$. For a measurable function $\varphi : \mathbb{R} \to \mathbb{R}$ and

$$Dom(\varphi(C)) := \left\{ u \in H \, : \, \sum_{i=1}^{\infty} |\varphi(\lambda_i)|^2 \langle u, e_i\rangle^2 < \infty \right\}$$

$$\varphi(C) : Dom(\varphi(C)) \to H\,, \qquad u \mapsto \sum_{i=1}^{\infty} \varphi(\lambda_i)\langle u, e_i\rangle e_i$$

defines a closed linear symmetric operator. Moreover, if φ is bounded on the set $(\lambda_i)_{i\in\mathbb{N}}$, then $\varphi(C)$ is everywhere defined and bounded. For measurable functions $\varphi, \psi : \mathbb{R} \to \mathbb{R}$ with ψ bounded, $(\varphi + \psi)(C) = \varphi(C) + \psi(C)$ and $(\varphi\psi)(C) = \varphi(C)\psi(C)$.

Assuming now that Q_1 is positive definite (not semi-!), we find that it has strictly positive eigenvalues $(\lambda_i)_{i\in\mathbb{N}}$, and, moreover, as it is a compact operator the eigenvalues converge to zero. Hence, from the above, we can therefore define the inverse Q_1^{-1} as the linear operator on H_1 given by

$$Q_1^{-1}u = \sum_{i=1}^{\infty} \lambda_i^{-1}\langle u, e_i\rangle_1 e_i \tag{2.9}$$

where $(e_i)_{i\in\mathbb{N}}$ is the ONB of eigenvectors of Q_1. Being positive definite, $\ker(Q_1) = \{0\}$, and hence $\overline{\text{Ran}(Q_1)} = \ker(Q_1)^{\perp} = H_1$. Hence, Q_1^{-1} is densely defined in H_1 as it is obviously defined on the range of Q_1. Also, it is an unbounded operator as $|Q_1^{-1}e_i|_1 = \lambda_i^{-1} \to \infty$ as $i \to \infty$. We wrap up in the following proposition:

Proposition 2.13 *Assume Q_1 is positive definite and $Q_{12} \in L(H_1, H_2)$ is such that $ran(Q_{12}^*) \subset ran(Q_1)$. Then the closure C of the densely defined operator $Q_{12}Q_1^{-1}$ is in $L(H_2, H_1)$ and with this choice Z in* (2.8) *will be uncorrelated from X_1.*

Proof Introduce $\mathcal{D} := Q_1^{-1}Q_{12}^*$. Then, since $\text{ran}(Q_{12}^*) \subset \text{ran}(Q_1)$ by assumption, we have that $\mathcal{D}$ is defined on H_2. Moreover, it is a closed operator, and it follows by the closed graph theorem that it is a bounded operator, i.e., $\mathcal{D} \in L(H_2, H_1)$. We observe, on the other hand, that $Q_{12}Q_1^{-1}$ is only densely defined on H_1 as Q_1^{-1} is. So, the dual of $\mathcal{D}$, must be the continuous linear continuation of $Q_{12}Q_1^{-1}$. Thus, $C = \mathcal{D}^*$ is the closure of $Q_{12}Q_1^{-1}$ as claimed.

As $C^* = Q_1^{-1}Q_{12}^*$, we find that $Q_1C^* = Q_{12}^*$, or $CQ_1 = Q_{12}$. Hence, by Lemma 2.7, the uncorrelation between Z and X_1 follows. □

We can apply these considerations in the modelling of $H = H_1 \times H_2$-valued Wiener processes. Let $W = (W_1, W_2)$ denote such a process with covariance operator Q, where we suppose that Q_1 is positive definite (not semi-!), and that $\text{ran}(Q_{12}^*) \subset \text{ran}(Q_1)$. Then we can define $C \in L(H_2, H_1)$ as in Prop. 2.13 above, and we have that $\widetilde{W} := W_1 - CW_2$ is an H_1-valued Wiener process independent of W_1. Thus, in modelling the time dynamics of a random phenomenon in H, we can use $W_1 = CW_2 + \widetilde{W}$ and W_2 as driving Wiener processes, much in align with the bivariate finite dimensional case.

2.6 Bibliographical Notes

The NIG distribution is a member of the generalized hyperbolic family of distributions, which has become a popular class to develop dynamical models for prices in various financial markets. Eberlein and Keller [91] model stock prices based on the hyperbolic Lévy process, while Eberlein and Stahl [92] apply similar dynamics for electricity prices in the German market. Eberlein with co-authors have also several papers on Lévy models for interest rate term structures which are relevant to commodity and power modeling, see for example Eberlein, Jacod and Raible [90]. In Frestad, Benth and Koekebakker [107] and Hinderks and Wagner [124] model power forwards in the Nordic and German markets, respectively, based on the NIG distribution. For a factorial HJM-model based on NIG, we refer to Piccirilli, Schmeck and Vargiolu [166]. A multivariate NIG model for forward prices in the Nordic market is found in Andersen, Koekebakker and Westgaard [5]. For power price models based on α-stable processes we refer to Garcia, Klüppelberg and Müller [109], and variance-gamma models has been applied in Goutte, Oudjane and Russo [118].

Subordination, which was first introduced by Bochner [61], has become a widely used tool to construct new Markov processes or C_0-semigroups. Barndorff-Nielsen, Pedersen and Sato [16] extended this approach to multivariate subordination of Lévy processes, i.e. subordination of d independent Lévy processes $U_1, \dots, U_d$ with d possibly dependent subordinators $\Theta_1, \dots, \Theta_d$. They proved that the resulting process $\mathbf{L}(t) := (U_1(\Theta_1(t)), \dots, U_d(\Theta_d(t)))$ is again a Lévy process and its characteristics as well as its Lévy exponent can be expressed easily in terms of properties of $\mathbf{U}$ and $\boldsymbol{\Theta}$. In the paper of Mendoza-Arriaga and Linetsky [156] multivariate subordination has been generalised to Markov processes with locally compact state spaces. Baeumer, Kovács and Meerschaert [10] treated multivariate subordination from an analytical point of view. Pérez-Abreu and Rocha-Arteaga [164] generalised the procedure to infinite-variate subordination of Hilbert space valued Lévy processes. General Lévy processes are treated in Pathasarathy [162], whereas Linde [148] treats the Banach space valued case.

Ané and Geman [6] model financial asset price returns by a stochastic time changed Brownian motion modeling the cumulative number of trades. A general study of time-changed Lévy processes in finance is found in Geman, Madan and Yor [112]. Borovkova and Schmeck [63] propose a time-changed Ornstein–Uhlenbeck model for the spot price dynamics of power in the German market, while Ladokhin, Schmeck and Borovkova [145] use time change to develop models for commodity forward curves which are applied to oil and gas. Kremer, Benth, Felten and Kiesel [142] analyse the effect of liquidity (number of trades) on the volatility of German power forward prices.

Chapter 3
The Filipović Space and Operators

Abstract We present the basic necessary properties of the state space of term structure curves, namely the Filipović space. Particular focus is on the Banach algebra structure and the quasi-contractivity of the shift semigroup. Next, we analyse operators on the Filipović space which are of interest in forward price modelling. This includes operators like Hilbert-Schmidt, nuclear and trace class to define covariance operators, integral operators to define the relation between forward prices with fixed delivery and delivery period, and multiplication operators to model the Samuelson effect. Several concrete examples are discussed.

In this chapter we introduce and analyse the Filipović space, a convenient separable Hilbert space for the realization of forward price curves. The material presented will be applied in later chapters when we study the stochastic dynamics of forward prices in energy markets. We also include an extensive analysis of different classes of operators on the Filipović space which is of interest for particular modeling in forward markets.

3.1 Basic Properties of the Filipović Space

This section is devoted to the study of basic properties of the Filipović space, where we focus our attention to those results relevant for our purposes. Some of the material of this section is collected from Filipović [99, Section 5.1], but as we dispense slightly with the assumptions in Filipović [99], we include some proofs of known results for the convenience of the reader (although they are mostly simple modifications of the arguments of Filipović [99]). We provide some new results beyond what can be found in Filipović [99] on the algebraic property of the space and further characterisations on the shift semigroup.

Let us define the Filipović space of absolutely continuous functions on the positive real line.

F. E. Benth, P. Krühner, *Stochastic Models for Prices Dynamics in Energy and Commodity Markets*, Springer Finance,
https://doi.org/10.1007/978-3-031-40367-5_3

Definition 3.1 Let $w : \mathbb{R}_+ \to [1, \infty)$ be a non-decreasing measurable function with $w(0) = 1$. The *Filipović space*, denoted H_w, is the space of absolutely continuous functions $f : \mathbb{R}_+ \to \mathbb{R}$ such that

$$|f|_w^2 := f^2(0) + \int_0^\infty w(x)|f'(x)|^2\,dx < \infty\,.$$

Here, f' is the weak derivative of f

Obviously, H_w is a linear (vector) space over the field $\mathbb{R}$ or $\mathbb{C}$. We shall use the field $\mathbb{R}$ in this book.

Remark 3.1 In Filipović [99] a standing assumption on the weight function w is that

$$\int_0^\infty w^{-1/3}(x)\,dx < \infty.$$

We will not assume this integrability property here, but frequently make use of the weaker assumption that $w^{-1} \in L^1(\mathbb{R}_+)$.

First, we prove that the fundamental theorem of calculus holds in H_w:

Lemma 3.1 *If $f \in H_w$, then $f' \in L^1([0, x])$ for any $x > 0$, $|f'|_{L^1([0,x])} \leq \sqrt{x}|f|_w$, and*

$$f(x) = f(y) + \int_y^x f'(z)\,dz$$

for any $y \in \mathbb{R}_+$, $x \geq y$.

Proof If $f \in H_w$, it has a weak derivative f'. Moreover, from Cauchy-Schwarz' inequality we find

$$\begin{aligned}\int_0^x |f'(y)|\,dy &= \int_0^x w^{-1/2}(y)w^{1/2}(y)|f'(y)|\,dy \\ &\leq (\int_0^x w^{-1}(y)\,dy)^{1/2}(\int_0^x w(y)|f'(y)|^2\,dy)^{1/2}\,.\end{aligned}$$

Since w is non-decreasing with $w(x) \geq 1$, we find $\int_0^x w^{-1}(y)\,dy \leq x$. Hence,

$$\int_0^x |f'(y)|\,dy \leq \sqrt{x}|f|_w < \infty.$$

Hence, $\int_y^x f'(z)\,dz$ is well defined for all $0 \leq y \leq x$, and the fundamental theorem of calculus for Lebesgue integrals (see Folland [104, Thm. 3.36]) implies that $f(x) = f(y) + \int_y^x f'(z)\,dz$. The result follows. □

Remark 3.2 In the case $w^{-1} \in L^1(\mathbb{R}_+)$, we get a stronger result. Indeed, as is seen from the proof of Lemma 3.1, we find that $f' \in L^1(\mathbb{R}_+)$ and

$$|f'|_{L^1(\mathbb{R}_+)} \leq \sqrt{|w^{-1}|_{L^1(\mathbb{R}_+)}} |f|_w .$$

Define for $f, g \in H_w$, the inner product

$$\langle f, g\rangle_w := f(0)g(0) + \int_0^\infty w(x) f'(x) g'(x)\, dx . \tag{3.1}$$

Naturally, $|f|_w^2 = \langle f, f\rangle_w$, and we have that $(H_w, |\cdot|_w)$ is a separable Hilbert space (Filipović [99, Thm. 5.1.1]).

A canonical choice of the weight function w of the Filipović space is $w(x) = \exp(\alpha x)$ with $\alpha > 0$. We immediately see that $w^{-1}(x) = \exp(-\alpha x) \in L^1(\mathbb{R}_+)$. With this choice of weight function, the Filipović space supports the Nelson-Siegel curves (see Nelson and Siegel [158]), which is defined as

$$f(x) = \beta_0 + \beta_1 e^{-\gamma x} + \beta_2 x e^{-\gamma x}, \tag{3.2}$$

for $x \geq 0$ and constant parameters $\beta_0, \beta_1, \beta_2 \in \mathbb{R}$ and $\gamma \in \mathbb{R}_+$. We know that H_w is a vector space supporting the constant functions. Moreover, as long as $\alpha < 2\gamma$, we have that

$$\int_0^\infty w(x) \exp(-2\gamma x) dx < \infty$$

and

$$\int_0^\infty w(x) x^2 \exp(-2\gamma x) dx < \infty$$

which implies that $f \in H_w$. The Nelson-Siegel curves are popular in parametric representations of market forward rates in fixed-income theory (see, e.g., Nelson and Siegel [158] and Filipović [100]). However, the curves may also be applied in the smooth representation of forward prices in energy and commodity markets (see for example Grønborg and Lunde [119] for a dynamic Nelson-Siegel model applied to oil futures). In Figure 3.1 we illustrate two possible shapes of the Nelson-Siegel curves. If we think of x as measuring days, then these two curves mimic the forward price of a commodity that flattens when maturity is in about one year. The forward price becomes constant equal to 50 (blue) and 40 (red) in the long end of the market, which is equal to the parameter β_0. The spot price of this commodity is found for $x = 0$, being equal to $\beta_0 + \beta_1$ in the Nelson-Siegel curve, here becoming 20 (blue) and 100 (red). Both curves have a hump in the shorter end of the market, before decaying towards the long-term level. The blue curve shows prices in the short end which is lower than the long end, while the red is opposite. A forward market is

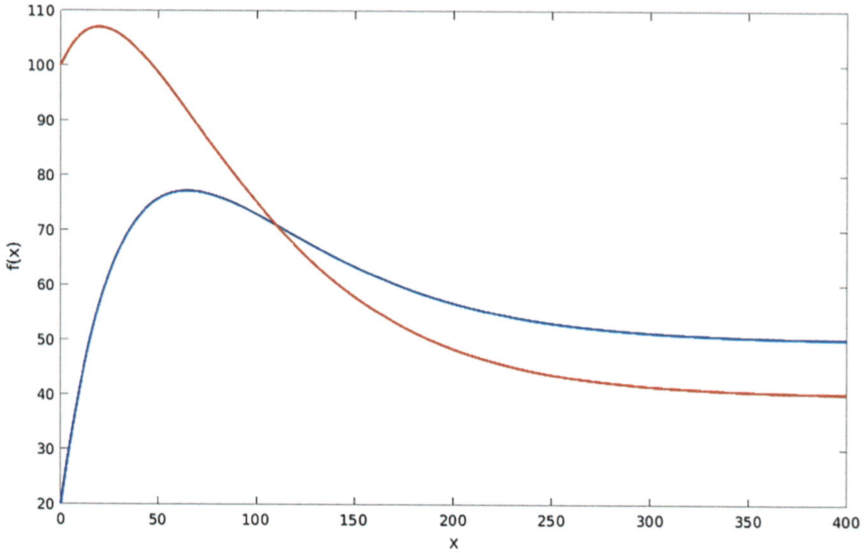

Fig. 3.1 Examples of Nelson-Siegel curves with parameters $\beta_0 = 40$, $\beta_1 = 60$. $\beta_2 = 2$, $\gamma = 0.02$ (red) and $\beta_0 = 50$, $\beta_1 = -30$, $\beta_2 = 2$, $\gamma = 0.02$ (blue)

said to be in *backwardation* when the spot price is higher than the forward prices. Apart from the small hump shape up to about $x = 40$, the red curve depicts a market in backwardation. When forward prices are higher than the spot, the market is said to be in *contango*. We refer to Geman [111] for more on this. In Section 6.8 of Chapter 6 we discuss further the Nelson-Siegel curve along with other methods to obtain a smooth forward curve. There, we consider smoothing of the term structure of crude oil prices in a case study.

We continue the theoretical exposition with some elementary properties of elements in the Filipović space. Any function in the Filipović space is bounded whenever w^{-1} is integrable, as the next result shows (see proof of Thm. 5.1.1 in Filipović [99] for this result).

Lemma 3.2 *Assume* $w^{-1} \in L^1(\mathbb{R}_+)$. *Then,*

$$\sup_{x \geq 0} |f(x)| \leq (1 + \sqrt{|w^{-1}|_{L^1(\mathbb{R}_+)}})|f|_w \,.$$

for any $f \in H_w$.

Proof If $f \in H_w$, it follows from Lemma 3.1 that

$$f(x) = f(0) + \int_0^x f'(y)\,dy$$

for any $x \geq 0$. By the Cauchy-Schwarz inequality it follows that

$$|f(x)| \leq |f(0)| + \int_0^x |f'(y)|\,dy \leq |f|_w + (\int_0^x w^{-1}(y)\,dy)^{1/2}|f|_w\,.$$

Thus, from the integrability assumption on w^{-1}, the claim follows. □

We next study the evaluation operator on H_w which will play an important role in our analysis in this book. To this end, introduce for $x \in \mathbb{R}_+$ the notation $\delta_x : H_w \to \mathbb{R}$ for the linear operator

$$\delta_x(f) = f(x)\,, \qquad f \in H_w\,. \tag{3.3}$$

Appealing to the fundamental theorem of calculus in H_w and the Cauchy-Schwarz inequality, we find that

$$\begin{aligned}|\delta_x(f)| &\leq |f(0)| + (\int_0^x w^{-1}(y)\,dy)^{1/2}(\int_0^x w(y)|f'(y)|^2\,dy)^{1/2} \\ &\leq \left(1 + (\int_0^x w^{-1}(y)\,dy)^{1/2}\right)|f|_w\,.\end{aligned}$$

We recall that w is non-decreasing with $w(0) = 1$, and thus $w^{-1}(y) \leq 1$. Therefore, δ_x is a linear functional on H_w, i.e., $\delta_x \in H_w^*$ for all $x \geq 0$.

In applications to energy and commodity forward contracts we will use H_w as the state space of the forward price dynamics. Hence, as δ_x is a linear functional on H_w, we can make sense out of the forward price dynamics for a particular contract with given maturity, by simply applying the evaluation operator. This also includes the spot prices dynamics as a particular choice of the evaluation operator.

For $x \in \mathbb{R}_+$, define the function $h_x : \mathbb{R}_+ \to \mathbb{R}$ by

$$h_x(y) = 1 + \int_0^{x \wedge y} w^{-1}(z)\,dz\,. \tag{3.4}$$

Note that since w is non-decreasing with $w(0) = 1$, $w^{-1}(z) \leq 1$ for all $z \geq 0$. Thus, h_x is well-defined for all $y \in \mathbb{R}_+$. Moreover, it is a non-decreasing function, being constant equal to $h_x(x)$ for $y > x$. We have:

Lemma 3.3 *For every fixed $x \in \mathbb{R}_+$, $h_x \in H_w$ and $\delta_x(f) = \langle f, h_x \rangle_w$ for any $f \in H_w$. Moreover, the adjoint of δ_x, $\delta_x^* : \mathbb{R}_+ \to H_w$ is the linear operator $\delta_x^*(a) = ah_x$ for $a \in \mathbb{R}_+$ and $\|\delta_x\|_{op}^2 = h_x(x)$.*

Proof First, h_x is an absolutely continuous function on $\mathbb{R}_+$ for every $x \in \mathbb{R}_+$, with derivative $h_x'(y) = w^{-1}(y)$ for $y < x$ and $h_x'(y) = 0$ for $y > x$. Moreover, since

$h_x(0) = 1$, we find

$$|h_x|_w^2 = 1 + \int_0^\infty w(y)|h_x'(y)|^2\,dy = 1 + \int_0^x w^{-1}(y)\,dy = h_x(x) < \infty\,,$$

and thus $h_x \in H_w$.

If $f \in H_w$, we find from the fundamental theorem of calculus of Lemma 3.1 that

$$\delta_x(f) = f(x) = f(0) + \int_0^x f'(y)\,dy\,,$$

for any $f \in H_w$. But $h_x(0) = 1$ and $h_x'(y) = w^{-1}(y)$ for $y < x$ and zero otherwise, and thus

$$\delta_x(f) = f(0)h_x(0) + \int_0^\infty w(y)f'(y)h_x'(y)\,dy = \langle f, h_x\rangle_w\,.$$

This shows the first part of the Lemma.

By definition of an adjoint operator, it holds for any $a \in \mathbb{R}_+$ and $f \in H_w$ that $\langle \delta_x^*(a), f\rangle_w = a\delta_x(f) = a\langle h_x, f\rangle_w$, and hence $\delta_x^*(a) = ah_x$.

Since $\|\delta_x\|_{\text{op}} = \|\delta_x^*\|_{\text{op}}$, we find

$$\|\delta_x\|_{\text{op}} = \sup_{|a|=1} |\delta_x^* a|_w = \sup_{|a|=1} |a||h_x|_w = |h_x|_w\,.$$

Thus, from the above, $\|\delta_x\|_{\text{op}}^2 = |h_x|_w^2 = h_x(x)$, and the proof is complete. □

We use this to show that any element in the Filipović space has an asymptotic limit when $x \to \infty$:

Lemma 3.4 *Assume $w^{-1} \in L^1(\mathbb{R}_+)$. Then, with h_∞ defined as*

$$h_\infty : \mathbb{R}_+ \to \mathbb{R}, \qquad x \mapsto 1 + \int_0^x w^{-1}(y)dy, \tag{3.5}$$

it holds that $h_\infty \in H_w$ and

$$\lim_{x\to\infty} f(x) = \langle h_\infty, f\rangle_w$$

for any $f \in H_w$.

Proof By integrability of w^{-1}, we find

$$|h_\infty|_w^2 = 1 + \int_0^\infty w^{-1}(y)dy < \infty$$

and therefore $h_\infty \in H_w$. It also holds that $h_x \to h_\infty$ in H_w since

$$|h_x - h_\infty|_w^2 = \int_x^\infty w^{-1}(y)dy \to 0$$

as $x \to \infty$. Hence,

$$\lim_{x\to\infty} f(x) = \lim_{x\to\infty} \delta_x(f) = \lim_{x\to\infty} \langle h_x, f\rangle_w = \langle h_\infty, f\rangle_w$$

for any $f \in H_w$. The claim follows. □

It turns out that the the geometric structure of H_w implies a strong connection between point evaluations at different spatial points $x \in \mathbb{R}_+$. The following result from Benth and Krühner [34, Lemma 3.11] is useful for example when we want to analyse the correlation structure of random elements in H_w interpreted as random fields in $x \in \mathbb{R}_+$ (see Section 3.2 in this Chapter).

Lemma 3.5 *For the function $h_x \in H_w$ defined in* (3.4)*, it holds that the map*

$$h : \mathbb{R}_+ \to H_w, x \mapsto h_x$$

is 0.5*-Hölder continuous with constant 1. Moreover,*

$$|h_y - h_x|_w \geq \sqrt{\frac{y-x}{w(y)}}$$

for any $0 \leq x \leq y < \infty$.

Proof By the definition of h_x, we find for any $0 \leq x \leq y < \infty$ that

$$|h_y - h_x|_w^2 = \int_x^y w^{-1}(z)dz.$$

Since w is non-decreasing, $w^{-1}(y) \leq w^{-1}(z) \leq w^{-1}(x)$ for $x \leq z \leq y$. We recall $w(0) = 1$, and therefore $w^{-1}(y) \leq w^{-1}(z) \leq 1$. Hence

$$\frac{y-x}{w(y)} \leq |h_y - h_x|_w^2 \leq y - x$$

and the result follows. □

As is shown in Benth and Krühner [34, Prop. 4.18], the Filipović space can be turned into a Banach algebra after a simple re-scaling of the norm. This is an attractive property when analysing geometric term structure models.

Proposition 3.1 *Assume $w^{-1} \in L^1(\mathbb{R}_+)$. Then, H_w is a Banach algebra with respect to the norm $\|\cdot\|_w := c|\cdot|_w$, where $c = \sqrt{1 + 4(1 + \sqrt{|w^{-1}|_{L^1(\mathbb{R}_+)}})^2}$.*

Proof If $f, g \in H_w$, then it follows from Lemma 3.2 that fg is absolutely continuous. Indeed, as f and g are absolutely continuous, if holds by definition that for a given $\epsilon > 0$, there exists a $\delta > 0$ such that any for disjoint sequence of subintervals $\{(a_i, b_i)\}_{i=1}^n$ of $\mathbb{R}_+$ for which $\sum_{i=1}^n (b_i - a_i) < \delta$ it holds that $\sum_{i=1}^n |f(b_i) - f(a_i)| < \epsilon/2\max(|f|_\infty, |g|_\infty)$ and $\sum_{i=1}^n |g(b_i) - g(a_i)| < \epsilon/2\max(|f|_\infty, |g|_\infty)$. Here, $|f|_\infty := \sup_{x\geq 0}|f(x)|$. By the triangle inequality, we find

$$
\begin{aligned}
\sum_{i=1}^n |f(b_i)g(b_i) - f(a_i)g(a_i)| &\leq \sum_{i=1}^n |g(b_i)||f(b_i) - f(a_i)| \\
&\quad + \sum_{i=1}^n |f(a_i)||g(b_i) - g(a_i)| \\
&\leq |g|_\infty \sum_{i=1}^n |f(b_i) - f(a_i)| \\
&\quad + |f|_\infty \sum_{i=1}^n |g(b_i) - g(a_i)| < \epsilon\,.
\end{aligned}
$$

Absolute continuity of fg follows. From the product rule its weak derivative becomes

$$
(fg)'(x) = f'(x)g(x) + f(x)g'(x)\,.
$$

Hence, from the elementary inequality $(a+b)^2 \leq 2a^2 + 2b^2$,

$$
\begin{aligned}
|fg|_w^2 &= f^2(0)g^2(0) + \int_0^\infty w(x)|f'(x)g(x) + f(x)g'(x)|^2\,dx \\
&\leq f^2(0)g^2(0) + 2\int_0^\infty w(x)|g(x)|^2|f'(x)|^2\,dx \\
&\quad + 2\int_0^\infty w(x)|f(x)|^2|g'(x)|^2\,dx \\
&\leq f^2(0)g^2(0) + 2k^2|g|_w^2 \int_0^\infty w(x)|f'(x)|^2\,dx \\
&\quad + 2k^2|f|_w^2 \int_0^\infty w(x)|g'(x)|^2\,dx \\
&\leq |f|_w^2|g|_w^2 + 4k^2|f|_w^2|g_w^2\,,
\end{aligned}
$$

where we appealed to Lemma 3.2 in the third step, setting $k = 1 + \sqrt{|w^{-1}|_{L^1(\mathbb{R}_+)}}$. It follows that $fg \in H_w$ and hence H_w is closed under multiplication. Moreover,

$$|fg|_w \leq \sqrt{1+4k^2}|f|_w|g|_w\,.$$

Defining $\|\cdot\|_w := \sqrt{1+4k^2}|\cdot|_w$, $\|fg\|_w \leq \|f\|_w\|g\|_w$ as claimed. □

Let $g \in H_w$ with $w^{-1} \in L^1(\mathbb{R}_+)$. Define $f(x) := \exp g(x)$, which is a measurable function on $\mathbb{R}_+$. By Lemma 3.2, $\sup_{x\geq 0}|g(x)| < \infty$ so $f(\mathbb{R}_+) \subset [a,b]$ for some $-\infty < a \leq b < \infty$. But the exponential function is Lipschitz continuous on any bounded interval $[a,b]$, and it follows that f is absolutely continuous on $\mathbb{R}_+$. Invoking Prop. 3.1, we have $|g|_w^n \leq c^{n-1}|g|_w^n$ for any $n \in \mathbb{N}$ and c given in the proposition. Thus,

$$|\exp g|_w \leq c^{-1}\exp(c|g|_w) < \infty\,,$$

and it follows that $\exp g \in H_w$. Indeed, the Filipović space with $w^{-1} \in L^1(\mathbb{R}_+)$ is stable with respect to exponentiating.

Remark 3.3 A more direct argument to show that $\exp g \in H_w$ whenever $g \in H_w$ is to observe from Prop. 3.1 that $g^k \in H_w$ and $\|g^k\|_w = \|g\|_w^k$ for any $k \in \mathbb{N}$ by the Banach algebra structure. Hence, $(f_n)_{n\in\mathbb{N}}$ with $f_n := \sum_{k=0}^{n}\frac{1}{k!}g^k$ is a Cauchy sequence in H_w, and by completeness, there exists an $f \in H_w$ such that $f_n \to f$.

In the remainder of this Section we focus on the family of shift semigroup operators $(\mathcal{S}_t)_{t\geq 0}$ that plays an important role in the study of the dynamics of the forward curve dynamics. For any $t \geq 0$ and $f \in H_w$, we define the function $\mathcal{S}_t f : \mathbb{R}_+ \to \mathbb{R}$ as

$$\mathcal{S}_t f(x) = f(x+t)\,. \tag{3.6}$$

Under the condition that w is a continuously differentiable function on $\mathbb{R}_+$, $(\mathcal{S}_t)_{t\geq 0}$ is a C_0-semigroup on H_w (see Thm. 5.1.1 in Filipovic [99]). At an informal level, for any sufficiently smooth function in $f \in H_w$ we find that $\lim_{t\downarrow 0}\frac{1}{t}(\mathcal{S}_t - f)f = f'$. Hence, the generator of $(\mathcal{S}_t)_{t\geq 0}$ is expected to be the first-order derivative operator d/dx. Indeed, from Filipović [99, Lemma 4.2.2 and Cor. 5.1.1] we find that the C_0-semigroup $(\mathcal{S}_t)_{t\geq 0}$ has generator $(d/dx, \mathrm{Dom}(d/dx))$, where

$$\mathrm{Dom}(d/dx) = \left\{f \in H_w \mid f' \in H_w\right\},$$

is its dense domain.

In this book we will make use of the Filipović space as the state space for the stochastic dynamics of forward price curves. To have available all the convenient properties of the Filipović space, like its Banach algebra structure and the C_0-

semigroup $(\mathcal{S}_t)_{t\geq 0}$, we assume from now on that the weight function w satisfies the following properties:

Assumption 3.1 The function $w : \mathbb{R}_+ \to \mathbb{R}_+$ is a non-decreasing continuously differentiable function with $w(0) = 1$ and $w^{-1} \in L^1(\mathbb{R}_+)$. □

Throughout the rest of this book we will work under Assumption 3.1 for w.

We next show some convenient bounds on the operator norm of the shift semigroup. The first is extracted from the proof of Thm. 5.1.1 in Filipović [99]:

Lemma 3.6 *The shift operator $(\mathcal{S}_t)_{t\geq 0}$ is uniformly bounded, that is, there exists a constant $K > 0$ such that*

$$\|\mathcal{S}_t\|_{op} \leq K$$

for any $t \geq 0$.

Proof For any $f \in H_w$, we have

$$|\mathcal{S}_t f|_w^2 = f^2(t) + \int_0^\infty w(y) f'(y+t)^2 \, dy \, .$$

But from the fundamental theorem of calculus and an elementary inequality,

$$\begin{aligned} f^2(t) &\leq 2f^2(0) + 2(\int_0^t f'(x)\,dx)^2 \\ &\leq 2f^2(0) + 2\int_0^t w^{-1}(x)\,dx \int_0^t w(x) f'(x)^2\,dx \\ &\leq 2(1 + \int_0^t w^{-1}(y)\,dy)|f|_w^2 \, . \end{aligned}$$

In the second estimate we applied the Cauchy-Schwarz inequality. Further, since w is a non-decreasing function,

$$\int_0^\infty w(y) f'(y+t)^2\,dy \leq \int_t^\infty w(x-t) f'(x)^2\,dx \leq \int_t^\infty w(x) f'(x)^2\,dx \, .$$

We conclude that $|\mathcal{S}_t f|_w^2 \leq (3 + 2|w^{-1}|_{L^1(\mathbb{R}_+)})|f|_w^2$, and the result follows. □

A similar argument shows that the shift semigroup is quasi-contractive, see Benth and Krühner [34, Lemma 3.5]:[1]

[1] At the end of the proof of Lemma 3.5 in Benth and Krühner [34] there is a flaw that has been corrected here.

Lemma 3.7 *The shift semigroup $(\mathcal{S}_t)_{t\geq 0}$ satisfies the exponential growth bound*

$$\|\mathcal{S}_t\|_{op} \leq e^{t/2},$$

for any $t \geq 0$, i.e., it is quasi-contractive.

Proof Fix $t \geq 0$ and $f \in H_w$. Define the functions $g(x) = f(t \wedge x)$ and $\widetilde{g}(x) = 1_{t\leq x}(f(x) - f(t))$. Then it is easy to see that $g, \widetilde{g} \in H_w$ are orthogonal and $f = g + \widetilde{g}$. Moreover, $|f|_w^2 = |g|_w^2 + |\widetilde{g}|_w^2$. Since $\mathcal{S}_t g(x) = g(x+t) = f(t \wedge (x+t)) = f(t) = g(t)$, we find

$$|\mathcal{S}_t g|_w^2 = g^2(t) = |\delta_t g|^2 \leq \|\delta_t\|_{\mathrm{op}}^2 |g|_w^2 .$$

But from Lemma 3.3 it holds that $\|\delta_t\|_{\mathrm{op}}^2 = h_t(t)$, and since w is non-decreasing with $w(0) = 1$,

$$\|\delta_t\|_{\mathrm{op}}^2 = h_t(t) = 1 + \int_0^t w^{-1}(x)\, dx \leq 1 + t .$$

Hence, $|\mathcal{S}_t g|_w^2 \leq (1+t)|g|_w^2$. On the other hand, it follows from the non-decreasing property of w and $\widetilde{g}(t) = 0$ that,

$$\begin{aligned}
|\mathcal{S}_t\widetilde{g}|_w^2 &= (\mathcal{S}_t\widetilde{g})(0)^2 + \int_0^\infty w(x)(\mathcal{S}_t\widetilde{g})'(x)^2\, dx \\
&= \widetilde{g}^2(t) + \int_0^\infty w(x)\widetilde{g}'(x+t)^2\, dx \\
&= \int_t^\infty w(y-t)\widetilde{g}'(y)^2\, dy \\
&\leq \int_t^\infty w(y)\widetilde{g}'(y)^2\, dy \\
&\leq |\widetilde{g}|_w^2 .
\end{aligned}$$

The constancy of $\mathcal{S}_t g$ and $\mathcal{S}_t\widetilde{g}(0) = \widetilde{g}(t) = 0$ yield orthogonality of $\mathcal{S}_t g$ and $\mathcal{S}_t\widetilde{g}$:

$$\langle \mathcal{S}_t g, \mathcal{S}_t\widetilde{g}\rangle_w = g(t)\widetilde{g}(t) + \int_0^\infty w(x)(\mathcal{S}_t g)'(x)(\mathcal{S}_t\widetilde{g})'(x)\, dx = 0 .$$

We therefore find,

$$|\mathcal{S}_t f|_w^2 = |\mathcal{S}_t g + \mathcal{S}_t\widetilde{g}|_w^2 = |\mathcal{S}_t g|_w^2 + |\mathcal{S}_t\widetilde{g}|_w^2 \leq (1+t)|g|_w^2 + |\widetilde{g}|_w^2$$

But as $t \geq 0$, $(1+t)|g|_w^2 + |\widetilde{g}|_w^2 \leq (1+t)(|g|_w^2 + |\widetilde{g}|_w^2) = (1+t)|f|_w^2$, and $(1+t) \leq \exp(t)$. Hence, $|\mathcal{S}_t f|_w^2 \leq \exp(t)|f|_w^2$, and we conclude that $\|\mathcal{S}_t\|_{\mathrm{op}} \leq \exp(t/2)$. □

Notice that we assume $w^{-1} \in L^1(\mathbb{R}_+)$ for the uniform boundedness, while this is not necessary for the quasi-contractive property of the shift semigroup. By the Hille-Yoshida Theorem (see, e.g., Engel and Nagel [94, Prop. 5.5, Ch. I]), any C_0-semigroup satisfies an exponential growth bound, which means that there exist positive constants M, C such that $\|\mathcal{S}_t\|_{\mathrm{op}} \leq Me^{Ct}$. The uniform boundedness and quasi-contractivity provide us with sharper estimates on the operator norm.

We end this section with remarking that H_w is a natural choice as state space for forward curves. The absolute continuity of elements in the Filipović space prevents discontinuities in the term structure of forward prices. A discontinuity at some maturity $x > 0$ would cause a possible arbitrage opportunity as one could enter a short and long position at either side of x. Further, it is reasonable to have flat forward curves in the long end of the market, as it should not be any difference in buying or selling a commodity maturing 100 years or 101 years from now (or any other maturities far into the future). As the evaluation functional is continuous in H_w, we can speak of the forward price with time to maturity x for any $x \in \mathbb{R}_+$. In particular, choosing $x = 0$, we recover the spot price. The neat properties of the shift semigroup is also very convenient when analysing arbitrage-free stochastic dynamical models for the temporal evolution of forward term structures. Being a separable Hilbert space, the Filipović space provides an attractive framework for infinite dimensional stochastic calculus and analysis of operators.

3.2 Operators on the Filipović Space

In specifying a Wiener or Lévy field, as defined in Chapter 2, we need to pin down a covariance operator $\mathcal{Q}$. Further, when defining a dynamics for forward prices, the volatility becomes naturally a linear bounded operator (or some process with values in this class). In this section we further explore the structure of linear operators on the Filipović space, with a particular focus on integral operators. As it turns out, integral operators play a key role in defining Hilbert-Schmidt operators, which is at the core of positive definite trace class operators (that is, the covariance operators, which are squares of symmetric Hilbert-Schmidt operators). Integral operators are also relevant in defining the dynamics of forwards on power and temperature, as we will see in Chapter 6 (Section 6.5). Finally, multiplication operators provides a rich class of operators for simple specifications of the terms structure of volatility, modelling for example the Samuelson effect.

In the analysis of operators on Filipović space, we will make frequent use of the map

$$\mathcal{W} : H_w \to L^2(\mathbb{R}_+), \quad f \mapsto \sqrt{w} f'. \tag{3.7}$$

As we show in the proof of Prop. A.1, the mapping

$$\mathcal{T} := (\delta_0, \mathcal{W}) : H_w \to \mathbb{R} \times L^2(\mathbb{R}_+), \quad f \mapsto (f(0), \mathcal{W}f)$$

is an isometric isomorphism of the Hilbert spaces H_w and $\mathbb{R} \times L^2(\mathbb{R}_+)$.

3.2.1 Integral Operators

A natural starting point for integral operators on the Filipović space would be to consider expressions like

$$\mathcal{I}f(x) = \int_0^\infty c(x, y) f(y) dy$$

for $f \in H_w$ and some suitable kernel function c. However, recalling Lemma 3.1 which says that any $f \in H_w$ satisfies the the fundamental theorem of calculus

$$f(y) = f(0) + \int_0^y f'(z) dz,$$

we derive

$$\begin{aligned} \mathcal{I}f(x) &= \int_0^\infty c(x, y) dy f(0) + \int_0^\infty c(x, y) \int_0^y f'(z) dz dy \\ &= \int_0^\infty c(x, y) dy \delta_0(f) + \int_0^\infty \left\{ \int_z^\infty c(x, y) dy \right\} f'(z) dz. \end{aligned}$$

The first term above is simply a function multiplied by the evaluation functional δ_0 applied to f. The second term is an integral operator applied on f', the weak derivative of f. This representation of $\mathcal{I}$ shows that on the Filipović space it is appropriate to focus the attention to integral operators being defined as

$$\mathcal{J}f(x) = \int_0^\infty k(x, y) f'(y) dy. \tag{3.8}$$

In the remainder of this subsection we will discuss conditions on the kernel function $k : \mathbb{R}_+^2 \to \mathbb{R}$ ensuring that $\mathcal{J}$ is well-defined, defined on H_w and becomes a bounded (i.e., continuous) operator.

We first show that under minimal conditions on k, the integral operator becomes a measurable function:

Lemma 3.8 *Assume that the integral operator $\mathcal{J}$ in* (3.8) *with $f \in H_w$ satisfies*

$$\int_0^\infty |k(x, y)f'(y)|dy < \infty$$

for any $x \in \mathbb{R}_+$. Then $\mathcal{J}f$ is a measurable function.

Proof According to Folland [104, Thm. 2.10(b)], we can select a sequence of elementary functions $k_n : \mathbb{R}_+^2 \to \mathbb{R}$ such that $k_n \to k$ pointwise and $|k_n(x, y)| \leq |k(x, y)|$. Similarly, as f' is measurable as well, we can choose a sequence of elementary functions $g_n : \mathbb{R}_+ \to \mathbb{R}$ such that $g_n \to f'$ pointwise and $|g_n(x)| \leq |f'(x)|$. It follows that $k_n g_n \to kf'$ pointwise and $|k_n(x, y)g_n(y)| \leq |k(x, y)f'(y)|$, and the dominated convergence theorem (see Folland [104, Thm. 2.24]) yields,

$$\mathcal{J}f(x) = \lim_{n\to\infty} \int_0^\infty k_n(x, y)g_n(y)dy.$$

But the functions $x \mapsto \int_0^\infty k_n(x, y)g_n(y)dy$ are elementary on $\mathbb{R}_+$ and therefore measurable, and the Lemma follows as $\mathcal{J}f$ is the pointwise limit of measurable functions. □

Under an explicit condition on integrability of the kernel function scaled by the weight function w of the Filipović space, we can show that $\mathcal{J}$ in (3.8) is a closed linear operator in H_w:

Lemma 3.9 *Suppose that for any $x \in \mathbb{R}_+$, $k(x, \cdot)/\sqrt{w} \in L^2(\mathbb{R}_+)$. Then the integral operator $\mathcal{J}$ in* (3.8) *is a closed linear operator on H_w.*

Proof For any $f \in H_w$, we find by the Cauchy-Schwarz inequality

$$\begin{aligned}\int_0^\infty |k(x, y)f'(y)|dy &= \int_0^\infty |k(x, y)|(\sqrt{w(y)})^{-1}\sqrt{w(y)}|f'(y)|dy \\ &\leq \left(\int_0^\infty k^2(x, y)w^{-1}(y)dy\right)^{1/2}\left(\int_0^\infty w(y)(f'(y))^2dy\right)^{1/2}\end{aligned}$$

which is finite by assumption. Hence, the operator $\mathcal{J}$ is defined for any $f \in H_w$. Furthermore, it is obviously linear on H_w.

Now, take a sequence $(f_n)_{n\in\mathbb{N}} \subset H_w$ such that $(\mathcal{J}f_n)_{n\in\mathbb{N}} \subset H_w$, $f_n \to f$ in H_w and $\mathcal{J}f_n \to g$ in H_w. We aim at proving $g = \mathcal{J}f$.

To this end, recall the continuous linear operator $\mathcal{W} : H_w \to L^2(\mathbb{R}_+)$ defined in (3.7). Hence, $\mathcal{W}f_n \to \mathcal{W}f$ in $L^2(\mathbb{R}_+)$, and for each $x \in \mathbb{R}_+$ we have

$$k(x, \cdot)f_n' = \frac{k(x, \cdot)}{\sqrt{w}}\mathcal{W}f_n \to \frac{k(x, \cdot)}{\sqrt{w}}\mathcal{W}f = k(x, \cdot)f'$$

where the convergence is in $L^1(\mathbb{R}_+)$ (which follows by again using the Cauchy-Schwarz inequality). By the dominated convergence theorem (see, e.g., Folland [104, Thm. 2.24]), we obtain,

$$\mathcal{J} f_n(x) = \int_0^\infty k(x,y) f_n'(y)dy \to \int_0^\infty k(x,y) f'(y)dy = \mathcal{J} f(x)$$

On the other hand, Lemma 3.3 yields that $\mathcal{J} f_n(x) \to g(x)$. Therefore, $g(x) = \mathcal{J} f(x)$ and $\mathcal{J} f \in H_w$. The result follows. □

We apply this result to give a characterisation of continuous integral operators $\mathcal{J}$ on H_w:

Proposition 3.2 *The integral operator $\mathcal{J}$ in* (3.8) *is continuous if and only if*

(1) $k(x,\cdot)/\sqrt{w} \in L^2(\mathbb{R}_+)$ for any $x \in \mathbb{R}_+$, and
(2) the function $(x \mapsto \int_0^\infty k(x,y) f'(y)dy) \in H_w$ for any $f \in H_w$

Proof Assume that (1) and (2) hold. Then, by (1), $\mathcal{J}$ is a closed linear operator on H_w from Lemma 3.9. Furthermore, by (2), its domain is H_w, i.e., it is everywhere defined. By the closed graph theorem (see, e.g., Folland [104, Thm. 5.11]) it follows that $\mathcal{J}$ is bounded on H_w.

Opposite, suppose $\mathcal{J}$ is continuous. Then it is everywhere defined and (2) holds. Furthermore, as $\mathcal{J}$ is a bounded operator on H_w, we find that

$$\|\mathcal{J}\|_{\text{op}} = \sup_{|f|_w = 1} \left| \int_0^\infty k(x,y) f'(y)dy \right| < \infty$$

Since,

$$\int_0^\infty k(x,y) f'(y)dy = \int_0^\infty \frac{k(x,y)}{\sqrt{w(y)}} \mathcal{W} f(y)dy$$

and $(\delta_0, \mathcal{W})$ is an isometric isomorphism of H_w onto $\mathbb{R} \times L^2(\mathbb{R}_+)$, it follows that for each $x \in \mathbb{R}_+$, $L^2(\mathbb{R}_+) \ni g \mapsto \int_0^\infty (k(x,y)/\sqrt{w(y)}) g(y)dy \in \mathbb{R}$ is a bounded linear functional. Property (1) then follows from the converse of Hölder's inequality (see Folland [104, Thm. 6.14]). □

Interestingly, all bounded linear operators on H_w has an adjoint which is the sum of an integral operator and a multiplication operation. We precise this in the next proposition (see Benth and Krühner [37, Prop. 3.9]):

Proposition 3.3 *Let $\mathcal{K} \in L(H_w)$. Then, for any $g \in H_w$*

$$\mathcal{K}^* g(x) = g(0)\eta(x) + \int_0^\infty q(x,y) g'(y)dy$$

where

$$\eta(x) := (\mathcal{K}h_x)(0)$$

and

$$q(x, y) := (\mathcal{K}h_x)'(y)w(y)$$

for any $x, y \in \mathbb{R}_+$ *and* h_x *defined in* (3.4).

Proof We know that $g(x) = \delta_x(g) = \langle g, h_x\rangle_w$ for any $g \in H_w$. Hence, by definition of the adjoint operator

$$\begin{aligned}\mathcal{K}^* g(x) &= \langle \mathcal{K}^* g, h_x\rangle_w \\ &= \langle g, \mathcal{K}h_x\rangle_w \\ &= g(0)\mathcal{K}h_x(0) + \int_0^\infty w(y)g'(y)(\mathcal{K}h_x)'(y)dy\end{aligned}$$

for any $x \geq 0$. The result follows. □

As an example, let us consider the shift semigroup $(\mathcal{S}_t)_{t\geq 0}$. In this case, it follows from Prop. 3.3 that

$$\eta(x) = (\mathcal{S}_t h_x)(0) = h_x(t) = h_t(x)$$

and

$$q(x, y) = (\mathcal{S}_t h_x)'(y) = \frac{w(y)}{w(y+t)} 1(y + t < x).$$

Therefore, for $g \in H_w$,

$$\mathcal{S}_t^* g(x) = g(0)h_t(x) + \int_0^\infty \frac{w(y)}{w(y+t)} 1(y + t < x)g'(y)dy \tag{3.9}$$

is an explicit representation of the adjoint of the shift semigroup.

3.2.2 Hilbert-Schmidt Operators

In a separable Hilbert space H with ONB $(e_n)_{n\in\mathbb{N}}$, a Hilbert-Schmidt operator is an operator $\mathcal{J} \in L(H)$ such that

$$\|\mathcal{J}\|_{\text{HS}}^2 := \sum_{n=1}^\infty |\mathcal{J}e_n|_H^2 < \infty$$

The space of Hilbert-Schmidt operators on H forms a separable Hilbert space with the inner product

$$\langle \mathcal{J}, \mathcal{K} \rangle_{\mathrm{HS}} := \sum_{n=1}^{\infty} \langle \mathcal{J} e_n, \mathcal{K} e_n \rangle_H$$

In this Subsection we will classify the Hilbert-Schmidt operators on H_w. These operators are closely linked with the covariance operators of H_w-valued Lévy processes. As we recall from Chapter 2 (see Section 2.1, say), covariance operators are positive semidefinite trace class operators. Hence, they can be expressed as the square of a Hilbert-Schmidt operator, since any positive semidefinite operator has a square-root, and this is Hilbert-Schmidt since the operator itself is trace class.

For $H = L^2(\mathbb{R}_+)$, any integral operator,

$$\mathcal{K} f(x) = \int_{\mathbb{R}} \kappa(x, y) f(y) dy$$

where the kernel function κ is square-integrable on $\mathbb{R}_+^2$, becomes a bounded linear operator. By monotone convergence and Parseval's identity

$$\begin{aligned} \|\mathcal{K}\|_{\mathrm{HS}}^2 &= \sum_{n=1}^{\infty} \int_{\mathbb{R}_+} (\int_{\mathbb{R}_+} \kappa(x, y) e_n(y) dy)^2 dx \\ &= \int_{\mathbb{R}_+} (\sum_{n=1}^{\infty} (\int_{\mathbb{R}_+} \kappa(x, y) e_n(y) dy)^2) dx \\ &= \int_{\mathbb{R}_+} \int_{\mathbb{R}_+} \kappa^2(x, y) dy dx \end{aligned}$$

for an ONB $(e_n)_{n \in \mathbb{N}}$ in $L^2(\mathbb{R}_+)$. Thus,

$$\|\mathcal{K}\|_{\mathrm{HS}}^2 = \int_{\mathbb{R}_+^2} \kappa^2(x, y) dx dy$$

and we can identify Hilbert-Schmidt operators on $L^2(\mathbb{R}_+)$ with $L^2(\mathbb{R}_+^2)$ (both spaces equipped with the Lebesgue measure). We formalise the result in a Lemma:

Lemma 3.10 *Let $\mathcal{K}$ be a Hilbert-Schmidt operator on $L^2(\mathbb{R}_+)$. Then there exists a kernel function $\kappa \in L^2(\mathbb{R}_+^2)$ such that $\mathcal{K} f(x) = \int_{\mathbb{R}_+} \kappa(x, y) f(y) dy$ for any $f \in L^2(\mathbb{R}_+)$ and $\|\mathcal{K}\|_{HS}^2 = \int_{\mathbb{R}_+^2} \kappa^2(x, y) dx dy$.*

Opposite, if $\kappa \in L^2(\mathbb{R}_+^2)$, then $\mathcal{K} f(x) = \int_{\mathbb{R}_+} \kappa(x, y) f(y) dy$ defines a Hilbert-Schmidt operator on $L^2(\mathbb{R}_+)$.

Proof From the discussion prior to this Lemma, we see that there exist isomorphisms between the space of Hilbert-Schmidt operators and $L^2(\mathbb{R}_+) \otimes L^2(\mathbb{R}_+)$, and between $L^2(\mathbb{R}_+) \otimes L^2(\mathbb{R}_+)$ and $L^2(\mathbb{R}_+^2)$. □

In our analysis to come, we are going to make use of Hilbert-Schmidt operators on $L^2(\mathbb{R}_+)$ along with the isometry between H_w and $L^2(\mathbb{R}_+)$ to study Hilbert-Schmidt operators on H_w. We start by an identification of Hilbert-Schmidt operators on a subspace of H_w as integral operators.

Lemma 3.11 *Let $\mathcal{J}$ be a closed linear operator on $H_w^0 := \{f \in H_w : f(0) = 0\}$. Then the following are equivalent:*

1. *$\mathcal{J}$ is a Hilbert-Schmidt operator,*
2. *$\mathcal{J}$ is an integral operator and there is a function $\kappa \in L^2(\mathbb{R}_+^2)$ such that the kernel of $\mathcal{J}$ is given by*

$$k(x, y) := \int_0^x \sqrt{\frac{w(y)}{w(z)}}\kappa(z, y)dz.$$

If (2) holds, then the Hilbert-Schmidt norm of $\mathcal{J}$ is equal to the $L^2(\mathbb{R}_+^2)$-norm of κ.

Proof Recall the isometric isomorphism $\mathcal{W} : H_w \to L^2(\mathbb{R}_+)$ in (3.7), and consider its restriction to H_w^0. Assume $\mathcal{J}$ is a Hilbert-Schmidt operator. Then $\mathcal{K} := \mathcal{W}\mathcal{J}\mathcal{W}^{-1}$ is a Hilbert-Schmidt operator on $L^2(\mathbb{R}_+)$ with the same Hilbert-Schmidt norm as $\mathcal{J}$. Furthermore, by Lemma 3.10, there exists a kernel function $\kappa \in L^2(\mathbb{R}_+^2)$ such that $\mathcal{K}f(x) = \int_{\mathbb{R}_+} \kappa(x, y)f(y)dy$ and such that the Hilbert-Schmidt norm coincides with the $L^2(\mathbb{R}_+^2)$-norm of κ. Since $\mathcal{W}^{-1}g(x) = \int_0^x g(y)/\sqrt{w(y)}dy$, we find for $f \in H_w^0$

$$\begin{aligned}
\mathcal{J}f(x) &= \mathcal{W}^{-1}\mathcal{K}\mathcal{W}f(x) \\
&= \int_0^x \frac{1}{\sqrt{w(y)}}\mathcal{K}\mathcal{W}f(y)dy \\
&= \int_0^x \frac{1}{\sqrt{w(y)}}\int_0^\infty \kappa(y, z)\mathcal{W}f(z)dzdy \\
&= \int_0^x \frac{1}{\sqrt{w(y)}}\int_0^\infty \kappa(y, z)\sqrt{w(z)}f'(z)dzdy \\
&= \int_0^\infty k(x, z)f'(z)dz.
\end{aligned}$$

In the last equality we appealed to the Fubini Theorem (see, e.g., Folland [104, Thm. 2.37]) and the definition of k. Statement (2) follows.

Suppose that Statement (2) holds for the operator $\mathcal{J}$. For the function $\kappa \in L^2(\mathbb{R}_+^2)$ in the kernel k of $\mathcal{J}$, define the operator $\mathcal{K}f(x) = \int_{\mathbb{R}_+} \kappa(x, y)f(y)dy$ on

$L^2(\mathbb{R}_+)$, which becomes a Hilbert-Schmidt operator by Lemma 3.10. Following the above calculations we find $\mathcal{J} = \mathcal{W}^{-1}\mathcal{K}\mathcal{W}$, which shows that $\mathcal{J}$ is Hilbert-Schmidt. Thus Statement (1) holds. □

With this result at hand, we are able to completely characterise the Hilbert-Schmidt operators on H_w. This is stated in the subsequent Theorem, which will be of fundamental importance in Subsection 3.2.3, where we analyse covariance operators on H_w.

Theorem 3.1 *Assume C is a Hilbert-Schmidt operator on H_w. Then there exist a constant $c \in \mathbb{R}$ and functions $g, h \in H_w$ and $\kappa \in L^2(\mathbb{R}_+)$ where $g(0) = 0 = h(0)$, such that for any $f \in H_w$,*

$$Cf(x) = (c + h(x))f(0) + \langle g, f\rangle_w + \int_0^\infty k(x,z)f'(z)dz,$$

where $k(x,z) = \int_0^x \sqrt{w(z)/w(y)}\kappa(y,z)dy$. Moreover, the Hilbert-Schmidt norm of C is

$$\|C\|_{HS}^2 = c^2 + |g|_w^2 + |h|_w^2 + \int_{\mathbb{R}_+^2} \kappa^2(x,y)dxdy.$$

The dual operator of C is given by

$$C^* f(x) = (c + g(x))f(0) + \langle f, h\rangle_w + \int_0^\infty k^*(x,z)f'(z)dz$$

where $k^*(x,y) = \int_0^x \sqrt{w(y)/w(z)}\kappa(y,z)dz$. In particular, C is symmetric if and only if $g = h$ and κ is symmetric.

Proof Recall that H_w^0 is the subspace of H_w consisting of all functions $f \in H_w$ such that $f(0) = 0$. Let H_w^c be the subspace of H_w of constant functions, and we observe that $H_w = H_w^c \oplus H_w^0$, i.e., the orthogonal sum of H_w^0 and H_w^c. Hence,

$$H_w \otimes H_w = (H_w^c \otimes H_w^c) \oplus (H_w^c \otimes H_w^0) \oplus (H_w^0 \otimes H_w^c) \oplus (H_w^0 \otimes H_w^0).$$

Let $g \in H_w^0$ and $u \in H_w^0$, which we can take as $u(x) = 1$ without loss of generality. Then $u \otimes g(f)(x) = \langle u, f\rangle_w g(x) = f(0)g(x)$ for any $f \in H_w$ with $u \otimes g \in H_w^c \otimes H_w^0$. For $h \in H_w^0$, we find that $h \otimes u \in H_w^0 \otimes H_w^c$ and $h \otimes u(f)(x) = \langle h, f\rangle_w u(0) = \langle h, f\rangle_w$. Also, $u, v \in H_w^c$, $u \otimes v \in H_w^c \otimes H_w^c$ and $u \otimes v(f)(x) = \langle u, f\rangle_w v(x) = u(0)v(0)f(0) = cf(0)$ for some constant $c \in \mathbb{R}$. This proves the first terms of C. The integral term comes from Lemma 3.11 characterising all Hilbert-Schmidt operators on H_w^0. The Hilbert-Schmidt norm of C follows by the orthogonal representation.

Consider the operator C^* defined in the Proposition. It follows that for $f \in H_w$,

$$C^* f(0) = cf(0) + \langle f, h \rangle_w$$

and

$$(C^* f)'(x) = g'(x) f(0) + \int_0^\infty \sqrt{\frac{w(y)}{w(z)}} \kappa(y, x) f'(y) dy$$

But then we see that $\langle C^* f, v \rangle_w = \langle f, Cv \rangle_w$ for any $f, v \in H_w$, and C^* is the dual of C. The claimed result on symmetry follows by reading off the definitions of C and C^*. □

With this result at hand, we are ready to analyse covariance operators in the Filipovic space, a very important topic in connection with modeling the forward curve dynamics.

3.2.3 Covariance Operators

To this end, consider a zero-mean square-integrable random variable W on H_w, where we denote its covariance operator by Q. Hence, we recall a covariance operator Q to be a symmetric positive semidefinite trace class operator with the property that for any $f, g \in H_w$,

$$\mathbb{E}[\langle W, f \rangle_w \langle W, g \rangle_w] = \langle Qf, g \rangle_w.$$

Now, any positive semidefinite operator has a symmetric square-root $Q^{1/2}$ again being positive semidefinite. Moreover, by the definition of the trace class and Hilbert-Schmidt norms, it follows that $Q^{1/2}$ is a Hilbert-Schmidt operator on H_w. Hence, we can appeal to Theorem 3.1 for a complete characterisation of the covariance operators on H_w, a result which we make precise here (see Benth and Krühner [34, Cor. 4.12]):

Corollary 3.1 *Let Q be a positive semidefinite trace class operator on H_w. Then there exists a $c \in \mathbb{R}_+$ and a measurable function $\ell : \mathbb{R}_+^2 \to \mathbb{R}$ such that ℓ is absolutely continuous in its first variable,*

1. $\ell(0, \cdot)/\sqrt{w} \in L^2(\mathbb{R}_+)$,
2. $(x, z) \mapsto \frac{w(z)}{w(x)}(\partial_x \ell(x, z))^2$ *is symmetric and integrable,*

3. *for any* $f \in H_w$ *and* $x \in \mathbb{R}_+$,

$$Qf(x) = \left(f(0)c + \int_0^\infty \ell(0,z) f'(z) dz \right) \left(c + \int_0^x \frac{\ell(0,z)}{w(z)} dz \right) + f(0) \int_0^\infty \ell(x,z) \frac{\ell(0,z)}{w(z)} dz + \int_0^\infty \int_0^\infty \ell(x,z) \partial_z \ell(z,y) f'(y) dz dy,$$

4. *for any* $f \in H_w$,

$$\langle Qf, f \rangle_w = \left(f(0)c + \int_0^\infty \ell(0,z) f'(z) dz \right)^2 + \int_0^\infty \left(f(0)\ell(0,x) + \int_0^\infty w(x) \partial_x \ell(x,z) f'(z) dz \right)^2 \frac{1}{w(x)} dx,$$

5. *the trace of* Q *is,*

$$Tr(Q) = c^2 + \int_0^\infty \frac{\ell^2(0,y)}{w(y)} dy + \int_0^\infty \frac{|\ell(\cdot,y)|_w^2}{w(y)} dy.$$

Moreover, if $c \in \mathbb{R}_+$ *and* $\ell : \mathbb{R}_+^2 \to \mathbb{R}$ *is measurable, absolutely continuous in its first variable,* ℓ *satisfies Statements (1) and (2), and* Q *is defined by Statement (3), then* Q *is a symmetric positive semidefinite trace class operator satisfying Statements (4) and (5).*

Proof From Pedersen [163, Prop. 3.2.11], there exists a unique positive semidefinite symmetric operator C such that $C^2 = Q$. I.e., C is the square-root of Q. We have that

$$\|C\|_{\text{HS}}^2 = \sum_{n=1}^\infty \langle Ce_n, Ce_n \rangle_w = \sum_{n=1}^\infty \langle Qe_n, e_n \rangle_w = \text{Tr}(Q) < \infty,$$

and therefore C is a Hilbert-Schmidt operator on H_w.

From Theorem 3.1 we have that there exist $c \in \mathbb{R}$, $g, h \in H_w$ and $\kappa \in L^2(\mathbb{R}_+^2)$ yielding a representation of the operator C. Moreover, due to the symmetry of C, $h = g$ and κ is symmetric. We have

$$Cf(x) = cf(0) + h(x)f(0) + \langle f, h \rangle_w + \int_0^\infty k(x,z) f'(z) dz$$

where $k(x,z) = \int_0^x \sqrt{w(z)/w(y)} \kappa(y,z) dy$. for all $f \in H_w$. If f is a constant, i.e., $f(x) = f(0)$, we find that $\langle Cf, f \rangle_w = cf^2(0)$, and as C is positive definite, $c \in \mathbb{R}_+$. Define

$$\ell(x,z) := w(z)h'(z) + k(x,z).$$

Then, $\ell(0,z) = w(z)h'(z)$, and the claim in Statement (1) follows as $h \in H_w$. Further, we find that

$$\frac{w(x)}{w(z)}(\partial_x \ell(x,z))^2 = \frac{w(x)}{w(z)}\left(\sqrt{\frac{w(z)}{w(x)}}\kappa(x,z)\right)^2 = \kappa^2(x,z).$$

Hence, the claim in Statement (2) follows. We find that

$$Cf(x) = (c + h(x))f(0) + \int_0^\infty \ell(x,z)f'(z)dz.$$

This implies that

$$Cf(0) = cf(0) + \int_0^\infty \ell(0,z)f'(z)dz$$

and

$$(Cf)'(z) = f(0)h'(z) + \int_0^\infty \partial_x \ell(z,y)f'(y)dy.$$

Thus,

$$\begin{aligned} Qf(x) &= C^2 f(x) = C(Cf)(x) \\ &= (c + h(x))(cf(0) + \int_0^\infty \ell(0,z)f'(z)dz) \\ &\quad + \int_0^\infty \ell(x,z)(f(0)h'(z) + \int_0^\infty \partial_x \ell(z,y)f'(y)dy)dz. \end{aligned}$$

Statement (3) follows after using the identity $\ell(0,z) = w(z)h'(z)$, the fundamental theorem of calculus in Lemma 3.1 and the Fubini theorem. Appealing to the definition of the inner product in the Filipović space along with the expression of $Qf(x)$, Statement (4) follows readily. For Statement (5), we get immediately from Theorem 3.1 that

$$\mathrm{Tr}(Q) = c^2 + 2|h|_w^2 + \int_{\mathbb{R}_+^2} \kappa^2(x,y)dxdy.$$

But

$$\begin{aligned} \int_{\mathbb{R}_+^2} \kappa^2(x,y)dxdy &= \int_{\mathbb{R}_+^2} \frac{w(x)}{w(y)}(\partial_x \ell(x,y))^2 dxdy \\ &= \int_0^\infty (\int_0^\infty w(x)(\partial_x \ell(x,y))^2 dx)\frac{1}{w(y)}dy \end{aligned}$$

$$= \int_0^\infty \frac{|\ell(\cdot, y)|_w^2 - \ell(0, y)}{w(y)} dy$$

$$= \int_0^\infty \frac{|\ell(\cdot, y)|_w^2}{w(y)} dy - |h|_w^2 .$$

Since $\ell(0, y) = w(y)h'(y)$ we find

$$|h|_w^2 = \int_0^\infty \frac{\ell^2(0, y)}{w(y)} dy$$

which yields the claimed trace of Q.

The opposite assertion follows from similar arguments. □

As we know, we can evaluate the elements in H_w to get random variables $W(x) := \delta_x(W)$ for $x \in \mathbb{R}_+$. Thus, $(W(x))_{x \in \mathbb{R}_+}$ defines a random field on $\mathbb{R}_+$. Let us study the covariance of this field at different spatial locations $x, y \in \mathbb{R}_+$. It holds,

$$\mathbb{E}[W(x)W(y)] = \mathbb{E}[\delta_x(W)\delta_y(W)] = \mathbb{E}[\langle W, h_x\rangle_w \langle W, h_y\rangle_w]$$
$$= \langle Qh_x, h_y\rangle_w = Qh_x(y) .$$

Thus, the covariance of the random field $(W(x))_{x \in \mathbb{R}_+}$ can be expressed through the covariance operator of W applied to the function h_x defined in (3.4), evaluated at $y \in \mathbb{R}_+$. Let us analyse this further.

In view of Corollary 3.1 above, choose ℓ such that $\ell(0, y) = 0$. Then we have that

$$Qf(x) = c^2 f(0) + \int_0^\infty \int_0^\infty \ell(x, u)\partial_u \ell(u, v)\, du f'(v)\, dv .$$

With $f = h_x$ we find

$$Qh_x(y) = c^2 + \int_0^x \left(\int_0^\infty \ell(y, u)\partial_u \ell(u, v)\, du\right) w^{-1}(v)\, dv$$

Moreover, as we recall from the proof of Corollary 3.1, $\partial_u \ell(u, v) = \sqrt{w(v)/w(u)}\kappa(u, v)$ for a symmetric function $\kappa \in L^2(\mathbb{R}_+^2)$. It follows,

$$Qh_x(y) = c^2 + \int_0^x \int_0^\infty \left(\int_0^y \sqrt{\frac{w(u)}{w(r)}} \kappa(r, u)\, dr\right) \sqrt{\frac{w(v)}{w(u)}} \kappa(u, v)\, du w^{-1}(v)\, dv$$
$$= c^2 + \int_0^x \int_0^y \left(\int_0^\infty \kappa(r, u)\kappa(u, v)\, du\right) w^{-1/2}(r) w^{-1/2}(v)\, dr\, dv .$$

Defining the function

$$\rho(u, v) := \int_0^\infty \kappa(u, z)\kappa(z, v)\, dz w^{-1/2}(u) w^{-1/2}(v)\,, \tag{3.10}$$

which becomes symmetric since κ is symmetric, we can conclude

$$\mathrm{Cov}(W(x), W(y)) = c^2 + \int_0^x \int_0^y \rho(u, v)\, du\, dv\,, \tag{3.11}$$

for any $x, y \in \mathbb{R}_+$. It is immediate from this expression that the random field W is not stationary in space in the sense that its covariance structure is not a function of $|x - y|$. Obviously, the spatial correlation structure of the random field $(W(x))_{x\in\mathbb{R}_+}$ will become

$$\mathrm{Corr}(W(x), W(y)) = \frac{c^2 + \int_0^x \int_0^y \rho(u, v)\, du\, dv}{\sqrt{c^2 + \int_0^x \int_0^x \rho(u, v)\, du\, dv}\sqrt{c^2 + \int_0^y \int_0^y \rho(u, v)\, du\, dv}}\,. \tag{3.12}$$

We find the following:

Lemma 3.12 *For fixed* $\Delta > 0$, $\lim_{x\to\infty} \mathrm{Corr}(W(x), W(x + \Delta)) = 1$.

Proof Let us first show that ρ is integrable on $\mathbb{R}_+^2$: from the symmetry of κ and Cauchy-Schwarz' inequality,

$$\begin{aligned}\int_0^\infty \int_0^\infty |\rho(u, v)| du dv &\leq \int_0^\infty \int_0^\infty w^{-1/2}(u) w^{-1/2}(v) \\ &\quad \int_0^\infty |\kappa(u, z)||\kappa(z, v)| dz du dv \\ &= \int_0^\infty (\int_0^\infty w^{-1/2}(u)|\kappa(u, z)| du)^2 dz \\ &\leq \int_0^\infty w^{-1}(u) du \int_0^\infty \int_0^\infty \kappa^2(u, z) du dz < \infty.\end{aligned}$$

By dominated convergence, we have that $\int_0^x \int_0^x \rho(u, v) du\, dv$, $\int_0^x \int_0^{x+\Delta} \rho(u, v)\, du\, dv$ and $\int_0^{x+\Delta} \int_0^{x+\Delta} \rho(u, v)\, du\, dv$ all converge to $\int_0^\infty \int_0^\infty \rho(u, v)\, du\, dv$ when $x \to \infty$. Thus, the claim holds. □

This result is not surprising in view of the fact that for any $f \in H_w$ we must have $f'(x) \to 0$ as $x \to \infty$, otherwise $|f|_w$ will not be finite. It follows that f asymptotically will become a constant. For a random field W, this means informally that $W(x) \approx W_\infty$ for x large, with W_∞ some square integrable H_w-valued random variable, and thus $\mathbb{E}[W(x)W(x+\Delta)] \approx \mathbb{E}[W_\infty^2]$. This explains the asymptotically perfect correlation structure of W.

Similarly to Lemma 3.12 we find:

Lemma 3.13 *Assume $x > 0$. Then,*

$$\lim_{\Delta\to\infty} \operatorname{Corr}(W(x), W(x+\Delta))$$
$$= \frac{c^2 + \int_0^x \int_0^\infty \rho(u,v)\,du\,dv}{\sqrt{c^2 + \int_0^x \int_0^x \rho(u,v)\,du\,dv}\sqrt{c^2 + \int_0^\infty \int_0^\infty \rho(u,v)\,du\,dv}}.$$

We find that as long as $c > 0$,

$$\lim_{\Delta\to\infty} \operatorname{Corr}(W(0), W(\Delta)) = \frac{c}{\sqrt{c^2 + \int_0^\infty \int_0^\infty \rho(u,v)\,du\,dv}} > 0.$$

Thus, even far apart $W(0)$ is positively correlated with $W(\Delta)$. This is explained intuitively by the fact that from the fundamental theorem of calculus in Lemma 3.1, we find $W(\Delta) = W(0) + \int_0^\Delta W'(x)\,dx$, linking $W(0)$ and $W(\Delta)$. In the case $c = 0$, it follows $\mathbb{E}[W^2(0)] = \operatorname{Cov}(W(0), W(0)) = 0$ from (3.11), and hence $W(0) = 0$.

As a particular case, consider $w(x) = \exp(\alpha x)$ and $\kappa(u,v) = \exp(-\beta(u+v))$ for $\alpha, \beta > 0$. Then,

$$\rho(u,v) = \frac{1}{2\beta} \exp\left(-(\beta + \frac{1}{2}\alpha)(u+v)\right),$$

and

$$\operatorname{Cov}(W(x), W(y)) = c^2 + \frac{1}{2\beta(\beta + \frac{1}{2}\alpha)^2}\left(1 - e^{-(\beta+\frac{1}{2}\alpha)x}\right)\left(1 - e^{-(\beta+\frac{1}{2}\alpha)y}\right). \tag{3.13}$$

In Figure 3.2 we have plotted an example of the correlation function derived from (3.13). The correlation is obviously 1 for $x = y$, whereas it is decreasing with increasing distance between x and y. But we also observe that the correlation surface flattens as x and y are getting bigger. We may interpret the maturity axes as measured in years, say.

We end our analysis on covariance operators in H_w with a general lower bound on the correlation between close points in space (see Benth and Krühner [34, Thm. 3.12]):

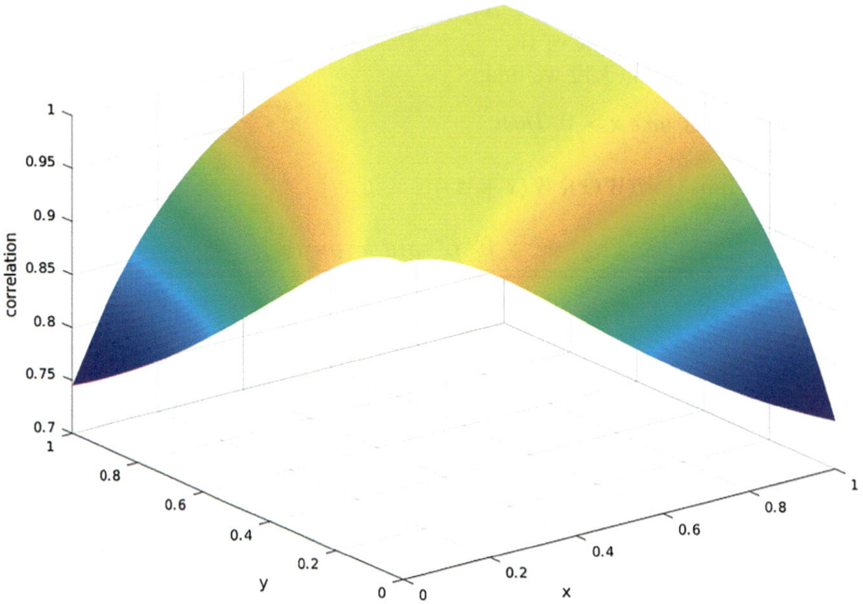

Fig. 3.2 Example of the correlation structure derived from the covariance function (3.13) with parameters $\alpha = 1$, $\beta = 0.5$ and $c = 1$.

Theorem 3.2 *Let W be a mean-zero square integrable random variable with values in H_w having covariance operator $\mathcal{Q}$. For any $x \in \mathbb{R}_+$, there exists an $\epsilon > 0$ such that*

$$\mathrm{Corr}(W(x), W(y)) \geq 1 - \frac{2\|\mathcal{Q}\|_{op}^{1/2}\sqrt{|x-y|}}{|\mathcal{Q}^{1/2}h_x|_w + \|\mathcal{Q}\|_{op}^{1/2}\sqrt{|x-y|}}$$

for any $|x - y| \leq \epsilon$. Here, $h_x \in H_w$ is the function such that $\delta_x(f) = \langle h_x, f\rangle_w$ for $f \in H_w$, defined in (3.4).

Proof Notice for any $x \in \mathbb{R}_+$

$$\mathbb{E}[W(x)^2] = \mathbb{E}[\delta_x(W)^2] = \langle \mathcal{Q}h_x, h_x\rangle_w = |\mathcal{Q}^{1/2}h_x|_w^2.$$

First, observe that if $|\mathcal{Q}^{1/2}h_x|_w = 0$ the result follows trivially. Hence, we suppose that $|\mathcal{Q}^{1/2}h_x|_w > 0$.

Now, by the triangle inequality

$$|\mathcal{Q}^{1/2}h_x|_w \leq |\mathcal{Q}^{1/2}h_y|_w + \|\mathcal{Q}^{1/2}\|_{\mathrm{op}}|h_x - h_y|_w.$$

Therefore, by Lemma 3.5,

$$|Q^{1/2}h_y|_w \geq |Q^{1/2}h_x|_w - \|Q^{1/2}\|_{\mathrm{op}}\sqrt{|x-y|}$$

Choose $\epsilon := |Q^{1/2}h_x|_w^2/\|Q\|_{\mathrm{op}}$. From Pedersen [163, Thm. 3.2.3] it holds that $\|Q\|_{\mathrm{op}} = \|Q^{1/2}\|_{\mathrm{op}}^2$, and we find

$$|Q^{1/2}h_y|_w \geq |Q^{1/2}h_x|_w - \|Q^{1/2}\|_{\mathrm{op}}\sqrt{|x-y|} > 0.$$

for all $|x-y| < \epsilon$.

Thus, for any $y \in \mathbb{R}_+$ such that $|y-x| < \epsilon$,

$$\mathrm{Corr}(W(x), W(y)) = \frac{\langle Qh_x, h_y\rangle_w}{|Q^{1/2}h_x|_w |Q^{1/2}h_y|_w}$$

From the triangle inequality it follows

$$|Q^{1/2}h_y|_w \leq |Q^{1/2}h_x|_w + \|Q^{1/2}\|_{\mathrm{op}}\sqrt{|x-y|}.$$

Furthermore, again by appealing to Lemma 3.5,

$$\begin{aligned}
\langle Qh_x, h_y\rangle_w &= \langle Q^{1/2}h_x, Q^{1/2}h_y\rangle_w \\
&= \langle Q^{1/2}h_x, Q^{1/2}h_x\rangle_w - \langle Q^{1/2}h_x, Q^{1/2}(h_x - h_y)\rangle_w \\
&\geq |Q^{1/2}h_x|_w^2 - |\langle Q^{1/2}h_x, Q^{1/2}(h_x - h_y)\rangle_w| \\
&\geq |Q^{1/2}h_x|_w^2 - \|Q^{1/2}\|_{\mathrm{op}}|Q^{1/2}h_x|_w\sqrt{|x-y|} \\
&= |Q^{1/2}h_x|_w\left(|Q^{1/2}h_x|_w - \|Q^{1/2}\|_{\mathrm{op}}\sqrt{|x-y|}\right).
\end{aligned}$$

Hence,

$$\begin{aligned}
\mathrm{Corr}(W(x), W(y)) &= \frac{\langle Qh_x, h_y\rangle_w}{|Q^{1/2}h_x|_w |Q^{1/2}h_y|_w} \\
&\geq \frac{|Q^{1/2}h_x|_w - \|Q^{1/2}\|_{\mathrm{op}}\sqrt{|x-y|}}{|Q^{1/2}h_x|_w + \|Q^{1/2}\|_{\mathrm{op}}\sqrt{|x-y|}},
\end{aligned}$$

and the claim follows. □

We see from the above lower bound that the correlation converges to 1 at the rate $\sqrt{|y-x|}$ when $y \to x$. Another phrasing of the result is that for $\sqrt{|x-y|} < \epsilon$,

$$1 - \mathrm{Corr}(W(x), W(y)) \leq K\sqrt{|x-y|}$$

for a constant $K > 0$. Hence, we have a local square-root behaviour in $x - y$ of $1 - \mathrm{Corr}(W(x), W(y))$. Notice that we did not use the explicit structure of Q in the above proof, only that Q was a covariance operator.

3.2.4 Multiplication Operators

Under the assumption of integrability of the reciprocal of the weight function (recall Assumption 3.1), we have that H_w is a Banach algebra according to Proposition 3.1. This gives rise to a class of linear operators defined as multiplication by a fixed element of H_w. Multiplication operators can provide a simple way to construct attractive models for the volatility term structure, say.

We define a multiplication operator as follows:

Definition 3.2 Let $a : \mathbb{R}_+ \to \mathbb{R}$ be a measurable function and

$$D_a := \{g \in H_w \ : \ ag \in H_w\}$$

Then the *multiplication operator with kernel a* is defined by

$$\mathcal{M}_a : D_a \to H_w, \ g \mapsto ag$$

As the next result shows, multiplication operators are closed operators on H_w.

Proposition 3.4 *The multiplication operator $\mathcal{M}_a$ with kernel a is closed.*

Proof Let $(g_n)_{n \in \mathbb{N}}$ be a sequence in D_a which converges to some $g \in H_w$ such that $M_a(g_n)$ converges to some $b \in H_w$. For $x \in \mathbb{R}_+$, we find by continuity of the evaluation functional

$$b(x) = \delta_x(b) = \lim_{n \to \infty} \delta_x(\mathcal{M}_a g_n) = a(x) \lim_{n \to \infty} \delta_x(g_n) = a(x)g(x).$$

We conclude that $ag = b \in H_w$, $g \in D_a$ and $\mathcal{M}_a g = b$. □

In our context, we have a much improved result than mere closedness of the multiplication operators when considering kernels belonging to H_w. Indeed, under the Assumption 3.1 we know $w^{-1} \in L^1(\mathbb{R}_+)$. This yields the not surprising result that *all* functions $a \in H_w$ defines linear bounded multiplication operators on H_w, as the following result proves (see Benth and Krühner [34, Thm. 4.17]):

Proposition 3.5 *Let $a : \mathbb{R}_+ \to \mathbb{R}$ be a measurable function. Then the following statements are equivalent,*

1. *$\mathcal{M}_a$ is a continuous linear operator on H_w,*
2. *$\mathcal{M}_a$ is everywhere defined,*
3. *$a \in H_w$.*

If $\mathcal{M}_a$ is a continuous linear operator on H_w, then its operator norm is bounded by

$$\|\mathcal{M}_a\|_{op} \leq |a|_w \sqrt{1 + 4(1 + \sqrt{|w^{-1}|_{L^1(\mathbb{R}_+)}})^2}$$

and $\mathcal{M}_a^* g(x) = \langle g, ah_x \rangle_w$ where h_x is defined in (3.4).

Proof As any continuous linear operator on a Hilbert space is everywhere defined, (1)⇒(2) is clear. Assuming Statement (2), we find that $a = \mathcal{M}_a 1$ where $1 \in H_w$ is the constant function with value 1. Hence, $a \in H_w$, yielding Statement (3). We prove (3)⇒(1): we know from Prop. 3.1 that H_w is a Banach algebra, so $ag \in H_w$ for any $g \in H_w$, and hence $D_a = H_w$. By inspection of the proof of Prop. 3.1, we find that

$$|ag|_w \leq c|a|_w|g|_w$$

for $c = \sqrt{1 + 4(1 + \sqrt{|w^{-1}|_{L^1(\mathbb{R}_+)}})^2}$. This proves boundedness of $\mathcal{M}_a$, and therefore Statement (1) follows.

The adjoint operator representation is derived as follows:

$$\mathcal{M}_a^* g(x) = \langle \mathcal{M}_a^* g, h_x \rangle_w = \langle g, \mathcal{M}_a h_x \rangle_w = \langle g, ah_x \rangle_w .$$

This concludes the proof. □

We have on several occasions discussed the Samuelson effect observed in the volatility term structure of commodity forwards. A simple model for the forward term structure in a commodity market could be

$$f(t, x) := f_0(x) + \Sigma(W(t))(x), \tag{3.14}$$

where $f_0 \in H_w$, $W(t)$ is an H_w-valued stochastic process (Wiener process, say) and $\Sigma \in L(H_w)$ is the volatility operator. The volatility operator may be defined as a multiplication operator $\Sigma(f) = \sigma f$, where $\sigma \in H_w$. Choosing $\sigma(x) = \exp(-\eta x)$ for some $\eta > 0$, we have a well-defined multiplication operator as long as $x \mapsto w(x)\exp(-2\eta x) \in L^1(\mathbb{R}_+)$, which is the case for $w(x) = \exp(\alpha x)$, $\alpha < 2\eta$. Choosing σ to be an exponentially decaying function aligns with the term structure of volatility resulting from classical one-factor Ornstein–Uhlenbeck spot price models, see Chapter 5 for elaborations on this. In Chapter 6 we will come back to forward prices models, where we also take no-arbitrage into consideration leading to different dynamics.

In Figure 3.3 we depict an example of a Samuelson effect in the term-structure of volatility. We show the function $\sigma(x) := a + b\exp(-\eta x)$, where the spot volatility is $\sigma(0) = a + b$. In the plot, we have set this equal to 100% on an annualized basis, which is not uncommon in power markets, say (see Benth, Šaltytė Benth and Koekebakker [49]). Moreover, the volatility flattens when the maturity becomes

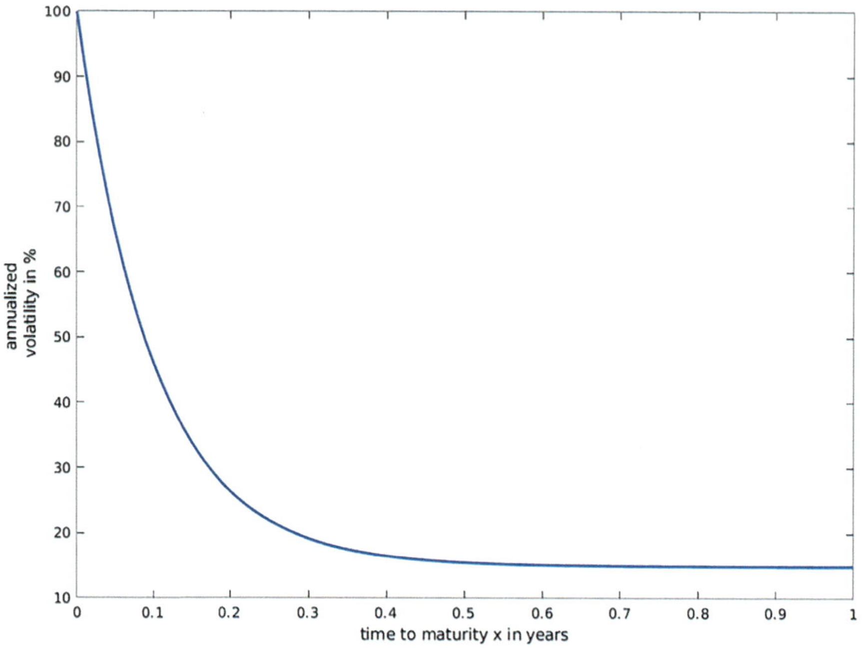

Fig. 3.3 A term-structure of volatility as a function of maturity x (measured in years).

large, to a constant level of 15% which is modelled by $\sigma(\infty) = a$. Choosing the scaling function w appropriately, this model of σ is an element in H_w.

3.3 Bibliographical Notes

The Filipović space was first introduced and analysed in the context of interest-rate models by Filipović [99]. An extension of this space to twice (weakly) differentiable functions has been suggested and applied by Filipović and Willems [103] in calibration of smooth term structure models.

Ekeland and Taflin [93] introduce a Sobolev space $H^s(\mathbb{R}_+)$, $s > 1/2$, as the state space for the term structure of zero-coupon prices. Their proposed state space consists of continuous functions vanishing at infinity. Bond prices tend to zero when time to maturity becomes large, however, this may not be a reasonable assumption for forward prices in commodity markets. This can be remedied by considering the state space $H^s(\mathbb{R}_+)\oplus\mathbb{R}$, which Ekeland and Taflin [93] propose as the state space for volatility modeling of bond prices. Barth and Benth [20] is making use of this state space when studying theoretically and numerically Heath-Jarrow-Morton models for forward contracts in the power markets. Vargiolu [184] uses an exponentially weighted Sobolev space of order 1 as state space of forward rates in fixed income theory when studying invariant measures.

The results in Section 3.2 of this Chapter on operators on Filipović spaces are to a large extent collected from Benth and Krühner [34].

Chapter 4
Stochastic Integration and Partial Differential Equations

Abstract We review the basic theory for stochastic integration in Hilbert space, and present results on the representation of multi-dimensional linear functionals applied to such integral. Our presentation includes Wiener processes and Lévy processes from subordination of Wiener processes as integrators. Then, an account on stochastic partial differential equations of parabolic type is given, with a focus on the existence and uniqueness of mild and weak solutions.

In this Chapter we start by recalling some basic theory and assumptions regarding stochastic integration with respect to Lévy processes in Hilbert space. As these integrals again become elements in the Hilbert space, it is of interest to analyse linear functionals applied to these. In the second Section we show that indeed such objects can be viewed as finite dimensional stochastic integrals with respect to multivariate Lévy processes. Finally, we recall some theory for time-inhomogeneous stochastic partial differential equations of parabolic type, relevant for a general model of forward prices.

4.1 Stochastic Integration in Hilbert Space

The aim of this Section is to give a brief account on stochastic integration with respect to Lévy processes in separable Hilbert spaces. Our discussion is heavily influenced by the theory presented in Peszat and Zabczyk [165], where a more general martingale integration theory is defined. We focus our attention to square-integrable Lévy processes as integrators.

Suppose that $(L(t))_{t\geq 0}$ is a square-integrable Lévy process with values in a separable Hilbert space K. Without loss of generality, we further assume that $(L(t))_{t\geq 0}$ is a martingale. By the Lévy-Kintchine decomposition (see Peszat and

F. E. Benth, P. Krühner, *Stochastic Models for Prices Dynamics in Energy and Commodity Markets*, Springer Finance,
https://doi.org/10.1007/978-3-031-40367-5_4

Zabczyk [165, Thm. 4.23], we know that for $t \geq 0$

$$L(t) = W(t) + \int_0^t \int_K z\widetilde{N}(dz, ds),$$

where $(W(t))_{t\geq 0}$ is a K-valued Wiener process and N is a Poisson random measure on K with Lévy measure ν such that $\widetilde{N}(dz, dt) = N(dz, dt) - \nu(dz)dt$ is the compensated Poisson random measure. Notice that by the assumption of $(L(t))_{t\geq 0}$ being square-integrable, we can represent the Lévy process as a continuous martingale and jump martingale, with the drift being equal to zero. Integration with respect to a drift will only be standard Lebesgue integration on the time line, and poses no additional difficulty.

We use the version of L which has paths being RCLL. As the expected value of $L(t)$ is zero, we have that the covariance operator $Q \in L(K)$ of L is defined by

$$\mathbb{E}[\langle L(t), u\rangle_K \langle L(t), v\rangle_K] = \langle Qu, v\rangle_K\, t$$

where $u, v \in K$. We find that Q can be expressed as $Q = \mathbb{E}[L(1)^{\otimes 2}]$, becoming a symmetric, positive-definite trace class operator. For $T > 0$, we denote by $\mathcal{P}_{[0,T]}$ the *predictable* σ-algebra, defined as the smallest σ-algebra on $[0, T] \times \Omega$ containing the sets $(s, t] \times A$ for $0 \leq s < t \leq T$ and $A \in \mathcal{F}_s$.

We now state conditions on the $L(K, H)$-valued stochastic process $(X(t))_{t\geq 0}$ such that the stochastic integral

$$\int_0^t X(s)dL(s)$$

makes sense. Here, H is a separable Hilbert space and $L(K, H)$ is a Banach space which we equip with its Borel σ-algebra. According to Peszat and Zabczyk [165, Section 8.6, Cor. 8.17], we make the following definition:

Definition 4.1 For fixed $T > 0$, if the $L(K, H)$-valued stochastic process $(X(t))_{t\geq 0}$ satisfies,

1. $(X(t))_{t\in[0,T]}$ is a predictable process, i.e., $[0, T] \times \Omega \ni (t, \omega) \mapsto X(t, \omega) \in L(K, H)$ is $\mathcal{P}_{[0,T]}$-measurable,
2. it holds that

$$\mathbb{E}\left[\int_0^T \|X(t)Q^{1/2}\|_{\mathrm{HS}}^2 dt\right] < \infty,$$

where $\|\cdot\|_{\mathrm{HS}}$ is the Hilbert-Schmidt norm for operators from K to H,

we say that $(X(t))_{t\geq 0}$ is *L-integrable* on the interval $[0, T]$. The space of all L-integrable processes is denoted $\mathcal{L}_L^2(\mathcal{F}, H)$.

Notice that if $(X(t))_{t\geq 0}$ is L-integrable on $[0,T]$ then it is L-integrable also on $[0,t]$ for any $t \leq T$ since $X(s)\mathbf{1}_{[0,t]}(s)$ is L-integrable on $[0,t]$. Also, we have emphasised the role of the filtration $(\mathcal{F}_t)_{t\geq 0}$ in the notation of the space of integrands $\mathcal{L}_L^2(\mathcal{F}, H)$. The reader is referred to Peszat and Zabczyk [165, Sects. 8.2 and 8.6] for further analysis and properties of stochastic integration with respect to Lévy processes in Hilbert space.

Let us investigate the integrability condition in Assumption (2) of Definition 4.1. Suppose that L is a pure-jump Lévy process, so that $L(t) = \int_0^t \int_K z\widetilde{N}(dz, ds)$. Then its covariance operator is given by the following calculation: for any $x, y \in K$,

$$\begin{aligned}\mathbb{E}[\langle L(1), x\rangle_K \langle L(1), y\rangle_K] &= \mathbb{E}\left[\int_0^1 \int_K \langle z, x\rangle_K \,\widetilde{N}(dz, ds) \int_0^1 \int_K \langle z, y\rangle_K \,\widetilde{N}(dz, ds)\right] \\ &= \int_K \langle z, x\rangle_K \langle z, y\rangle_K \,\nu(dz) \\ &= \langle \int_K z^{\otimes 2}\,\ell(dz)x, y\rangle_K .\end{aligned}$$

Thus, $Q = \int_K z^{\otimes 2}\,\nu(dz)$. According to Peszat and Zabczyk [165, Prop. A. 3], we have for an ONB $(e_n)_{n\in\mathbb{N}}$ in H,

$$\begin{aligned}\|X(s)Q^{1/2}\|_{\mathrm{HS}}^2 &= \|Q^{1/2}X(s)^*\|_{\mathrm{HS}}^2 \\ &= \sum_{n=1}^{\infty} |Q^{1/2}X(s)^*e_n|_K^2 \\ &= \sum_{n=1}^{\infty} \langle QX(s)^*e_n, X(s)^*e_n\rangle_K \\ &= \sum_{n=1}^{\infty} \int_K \langle z, X(s)^*e_n\rangle_K^2 \,\nu(dz) \\ &= \int_K |X(s)z|_H^2 \,\nu(dz),\end{aligned}$$

by using Parseval's identity in the last equality. Hence, we may re-express Condition (2) in Definition 4.1 as

$$\mathbb{E}\left[\int_0^T \int_K |X(t)z|_H^2 \,\nu(dz)dt\right] < \infty.$$

Indeed, this matches with the integrability condition posed by Filipović, Tappe and Teichmann [102, Sect. 2.4] in their definition of stochastic integration with respect to compensated Poisson random measures.

With this definition of L-integrable processes, one can define the stochastic integral by following the Itô-scheme by first use elementary processes, and then pass to the limit in an L^2-sense based on an Itô isometry (we refer to Peszat and Zabczyk [165, Sect. 8.2] for this construction). The resulting stochastic integral becomes a square-integrable H-valued martingale process on $[0, T]$ with mean zero, i.e.,

$$\mathbb{E}\left[\int_0^T X(s)dL(s)\right] = 0. \tag{4.1}$$

For any L-integrable process $(X(t))_{t\in[0,T]}$, the following Itô isometry holds:

$$\begin{aligned}\mathbb{E}\left[|\int_0^T X(t)dL(t)|_H^2\right] &= \mathbb{E}\left[\int_0^T \|X(t)Q^{1/2}\|_{\mathrm{HS}}^2 dt\right] \\ &= \mathbb{E}\left[\int_0^T \int_K |X(t)z|_H^2 \nu(dz)dt\right].\end{aligned} \tag{4.2}$$

As we know that Q is a bounded operator, using properties of the Hilbert-Schmidt operator norms show that a sufficient condition for L-integrability is that $(X(t))_{t\in[0,T]}$ is a process taking values in Hilbert-Schmidt operators and $(t, \omega) \mapsto \|X(t)\|_{\mathrm{HS}} \in L^2([0, T] \times \Omega)$. From the isometry and condition (2) in Def. 4.1, the stochastic integral is square-integrable (i.e., has finite variance). Moreover, the H-valued process $[0, T] \ni t \mapsto \int_0^t X(s)dL(s)$ is a martingale. Moreover, from Filipović, Tappe and Teichmann [102, Sect. 2.4], it follows that the stochastic integral has a version which has RCLL paths.

An important property of the stochastic integral is that it commutes with linear operators (see Peszat and Zabczyk [165, Thm. 8.7(v)]. If $A \in L(H, V)$, where V is a separable Hilbert space, then $(AX(t))_{t\geq 0}$ is L-integrable on $[0, T]$ whenever $(X(t))_{t\geq 0}$ is, and

$$A\int_0^t X(s)dL(s) = \int_0^t AX(s)dL(s)$$

for all $t \in [0, T]$.

The *angle bracket process* associated with two square-integrable H-valued martingales $(M(t))_{t\geq 0}$ and $(N(t))_{t\geq 0}$ is defined as the unique predictable real-valued process $(\langle\langle M, N\rangle\rangle(t))_{t\geq 0}$ such that $t \mapsto \langle M(t), N(t)\rangle_H - \langle\langle M, N\rangle\rangle(t)$ is a martingale. By Peszat and Zabczyk [165, Cor. 8.17],

$$\langle\langle\int_0^{\cdot} X(s)dL(s), \int_0^{\cdot} X(s)dL(s)\rangle\rangle(t) = \int_0^t \|X(s)Q^{1/2}\|_{\mathrm{HS}}^2 ds \tag{4.3}$$

where $(X(t))_{t\geq 0}$ is L-integrable. Using polarization identities, it is straightforward to derive that for two L-integrable processes $(X(t))_{t\geq 0}$ and $(Y(t))_{t\geq 0}$,

$$\langle\langle \int_0^{\cdot} X(s)dL(s), \int_0^{\cdot} Y(s)dL(s)\rangle\rangle(t) = \int_0^t \langle X(s)Q^{1/2}, Y(s)Q^{1/2}\rangle_{\mathrm{HS}} ds \tag{4.4}$$

where we notice that here $\langle\cdot,\cdot\rangle_{\mathrm{HS}}$ is the Hilbert-Schmidt inner product of operators from K to H.

The *operator angle bracket process* associated with a square-integrable H-valued martingale $(M(t))_{t\geq 0}$ is defined as the unique predictable operator-valued process $[M, M]$ such that $t \mapsto M(t)^{\otimes 2} - [M, M](t)$ is an integrable martingale (see Peszat and Zabczyk [165, Sect. 8.1]. From Peszat and Zabczyk [165, Cor. 8.17], it holds that

$$[\int_0^{\cdot} X(s)dL(s), \int_0^{\cdot} X(s)dL(s)](t) = \int_0^t X(s)QX^*(s)ds \tag{4.5}$$

for the operator angle bracket process.

We recall from Chapter 2 the definition of multivariate Lévy processes, where the state space is a product of Hilbert spaces. A product Hilbert space is again a Hilbert space with respect to the induced 2-norm, and thus we can use the definition above of stochastic integration also in this case. We will look closer at this in the next section. In Chapter 2 we constructed Lévy processes using subordination of Wiener processes. A Wiener process itself is a Lévy process, and can therefore be used as a stochastic integrator. If $L = W$, a Wiener process, we use the terminology W-integrable for the class of integrands satisfying Def.4.1. Moreover, the stochastic process $t \mapsto \int_0^t X(s)dW(s)$ is a square-integrable martingale with continuous paths in H (see Da Prato and Zabczyk [81, Sect. 4.2]).

4.2 Representations on $\mathbb{R}^n$ of Stochastic Integrals

We next study finite dimensional functionals of the infinite dimensional stochastic integrals introduced in the previous section. As we will see, for both Wiener and subordinated Wiener cases we can represent the functionals of these processes by finite dimensional Ito stochastic integrals, where the integrands are explicitly given from the infinite dimensional case. The interest for having available a dynamics for such finite dimensional functionals comes from the derivation of a forward price dynamics of contracts with fixed maturity time, in particular, the implied spot price dynamics which is obtained for time to maturity zero.

For stochastic integrals with respect to W, a Wiener process, we find the following result for $\mathbb{R}^n$-valued linear functionals:

Theorem 4.1 *Let $(X(t))_{t\geq 0}$ be a W-integrable stochastic process in $L(K, H)$, where W is a Wiener process in K with covariance operator Q. Assume that $A \in L(H, \mathbb{R}^n)$ and $\dim \operatorname{ran}(Q) \geq n$. Define the $\mathbb{R}^n$-valued stochastic process*

$$Y(t) := A \int_0^t X(s)dW(s).$$

Then there exists a n-dimensional standard Brownian motion B such that

$$Y(t) = \int_0^t \sigma(s)dB(s)$$

where $\sigma(s) := (AX(s)QX(s)^*A^*)^{1/2}$, $s \geq 0$, defines a B-integrable stochastic process.

If $\sigma(s)$ is invertible for $\lambda\otimes\mathbb{P}$-almost every $s \geq 0$, then it is integrable with respect to Y and

$$B(t) = \int_0^t \sigma(s)^{-1}dY(s).$$

Here, λ denotes the Lebesgue measure on the real line.

Proof We know from the discussion in Section 4.1 above that

$$Y(t) = \int_0^t AX(s)dW(s)$$

which has $a.s.$ continuous paths. Moreover, it is a martingale (see Peszat and Zabczyk [165, Thm. 8.7]) in $\mathbb{R}^n$, with bracket process

$$[Y, Y](t) = \int_0^t AX(s)QX(s)^*A^*ds.$$

Recall from Peszat and Zabczyk [165, Section 4.8] that

$$W(t) = \sum_{i=1}^{\infty} \sqrt{\lambda_i}\beta_i(t)e_i$$

for the ONB $(e_i)_{i\in\mathbb{N}}$ in K of eigenvectors of Q, with corresponding eigenvalues $\lambda_i > 0$, $i \in \mathbb{N}$. Here, $(\beta_i)_{i\in\mathbb{N}}$ are independent standard Brownian motions on the given filtered probability space, and by the dimensionality condition of the range of Q, we have at least n such. Thus, we can find a standard n-dimensional Brownian motion on our filtered probability space. Therefore, appealing to Jacod [127, Corollaire

14.47(b)], there exists an n-dimensional Brownian motion B such that the first claim follows.

To show the second claim, recall that Y is a square-integrable martingale. When $\sigma(s)$ is invertible for $\lambda \otimes \mathbb{P}$-almost every $s \geq 0$ we have

$$B(t) = \int_0^t I_n dB(s) = \int_0^t \sigma(s)^{-1}\sigma(s)dB(s) = \int_0^t \sigma(s)^{-1}dY(s)$$

as long as $(\sigma(s)^{-1})_{s\geq 0}$ is Y-integrable. Here, we appeal to the stochastic integration theory with respect to martingales as descibed in Peszat and Zabczyk [165, Section 8.2] for stochastic integration with respect to Y. We have also used the notation I_n for the $n \times n$-identity matrix. According to Definition 8.3 and Theorem 8.2 of Peszat and Zabczyk [165], we find that

$$Q_t^Y := 1_{\{\mathrm{Tr}(\sigma(t)\sigma(t)^*)\neq 0\}} \frac{\sigma(t)\sigma(t)^*}{\mathrm{Tr}(\sigma(t)\sigma(t)^*)}$$

is the martingale covariance of Y. It follows that for any $T < \infty$

$$\mathbb{E}\left[\int_0^T \mathrm{Tr}(\sigma(s)^{-1}Q_s^Y\sigma(s)^{-1*})d\langle\langle Y, Y\rangle\rangle(s)\right] = nT < \infty.$$

According to Eq. (8.6) on page 113 along with Theorem 8.7 in Peszat and Zabczyk [165], $(\sigma(s)^{-1})_{s\geq 0}$ is Y-integrable. This completes the proof. □

We now extend these results to the case when we have subordinated Wiener processes as integrators. The theory for subordinated Wiener processes in Hilbert space was presented in Chapter 2, and we recall this as an approach to define a rich class of Hilbert-valued Lévy processes.

To this end, we are interested to study stochastic processes in $\mathbb{R}^n$ of the form

$$Y(t) = A\int_0^t X(s)dL(s)$$

where $A \in L(H, \mathbb{R}^n)$, $L(t) := W(\Theta(t))$ for W a Wiener process in K and Θ a subordinator (on $\mathbb{R}_+$) with finite first moment and X being L-integrable. To analyse such processes, we need to study two different filtrations, as W is defined with respect to the given filtration $(\mathcal{F}_t)_{t\geq 0}$, whereas L is a Lévy process with respect to a "time-change" of this filtration. We make this precise in the next definition:

Definition 4.2 Assume $(\theta(t))_{t\geq 0}$ is a right-continuous increasing family of stopping times with respect to the filtration $(\mathcal{F}_t)_{t\geq 0}$. The time changed filtration $(\mathcal{F}_t^\theta)_{t\geq 0}$

is the filtration given by

$$\mathcal{F}_t^\theta := \bigcap_{s>t} \mathcal{F}_{\theta(s)}$$

for $t \geq 0$.

We can establish the following result for stochastic integrals with respect to subordinated Wiener processes:

Proposition 4.1 *Let W be a Wiener process in a separable Hilbert space K and Θ a non-zero subordinator with finite moment such that $\Theta(t)$ is a stopping time for each $t \geq 0$. Then $L(t) := W(\Theta(t)), t \geq 0$ is a K-valued square-integrable Lévy process (with respect to the time-changed filtration $(\mathcal{F}_t^\Theta)_{t\geq 0}$). Moreover, if X is a stochastic process in $L(K, H)$ where H is another separable Hilbert space which is L-integrable, then there exists an isometric embedding*

$$\Gamma : \mathcal{L}_L^2(\mathcal{F}^\Theta, H) \to \mathcal{L}_W^2(\mathcal{F}, H)$$

such that $\mathbb{P}$-a.s.

$$\int_0^t X(s)dL(s) = \int_0^{\Theta(t)} \Gamma(X)(s)dW(s). \tag{4.6}$$

The stochastic integral on the left-hand side is with respect to the filtration $(\mathcal{F}_t^\Theta)_{t\geq 0}$ while the one on the right is with respect to $(\mathcal{F}_t)_{t\geq 0}$.

Proof Since W is mean zero and square integrable and Θ has finite first moment, it follows from Lemma 2.3 that L is a square integrable Lévy process with mean zero. Its increments are independent with respect to the filtration $(\mathcal{F}_t^\Theta)_{t\geq 0}$, so the space of L-integrable processes will be $\mathcal{L}_L^2(\mathcal{F}^\Theta, H)$ as L is Lévy with respect to this filtration.

We first show the claimed result for elementary integrands $X \in \mathcal{L}_L^2(\mathcal{F}^\Theta, H)$. If X is elementary, then there are $m \in \mathbb{N}$, $0 \leq u_j < v_j < \infty$, $\mathcal{F}_{u_j}^\Theta$-measurable random variables Z_j and $\Phi_j \in L(K, H)$ such that

$$X(s) = \sum_{j=1}^m Z_j 1_{(u_j, v_j]}(s)\Phi_j.$$

Define Γ on such elementary integrands by

$$\Gamma(X)(s) := \sum_{j=1}^m Z_j 1_{(\Theta(u_j), \Theta(v_j)]}(s)\Phi_j$$

where we observe that the definition does not depend on the representation of X. It follows that $\Gamma(X) \in L_W^2(\mathcal{F}, H)$ and

$$\begin{aligned}\int_0^t X(s)dL(s) &= \sum_{j=1}^m Z_j\Phi_j\left(L(t\wedge v_j) - L(t\wedge u_j)\right)\\ &= \sum_{j=1}^m Z_j\Phi_j\left(W(\Theta(t)\wedge\Theta(v_j)) - W(\Theta(t)\wedge\Theta(u_j))\right)\\ &= \int_0^{\Theta(t)} \Gamma(X)(s)dW(s).\end{aligned}$$

To obtain the isometric property of Γ on the elementary integrands, we see that

$$\begin{aligned}\int_0^\infty \mathbb{E}\left[\|\Gamma(X)(s)Q_W^{1/2}\|_{\text{HS}}^2\right]ds &= \mathbb{E}\left[|\int_0^\infty \Gamma(X)(s)dW(s)|_H^2\right]\\ &= \mathbb{E}\left[|\lim_{k\to\infty}\int_0^{\Theta(k)} \Gamma(X)(s)dW(s)|_H^2\right]\\ &= \mathbb{E}\left[|\int_0^\infty X(s)dL(s)|_H^2\right]\\ &= \int_0^\infty \mathbb{E}\left[\|X(s)Q_L^{1/2}\|_{\text{HS}}^2\right]d\langle\langle L, L\rangle\rangle(s)\end{aligned}$$

for Q_L being the (martingale) covariance operator of L and Q_W the covariance operator of W. This establishes the result for elementary integrands. By a density argument, the proposition follows. □

With this proposition at hand, we are settled for the analogous result to Theorem 4.1 in the case of subordinated Wiener processes as integrators.

Theorem 4.2 *Let L be a mean-zero square-integrable Lévy process in K defined as $L(t) := W(\Theta(t)), t \geq 0$ for a K-valued Wiener process W and integrable non-zero subordinator Θ. Furthermore, let $(X(t))_{t\geq 0}$ be an L-integrable stochastic process in $L(K, H)$. Assume that $A \in L(H, \mathbb{R}^n)$ and* $\dim$ *$ran(Q_L) \geq n$ for Q_L being the martingale covariance operator of L. Define the $\mathbb{R}^n$-valued stochastic process*

$$Y(t) := A\int_0^t X(s)dL(s).$$

Then there exists an n-dimensional square-integrable mean-zero Lévy process N (again being a subordinated Brownian motion in $\mathbb{R}^n$) such that

$$Y(t) = \int_0^t \sigma(s) dN(s)$$

where $\sigma(s) := (AX(s)Q_L X(s)^* A^*)^{1/2}$, $s \geq 0$, defines an N-integrable stochastic process. If $\sigma(s)$ is invertible for $\lambda \otimes \mathbb{P}$-almost every $s \geq 0$, then it is integrable with respect to Y and

$$N(t) = \int_0^t \sigma(s)^{-1} dY(s).$$

Here, λ denotes the Lebesgue measure on the real line.

Proof We first prove the claim for elementary integrands X in $\mathcal{L}_L^2(\mathcal{F}^\Theta, H)$. I.e., there is $m \in \mathbb{N}$, $0 \leq u_j < v_j < \infty$, Z_j $\mathcal{F}_{u_j}^\Theta$-measurable random variables and operators $\Phi_j \in L(K, H)$ such that

$$X(s) = \sum_{j=1}^m Z_j 1_{(u_j, v_j]}(s) \Phi_j$$

for $s \geq 0$. For the isometric embedding Γ in Prop. 4.1 we see that $\Gamma(AX)(s) = A\Gamma(X)(s)$. This is true for all L-integrable processes X and not only elementary ones. Prop. 4.1 and Thm. 4.1 yield,

$$\begin{aligned} A \int_0^t X(s) dL(s) &= \int_0^t AX(s) dL(s) \\ &= \int_0^{\Theta(t)} \Gamma(AX)(s) dW(s) \\ &= \int_0^{\Theta(t)} A\Gamma(X)(s) dW(s) \\ &= \int_0^{\Theta(t)} \sigma_1(s) dB(s). \end{aligned}$$

Here, B is a standard Brownian motion in $\mathbb{R}^n$. Moreover, we find

$$\sigma_1(s) = \left(A\Gamma(X)(s) Q_W \Gamma(X)(s)^* A^*\right)^{1/2} = \Gamma\left(\left(AXQ_W X^* A^*\right)^{1/2}\right)(s)$$

for $s \geq 0$ and Q_W being the covariance operator of W. Applying again Prop. 4.1, we find

$$A\int_0^t X(s)dL(s) = \int_0^t \sigma(s)dN(s)$$

where $N(s) := B(\Theta(s))$ and

$$\sigma(s) := \left(AX(s)Q_W X(s)^* A^*\right)^{1/2}$$

for $s \geq 0$. Now, according to Sato [175, Thm. 30.1], N is an n-dimensional mean-zero square integrable Lévy process with the desired form of the integrand σ as Q_W coincides with Q_L. The second claim follows analogously.

We can conclude the proof by a density argument to extend the result to general integrand X in $\mathcal{L}_L^2(\mathcal{F}^\Theta, H)$. □

We remark that if L is a square-integrable K-valued Lévy process with characteristic triplet (b, C, ν), then the covariance operator Q of $L(1)$ is

$$Qx = Cx + \int_K y\langle x, y\rangle_K \nu(dy)$$

for $x \in K$. Moreover, the mean $m \in K$ is given by

$$m = b + \int_K y1(|y| \geq 1)\nu(dy).$$

According to Peszat and Zabczyk [165, Def. 4.45, Thm. 8.2], the angle bracket of L is $\langle\langle L, L\rangle\rangle(t) = t\mathrm{Tr}(Q)$ and the operator angle bracket process is $[L, L](t) = tQ$. Peszat and Zabczyk [165, Def. 8.3] provides us with the martingale covariance operator of L, which becomes $Q_L = Q/\mathrm{Tr}(Q)$.

4.3 A Brief Account on Parabolic Stochastic Partial Differential Equations

In this section we discuss existence and uniqueness of solutions of a class of stochastic partial differential equations (SPDEs) of parabolic type with time-inhomogeneous coefficients. The SPDEs we discuss are of a form relevant for modeling forward and futures prices, and our presentation is heavily influenced by the analysis presented in Filipović, Tappe and Teichmann [102] (see also Peszat and Zabczyk [165, Ch. 9] and Tappe [182]) Before giving a survey on the relevant parts, we notice in passing that we will focus on SPDEs driven by infinite dimensional Lévy processes, and not more general martingales as done in Peszat and Zabczyk [165, Ch. 9]. On the other hand, we split the Wiener part from the pure-jump component of the Lévy process in the formulation of the dynamics, following the set-up of Filipović, Tappe and Teichmann [102]:

To this end, let $(W(t))_{t\geq 0}$ be a Wiener process with values in a separable Hilbert space K. We recall the homogeneous Poisson random measure $N(dt, dz)$ on $\mathbb{R}_+ \times K$ with compensator $dt \otimes \nu(dz)$ for a σ-finite Lévy measure ν on K. We remark in passing in Filipović, Tappe and Teichmann [102] they consider a Blackwell space as the mark space for the Poisson random measure. We focus on a separable Hilbert space K being the same as the state space of W here to reduce notational burden and also to have a presentation which starts out with Lévy processes taking values in K. Notice that any Polish space equipped with its Borel σ-algebra is a Blackwell space, and thus K is. Furthermore, denote by $A : \mathrm{Dom}(A) \subset H \to H$ a linear, densely defined and unbounded operator on H, a separable Hilbert space. We assume that A generates a *quasi-contractive* C_0-semigroup $(S_t)_{t\geq 0}$. Consider the following SPDE,

$$
\begin{aligned}
dX(t) &= (AX(t) + \beta(t, X(t)))\, dt + \sigma(t, X(t))\, dW(t) \\
&\quad + \int_K \gamma(t, X(t), z)\, \widetilde{N}(dt, dz) \qquad (4.7)
\end{aligned}
$$

where $X(0) = X_0 \in H$ and measurable coefficient functions $\beta : \mathbb{R}_+ \times H \to H$, $\sigma : \mathbb{R}_+ \times H \to L_{\mathrm{HS}}(K, H)$ and $\gamma : \mathbb{R}_+ \times H \times K \to H$.

We introduce some local Lipschitz and growth conditions on the coefficient functions. First, let for $p \geq 1$, $L^p_{\mathrm{loc}}(\mathbb{R}_+)$ denote space of real-valued functions on $\mathbb{R}_+$ which are locally p-integrable. We suppose the following:

$$
\begin{aligned}
t &\mapsto |\alpha(t, 0)|_H \in L^2_{\mathrm{loc}}(\mathbb{R}_+) \\
t &\mapsto \|\sigma(t, 0)\|_{\mathrm{HS}} \in L^2_{\mathrm{loc}}(\mathbb{R}_+) \\
t &\mapsto \int_K |\gamma(t, 0, z)|^2_H\, \nu(dz) \in L^1_{\mathrm{loc}}(\mathbb{R}_+)
\end{aligned}
$$

Additionally, there exists a function $K \in L^2_{\mathrm{loc}}(\mathbb{R}_+)$ such that for any $x, y \in H$,

$$
\begin{aligned}
|\beta(t, x) - \beta(t, y)|_H &\leq K(t)|x - y|_H \\
\|\sigma(t, x) - \sigma(t, y)\|_{\mathrm{HS}} &\leq K(t)|x - y|_H \\
\int_K |\gamma(t, x, z) - \gamma(t, y, z)|^2_H\, \nu(dz) &\leq K(t)^2 |x - y|^2_H .
\end{aligned}
$$

By a straightforward application of the triangle inequality, the integrability at zero and Lipschitz conditions on the coefficient functions imply linear growth,

$$|\beta(t,x)|_H \leq K(t)(1+|x|_H)$$
$$\|\sigma(t,x)\|_{\mathrm{HS}} \leq K(t)(1+|x|_H)$$
$$\int_K |\gamma(t,x,z)|_H^2\,\nu(dz) \leq K^2(t)(1+|x|_H)^2,$$

where, possibly, the function $K \in L^2_{\mathrm{loc}}(\mathbb{R}_+)$ may have changed.

We remark that when modelling forward prices, the generator A is the derivative operator ∂_x. Recall that this is a densely defined operator on the Filipović space, and that its shift semigroup is quasi-contractive, see Lemma 3.7.

We define a mild solution of (4.7), following [102, Def. 7.3]:

Definition 4.3 An adapted RCLL H-valued process $(X(t))_{t\geq 0}$ is called a *mild solution* to (4.7) if

$$\begin{aligned} X(t) = \mathcal{S}_t X_0 + \int_0^t \mathcal{S}_{t-s}\beta(s,X(s))\,ds + \int_0^t \mathcal{S}_{t-s}\sigma(s,X(s))\,dW(s) \\ + \int_0^t\int_K \mathcal{S}_{t-s}\gamma(s,X(s-),z)\,\widetilde{N}(ds,dz) \end{aligned}$$

for all $t \geq 0$.

We remark that Filipović, Tappe and Teichmann [102] formulates mild solutions in terms of predictable processes in H (see [102, Def. 7.3]), but shows existence and uniqueness of an adapted RCLL solution of the SPDE. We prefer to include RCLL and adaptedness as part of the definition of a solution. Furthermore, in Peszat and Zabczyk [165, Def. 9.5], an additional integrability and boundedness condition on the solution is added to the definition of a mild solution. I.e., it is required that

$$\sup_{t\in[0,T]} \mathbb{E}[|X(t)|_H^2] < \infty$$

for $T < \infty$. However, Filipović, Tappe and Teichmann [102] obtain a slightly stronger bound on their solutions, so we formulate this in the existence and uniqueness result rather than including such a restriction in the definition of a mild solution.

According to Filipović, Tappe and Teichmann [102, Cor. 10.6], we have the following:

Theorem 4.3 *Suppose that X_0 is an H-valued $\mathcal{F}_0$-measurable square integrable random variable. Then there exists a unique mean-square continuous mild solution*

$(X(t))_{t\geq 0}$ *of* (4.7) *such that*

$$\mathbb{E}[\sup_{t\in[0,T]} |X(t)|_H^2] < \infty$$

for all $T > 0$.

Corollary 10.6 of Filipović, Tappe and Teichmann [102] also shows that the mild solution is a weak solution. Below we define a weak solution of the SPDE (4.7):

Definition 4.4 An adapted RCLL H-valued process $(X(t))_{t\geq 0}$ is called a *weak solution* to (4.7) if

$$\begin{aligned}\langle X(t), h\rangle_H = \langle X_0, h\rangle_H &+ \int_0^t \langle X(s), A^*h\rangle_H + \langle \beta(s, X(s)), h\rangle_H \, ds \\ &+ \int_0^t \langle \sigma(s, X(s)), h\rangle_H \, dW(s) \\ &+ \int_0^t \int_K \langle \gamma(s, X(s-), z), h\rangle_H \, \widetilde{N}(ds, dz)\end{aligned}$$

for all $t \geq 0$ and $h \in \mathrm{Dom}(A^*)$.

Notice that A^* is the adjoint of A, which also has a dense domain in H. The above definition is similar to [102, Def. 7.2], and it is shown in Lemma 7.6 of Filipović, Tappe and Teichmann [102] that any weak solution of (4.7) is a mild solution. In the formulation of Theorem 4.3, we can substitute *mild* with *weak*, according to Filipović, Tappe and Teichmann [102, Cor. 10.6]. We remark that in Tappe [182], the conditions on the coefficient functions are relaxed to be locally Lipschitz and linear growth, and mild, weak and strong solutions are shown under various assumptions.

4.4 Bibliographical Notes

Our main source for stochastic analysis in Hilbert space is Peszat and Zabczyk [165]. They define stochastic integration theory for general Hilbert-valued martingales, including Lévy processes as we have a focus on in our exposition. The Wiener case is extensively presented in Da Prato and Zabczyk [81]. Dinculeanu [87], Mandrekar and Rüdiger [154], and van Neerven, Veraar and Weis [183] present stochastic integration theory and calculus in Banach spaces.

Topics going beyond our exposition is the rough path integration theory which defines stochastic integration in a pathwise sense. An introduction to the rough path theory is given in Friz and Hairer [108]. Cylindrical Lévy processes is another topic that we do not touch. A stochastic calculus for cylindrical Lévy processes is found in Applebaum and Riedle [8] and Jakubowski and Riedle [129].

Both Da Prato and Zabczyk [81] and Peszat and Zabczyk [165] study existence and uniqueness of solutions of stochastic partial differential equations in Hilbert space driven by Wiener and Lévy noise, respectively. They also provide several applications of these including HJM-models of forward rates. A background on stochastic partial differential equations in infinite dimensions can also be found in Gawarecki and Mandrekar [110] and Prévôt and Röckner [168]. A white noise approach to stochastic partial differential equations is presented in Holden *et al.* [125], while a rough path approach is found in Friz and Hairer [108].

The theory and analysis of Section 4.2 in this Chapter is adopted from Benth and Krühner [34], where some further connections between the forward prices and spot dynamics are discussed, in particular Lévy semistationary spot price models (see Barndorff-Nielsen, Benth and Veraart [15]).

Part II
Modelling the Forward Price Dynamics and Derivatives Pricing

Chapter 5
Spot Models and Forward Pricing

Abstract Arithmetic and geometric factor models for the spot price dynamics in energy markets are reviewed, and the implied forward price dynamics from these models are derived. The forward price dynamics is analysed in the context of HJM-models under the Musiela parametrisation using convenient pricing measures based on the Girsanov and Esscher transforms. In particular, we derive a stochastic partial differential equation for the term structure dynamics and introduce the Filipović space as state space. Several particular examples are presented, including continuous-time autoregressive moving average processes applied to temperature data. We also establish a link between pricing measures and the classical theory of forward pricing which is based on storage costs and convenience yield. On the technical side, we prove a stochastic Fubini theorem tailored to our needs.

Forward prices in commodity markets are in many cases settled on the spot price of the commodity, or, like in weather markets, on an index established from a weather factor like temperature, say. A natural approach to model the forward price dynamics is thus to first introduce a stochastic dynamics for the underlying spot or index, and next derive the forward price based on no-arbitrage principles. In this Chapter we propose a general spot dynamics, which includes many of the popular factor models proposed and analysed theoretically and empirically in commodity and energy markets. The derived term structure of forward prices is next analysed as a dynamics taking values in a space of curves, conveniently expressed as a Hilbert space. We also show that the term structure is the solution of a certain stochastic partial differential equation, laying the foundation for the subsequent Chapters on the Heath-Jarrow-Morton modelling approach for the term structure. Finally, we also include a discussion relating the use of pricing measures to connect spot and forward prices with the classical theory of convenience yield and storage costs.

F. E. Benth, P. Krühner, *Stochastic Models for Prices Dynamics in Energy and Commodity Markets*, Springer Finance,
https://doi.org/10.1007/978-3-031-40367-5_5

5.1 Spot Price Models in Commodity Markets

In this Section we review some classical models for the spot price dynamics of commodity markets, with a particular view towards energy markets. We end the Section with defining a general spot price dynamics which we will use for further analysing the forward price dynamics and the evolution of the term structure in the forward market.

Lucia and Schwartz [149] proposed four stochastic processes to model the spot price dynamics in the NordPool electricity market. They divided into geometric and arithmetic dynamics, and further into one- and two-factor models. The geometric one-factor model is the so-called Schwartz dynamics, proposed by Schwartz [177] for modelling commodity spot prices with oil as the prime example. The dynamics is defined as

$$S(t) = \exp(Y(t)), \tag{5.1}$$

where $S(t)$ is the spot price at time $t \geq 0$ and $(Y(t))_{t\geq 0}$ is an Ornstein–Uhlenbeck process

$$dY(t) = (\mu - \alpha Y(t))\,dt + \sigma\,dB(t), \tag{5.2}$$

for $\mu, \alpha \geq 0, \sigma \geq 0$ constants and B a standard Brownian motion on $\mathbb{R}$. In the context of electricity, it is natural to include a measurable deterministic seasonal mean function $\Lambda : \mathbb{R}_+ \to \mathbb{R}$, and assume that the spot price is $S(t) = \exp(\Lambda(t) + Y(t))$. The logarithmic spot price dynamics will evolve as a stationary process around its mean Λ in this model.

The two-factor model proposed by Lucia and Schwartz [149] is motivated from Schwartz and Smith [178], and assumes that the spot prices evolve with stationary short term variations and non-stationary long-term variations:

$$S(t) = \exp(\Lambda(t) + X(t) + Y(t)), \tag{5.3}$$

where $X(t) = \beta t + \eta W(t)$ being a drifted Brownian motion, e.g., $\beta, \eta \geq 0$ are constants and W a standard Brownian motion on $\mathbb{R}$, possibly correlated with B. The corresponding arithmetic models take the form $S(t) = \Lambda(t) + Y(t)$ or $S(t) = \Lambda(t) + X(t) + Y(t)$.

Benth, Kallsen and Meyer-Brandis [30] suggest a multi-factor arithmetic spot price dynamics for electricity spot prices, suitable to derive analytical forward prices for contracts delivering over a period,

$$S(t) = \Lambda(t) + \sum_{i=1}^{n} w_i X_i(t), \tag{5.4}$$

where $w_i > 0, i = 1, \dots, n$ are weights and

$$dX_i(t) = -\alpha_i X_i(t)\,dt + dL_i(t)\,, \tag{5.5}$$

with $L_i, i = 1, \dots, n$ being n subordinators [1] and $\alpha_i > 0, i = 1, \dots, n$ constants. Whenever the seasonality function Λ is non-negative, this model is tailored to ensure positivity of the spot prices. However, negative prices may occur in power markets, with frequent examples in the EEX market, and we may use general Lévy processes L_i in the dynamics for X_i, for $i = 1, \dots, n$. We remark in passing that a non-Markovian dynamics explaining both negative and positive spikes as well as allowing for the possibility of negative prices was proposed by Kholodnyi [135].

At the EEX power exchange, one can trade in forward contracts written on a wind index for Germany. The wind index is defined as the relative production of wind power with respect to the total capacity, and hence a number between 0 and 1.[2] In Benth and Pircalabu [43], a dynamics for the wind index is given as an exponential model $P(t) = \exp(-X(t))$ where X follows an Ornstein–Uhlenbeck process of the form (5.5) driven by a subordinator. Since the state space of X is the positive real line, we find that $\exp(-X(t)) \in (0, 1)$. We remark in passing that Benth and Pircalabu [43] allow for a seasonality function in the dynamics of the wind index, and one must scale X with a mean-reversion level in order to ensure that the overall dynamics stay within the desired interval. This model of the wind index has been generalised to a multivariate dynamics and applied to German wind index data in a hedging study by Benth, Christiansen and Rohde [27] (cf. also Benth and Rohde [44]).

Benth *et al.* [31] use data from the EEX market to calibrate a two-factor arithmetic spot model which is extending (5.4) to $X_1 = L_1$ (e.g., $\alpha_1 = 0$) and X_2 following a so-called CARMA(2,1)-dynamics driven by an α-stable Lévy process L_2. In their empirical analysis, L_1 turns out to be a NIG Lévy process. In general, for $p, q \in \mathbb{N} \cup \{0\}, p > q$, a CARMA$(p, q)$-process is defined as $Y(t) = \int_0^t g(t-s)\,dL(s)$, where L is a Lévy process and $g : \mathbb{R}_+ \to \mathbb{R}$ the function

$$g(x) = \mathbf{b}^\top \exp(Ax)\mathbf{e}_p\,. \tag{5.6}$$

---

[1] Recall that a subordinator is a Lévy process with increasing paths, which means a pure-jump Lévy process with only positive jumps and a non-negative drift.

[2] Such a wind index is sometimes referred to as the *capacity factor* of wind.

Here, $(\mathbf{e}_i)_{i=1}^p$ is the canonical basis on $\mathbb{R}^p$, $A \in \mathbb{R}^{p\times p}$ given by

$$A = \begin{bmatrix} 0 & 1 & 0 & 0 & \cdots & 0 \\ 0 & 0 & 1 & 0 & \cdots & 0 \\ \cdot & \cdot & \cdot & \cdot & \cdots & \cdot \\ \cdot & \cdot & \cdot & \cdot & \cdots & \cdot \\ 0 & 0 & 0 & 0 & \cdots & 1 \\ -\alpha_p & -\alpha_{p-1} & -\alpha_{p-2} & -\alpha_{p-3} & \cdots & -\alpha_1 \end{bmatrix},$$

with $\alpha_k > 0, k = 1, \dots, p$ and $\mathbf{b} = (b_0, b_1, \dots, b_{p-1})^\top \in \mathbb{R}^p$ with $b_q = 1$, $b_k = 0, k = q+1, \dots, p-1$. CARMA-processes are stationary whenever the eigenvalues of the matrix A have negative real part.

In Benth and Šaltytė Benth [49, Chapter 3], temperature data series at different locations in Europe have been successfully modelled by CARMA-processes (in discrete time). We now make a small empirical case study of a data series of daily average temperatures collected in Vilnius, Lithuania, ranging from September 1, 2004 to September 30, 2019 (in total 5508 data points). The data were recorded by the Lithuanian Hydrometeorological Service,[3] where the average is defined as the mean between the daily maximum and minimum observed temperature. Temperature impacts the demand for electricity, due to consumer's need for heating (in the winter) and cooling (in the summer). Furthermore, there is a market for temperature futures contracts at the Chicago Mercantile Exchange (CME) based on different indices on temperatures measured in a selection of US cities. We refer to Benth and Šaltytė Benth [49] for more on this market.

In Figure 5.1 we have plotted the daily average temperature series along with a fitted seasonal function with trend (black line). We have assumed a seasonality function of the form

$$\Lambda(t) = a + bt + c\cos\left(\frac{2\pi}{365}(t-\varphi)\right),$$

for constants a, b, c and φ. The constants were fitted to the data by nonlinear regression.[4] In Figure 5.2 we have plotted the partial autocorrelation function for the temperature data less the seasonality function $\Lambda(t)$. Three lags are significant, pointing to an autoregressive time series model of order three. From Benth and Šaltytė Benth [49, Chapter 4] we know that a discretisation of CARMA(3,0)-dynamics leads to an autoregressive time series of order 3. A further analysis of the residuals of such a time series of temperature data would reveal a seasonality in the variance. We refer to Benth and Šaltytė Benth [49, Chapters 3 and 5] for

[3] www.meteo.lt/en/

[4] We used the `nlinfit`-function in Matlab. Estimated values are $a = 6.9594$, $b = 0.0002$, $c = 11.5332$, $\varphi = 321.8544$.

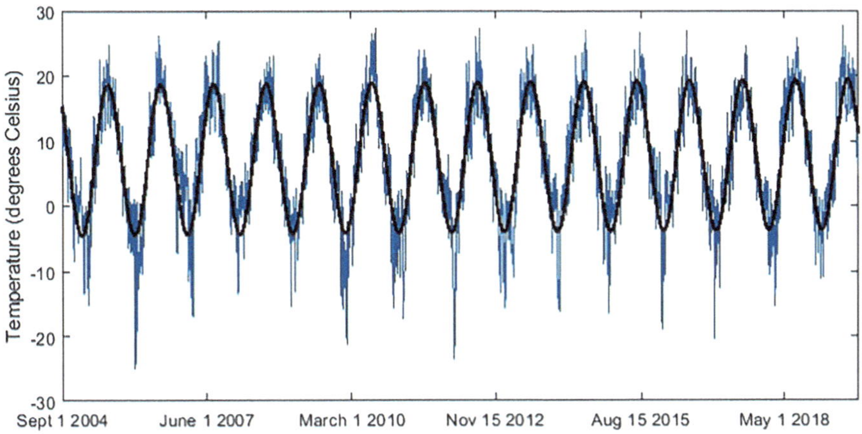

Fig. 5.1 Daily average temperature in Vilnius from September 1, 2004 to September 30, 2019. Black curve is the estimated seasonal function with linear trend

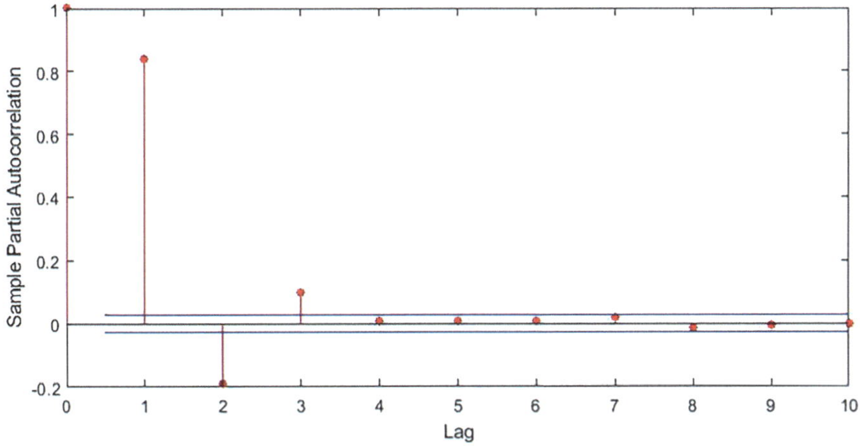

Fig. 5.2 Partial autocorrelation function for the daily average temperatures in Vilnius less its fitted seasonality function. The blue lines are 95% confidence bands

more details, where additionally temperature futures prices are analysed based on a CARMA(3,0)-dynamics. We further would like to mention that in the same monograph one can find empirical analysis of wind speed data from areas in New York and Texas which after a logarithmic transformation has been modelled using a CARMA(4,0)-dynamics (see also Benth and Šaltytė Benth [47] and Benth and Rohde [44]). In a recent study of solar irradiation (measured at noon) in German cities, Larsson, Green and Benth [146] found clear evidence of an autoregressive dynamics of order two.

CARMA-processes are special cases of Lévy semistationary processes, which again are special cases of volatility modulated Volterra processes. Volatility modulated Volterra processes is defined and applied in the context of energy markets by Barndorff-Nielsen *et al.* [13] (see also Barndorff-Nielsen *et al.* [14]), and takes the form

$$Y(t) = \int_0^t g(t,s)\sigma(s)\,dL(s)\,, \tag{5.7}$$

for a kernel function $(t,s) \mapsto g(t,s) \in \mathbb{R}$ with $0 \leq s \leq t$, L a real-valued Lévy process and $(\sigma(t))_{t\geq 0}$ a predictable stochastic volatility/intermittency process, assuming that $s \mapsto g(t,s)\sigma(s)$, $s \leq t$ is L-integrable. The volatility process is typically assumed to be independent of L, and in most cases L is a Brownian motion. Barndorff-Nielsen *et al.* [13] (see also Barndorff-Nielsen *et al.* [14]) discuss both arithmetic and geometric spot price models for energy prices based on volatility modulated Volterra processes.

We remark in passing that choosing $g(t,s) = \exp(-\alpha(t-s))$ and $\sigma(s) = 1$ in (5.7), we recover Y to be an Ornstein–Uhlenbeck process. With $L = B$, a Brownian motion and the kernel function being of Ornstein–Uhlenbeck type, various classical stochastic volatility models applied to energy markets are proposed and discussed in Eydeland and Wolyniec [97] and Geman [111]. For example, the Heston stochastic model proposed in [122] is defined as $\sigma^2(t) = V(t)$, where V is the Cox-Ingersoll-Ross process (see [74])

$$dV(t) = \kappa(\theta - V(t))dt + \eta\sqrt{V(t)}\,dU(t). \tag{5.8}$$

Here, U is a Brownian motion, possibly correlated with B. The parameters are assumed to satisfy the Feller condition $\eta^2 < 2\kappa\theta$, ensuring that V is strictly positive. The Barndorff-Nielsen and Shephard (BNS) stochastic volatility model (see [18]) is studied in Benth [21] for the UK gas spot market. The BNS stochastic volatility model is a weighted sum of Ornstein–Uhlenbeck processes driven by subordinators, as defined in (5.5).

We wrap up the discussion of this Section by introducing a general multi-factor spot model which includes the various special cases presented above. To this end, let $(\mathbf{L}(t))_{t\geq 0}$ be a square-integrable n-dimensional Lévy process with respect to $(\mathcal{F}_t)_{t\geq 0}$ and $\mathbf{g} : \mathbb{R}_+^2 \to \mathbb{R}^n$ being a (deterministic) measurable function. Furthermore, introduce the (matrix-valued) stochastic volatility process $(\boldsymbol{\Sigma}(t))_{t\geq 0}$, where $\boldsymbol{\Sigma}(t) \in \mathbb{R}^{n\times n}$ for all $t \geq 0$, being a predictable process. If $s \mapsto \mathbf{g}(t,s)^\top \boldsymbol{\Sigma}(s) \in \mathbb{R}^n$ is $\mathbf{L}$-integrable for $s \leq t$, we define the spot price at time $t \geq 0$ as

$$S(t) = \Lambda(t) + \int_0^t \mathbf{g}(t,s)^\top \boldsymbol{\Sigma}(s)\,d\mathbf{L}(s)\,. \tag{5.9}$$

Here, $\Lambda : \mathbb{R}_+ \to \mathbb{R}$ is a measurable function which models the seasonal mean of the spot price.

The dynamics in (5.9) defines an *arithmetic* spot price model. A *geometric* spot price model is conveniently defined as

$$\ln S(t) = \Lambda(t) + \int_0^t \mathbf{g}(t,s)^\top \mathbf{\Sigma}(s)\, d\mathbf{L}(s)\,. \tag{5.10}$$

In the remaining of this Chapter, we will discuss both models in view of futures pricing and the analysis of the term structure of the futures market.

5.2 Pricing of Forwards

Let $F(t,T)$ be the forward price at time $0 \leq t \leq T$ of a contract delivering the underlying spot commodity at time $T \geq t$. Introduce a probability $\mathbb{Q} \sim \mathbb{P}$, which we call the *pricing measure*. If the spot price dynamics $(S(t))_{t\geq 0}$ satisfies $S(t) \in L^1(\mathbb{Q})$ for all $t \geq 0$ we define

$$F(t,T) = \mathbb{E}_{\mathbb{Q}}[S(T)\,|\,\mathcal{F}_t]\,. \tag{5.11}$$

Strictly speaking, it is sufficient that $S(T) \in L^1(\mathbb{Q})$ for the given delivery time T, however, as we later will be interested in all delivery times $T \geq 0$ it is convenient to have $\mathbb{Q}$-integrability of the spot price at all times. Remark that the stochastic process $(F(t,T))_{t\in[0,T]}$ is a $\mathbb{Q}$-martingale which thus defines an arbitrage-free forward price dynamics. This is the standard approach to defining the forward price dynamics in commodity markets which are incomplete, that is, in markets for which the classical buy-and-hold strategy cannot be implemented due to lack of or limited storage possibilities and absence of a convenience yield (weather, electricity, energy). We refer the interested reader to Geman [111] and Eydeland and Wolyniec [97] as well as Section 5.5 at the end of this Chapter for more on this.

Let us consider a spot price dynamics $(S(t))_{t\geq 0}$ as defined in (5.9). In practice, the spot price is naturally formulated under the market probability $\mathbb{P}$ and a pricing measure $\mathbb{Q} \sim \mathbb{P}$ is introduced for the purpose of forward pricing, and thereby implicitly being a model for the *risk premium* in the commodity market. The risk premium is defined as the difference between the forward price and the predicted spot price at delivery, and we will come back to this important notion in Chapter 6. Typically, as discussed in detail in Benth, Šaltytė Benth and Koekebakker [25], an Esscher transform is applied to define a pricing measure, which has the convenient structure preserving property where $\mathbf{L}$ is a $\mathbb{Q}$-Lévy process. Benth and Ortiz-Latorre [40] extends the Esscher transform to Ornstein–Uhlenbeck factor processes, where also the speed of mean reversion can be changed. More precisely, they propose a state dependent pricing measure $\mathbb{Q} \sim \mathbb{P}$ where the speed of mean reversion of the Ornstein–Uhlenbeck process is slower with respect to $\mathbb{Q}$ than $\mathbb{P}$. The implied risk premium will in this case become stochastic, explicitly depending on the factors modelling the spot price dynamics (see Benth and Ortiz-Latorre [41] for an extension to a stochastic volatility model in commodity markets). We remark that the Esscher transform will only allow for a deterministic risk premium (see Benth and Sgarra [51] for an extensive discussion of the Esscher transform and the risk premium in energy markets).

We have the following expression for the forward price:

Proposition 5.1 *Assume that the arithmetic $\mathbb{P}$-dynamics of $(S(t))_{t\geq 0}$ given in* (5.9) *satisfies $S(t) \in L^1(\mathbb{Q})$ for all $t \geq 0$. Then,*

$$F(t,T) = \Lambda(T) + \mu_Q(T,t) + \int_0^t \mathbf{g}(T,s)^\top \mathbf{\Sigma}(s)\,d\mathbf{L}(s)\,,$$

for $0 \leq t \leq T$, where $\mu_Q(T,t) = \mathbb{E}_{\mathbb{Q}}[\int_t^T \mathbf{g}(T,s)^\top \mathbf{\Sigma}(s)\,d\mathbf{L}(s)\,|\,\mathcal{F}_t]$.

Proof This follows by the $\mathcal{F}_t$-measurability of $\int_0^t \mathbf{g}(T,s)^\top \mathbf{\Sigma}(s)\,\mathbf{L}(s)$. □

Remark that to compute the expectation defining μ_Q, we must have available the $\mathbb{Q}$-dynamics of $(\mathbf{\Sigma}(t))_{t\geq 0}$ and $(\mathbf{L}(t))_{t\geq 0}$. For example, if $\mathbf{\Sigma}$ is independent of the driving noise $\mathbf{L}$, and $\mathbf{L}$ is a $\mathbb{Q}$-Lévy process, then the independent increment property of $\mathbf{L}$ together with a double conditioning with respect to the σ-algebra generated by the path of $t \mapsto \mathbf{\Sigma}(t)$ for $t \leq T$, will yield

$$\begin{aligned}\mu_Q(T,t) &= \mathbb{E}_{\mathbb{Q}}[\int_t^T \mathbf{g}(T,s)^\top \mathbf{\Sigma}(s)\,d\mathbf{L}(s)\,|\,\mathcal{F}_t] \\ &= \int_t^T \mathbf{g}(T,s)^\top \mathbb{E}_{\mathbb{Q}}[\mathbf{\Sigma}(s)\,|\,\mathcal{F}_t]\mathbb{E}_{\mathbb{Q}}[\mathbf{L}(1)]\,ds\,.\end{aligned}$$

If we know the $\mathbb{Q}$-characteristics of $\mathbf{L}$ and $\mathbf{\Sigma}$, we can further compute the integral defining μ_Q.

Let us derive the forward price for the geometric spot dynamics.

Proposition 5.2 *Assume that the geometric $\mathbb{P}$-dynamics of $(S(t))_{t\geq 0}$ given in* (5.10) *satisfies $S(t) \in L^1(\mathbb{Q})$ for all $t \geq 0$. Then,*

$$F(t,T) = \exp\left(\Lambda(T) + \mu_Q(T,t) + \int_0^t \mathbf{g}(T,s)^\top \mathbf{\Sigma}(s)\,d\mathbf{L}(s)\right)\,,$$

for $0 \leq t \leq T$, where $\mu_Q(T,t) = \ln \mathbb{E}_{\mathbb{Q}}[\exp(\int_t^T \mathbf{g}(T,s)^\top \mathbf{\Sigma}(s)\,d\mathbf{L}(s))\,|\,\mathcal{F}_t]$.

Proof This follows by the $\mathcal{F}_t$-measurability of $\int_0^t \mathbf{g}(T,s)^\top \mathbf{\Sigma}(s)\,\mathbf{L}(s)$. □

We observe that the logarithmic forward price for the geometric spot dynamics has the same form as the forward price in the arithmetic case, except for a different expression for μ_Q.

Let us consider an example: Assume that the spot price dynamics is given by a two-factor model under the pricing measure $\mathbb{Q}$, extending the model of Lucia and Schwartz [149] in (5.3),

$$S(t) = \exp\left(X(t) + Y(t)\right) ,$$

where $X(t) = \beta t + \eta W(t)$, a drifted Brownian motion, and

$$dY(t) = \alpha(\theta - Y(t))\, dt + dL(t) ,$$

an Ornstein–Uhlenbeck process driven by a square integrable Lévy process L with zero mean. Here, α, β, θ and η are constants with $\alpha, \eta > 0$. Note that the assumption of L having zero mean is not restrictive, as we can always write $L(t) = \widetilde{L}(t) - t\mathbb{E}_{\mathbb{Q}}[L(1)]$, and redefine θ. To simplify matters, we let L be independent of W. In the framework of our general model, we have $n = 2$, and $\mathbf{L}(t) = (\eta W(t), L(t))^\top$ a two-dimensional Lévy process under $\mathbb{Q}$. Furthermore, the stochastic volatility process is the identity matrix $I_2 \in \mathbb{R}^{2\times 2}$, i.e., $\mathbf{\Sigma}(t) = I_2$ and $\mathbf{g}(t, s) = (1, \exp(-\alpha(t - s)))^\top$. The "seasonality function" is defined to be $\Lambda(t) = \beta t + \theta(1 - \exp(-\alpha t))$. Note that we have not introduced any seasonality function in the spot price dynamics, but use the Λ here to include the drift from X and the level towards which Y mean reverts. If $L(1)$ has finite exponential moment, $S(t) \in L^1(\mathbb{Q})$ for all $t \geq 0$, and we find the forward price dynamics from Prop. 5.1 as

$$F(t, T) = \exp\left(\beta T + \theta(1 - e^{-\alpha T}) + \mu_Q(t, T) + \eta W(t) + \int_0^t e^{-\alpha(T-s)}\, dL(s)\right) ,$$

with

$$\begin{aligned}\mu_Q(t, T) &= \ln \mathbb{E}_{\mathbb{Q}}\left[\exp\left(\int_t^T \mathbf{g}(T, s)^\top\, d\mathbf{L}(s)\right) \mid \mathcal{F}_t\right] \\ &= \ln \mathbb{E}_{\mathbb{Q}}\left[\exp\left(\eta(W(T) - W(t)) + \int_t^T e^{-\alpha(T-s)}\, dL(s)\right)\right] \\ &= \frac{1}{2}\eta^2(T - t) + \int_0^{T-t} \varphi(e^{-\alpha s})\, ds .\end{aligned}$$

In the second equality we used the independent increment property of Lévy processes, while in the third we used that L has finite exponential moment along with the Lévy-Kintchine representation. The function φ is the logarithm of the moment generating function of $L(1)$ (with respect to the measure $\mathbb{Q}$). In conclusion, we obtain

$$F(t, T) = \exp\left(\Phi(T - t) + X(t) + e^{-\alpha(T-t)} Y(t)\right) \tag{5.12}$$

where

$$\Phi(x) = (\beta + \frac{1}{2}\eta^2)x + \theta(1 - e^{-\alpha x}) + \int_0^x \varphi(e^{-\alpha s})\, ds\,. \tag{5.13}$$

We observe that the futures price at time t is a function of the states $X(t)$ and $Y(t)$ defining the spot price, but not a function of the spot price $S(t)$ itself. Moreover, F depends on time to maturity $T - t$ explicitly.

5.3 Analysis of the Term Structure

At each time t, the market will trade in forward contracts with different delivery times, and one refers to $T \mapsto F(t, T)$, $T \geq t$ as the *term structure*. Altough in practice we only have a finite number of maturity times T observable, theoretically this is a curve in the variable T. It is convenient to study the term structure as a curve in *time to delivery* rather than as a function of *time of delivery*. Hence, for $x \in \mathbb{R}_+$, define

$$f(t, x) := F(t, t + x)\,. \tag{5.14}$$

With $x = T - t$, $f(t, T - t) = F(t, T)$, and we recover the forward price for a contract with delivery time T. One says that $f(t, x)$ is the *Musiela parametrisation* of the forward price, and we will refer to $\mathbb{R}_+ \ni x \mapsto f(t, x) \in \mathbb{R}$ as the term structure.

We can view $(f(t, \cdot))_{t\geq 0}$ as stochastic process which takes values in a space of real-valued functions on $\mathbb{R}_+$. Assuming H is a separable Hilbert space of real-valued functions on $\mathbb{R}_+$, then $(f(t, \cdot))_{t\geq 0}$ is an H-valued stochastic process.

From Prop. 5.1, we find that the forward price for an arithmetic spot dynamics price given in (5.9) follows

$$f(t, x) = \Lambda(t+x)+\mu_Q(t+x, t)+\int_0^t \mathbf{g}(t+x, s)^\top \mathbf{\Sigma}(s)\, d\mathbf{L}(s)\,, \quad x \geq 0\,. \tag{5.15}$$

This is the arithmetic term structure of forward prices. In the next Proposition we show that $x \mapsto f(t, x)$ is an element of H for every $t \geq 0$ under rather mild conditions.

Proposition 5.3 *Assume $\Lambda(t + \cdot) \in H$, $\mu_Q(t + \cdot, t) \in L^1(H; \mathbb{P})$, and $\int_0^t |\mathbf{g}(t + \cdot, s)|_H^2 \mathbb{E}[\|\mathbf{\Sigma}(s)\|^2]\, ds < \infty$ for every $t \geq 0$, where $\|\cdot\|$ is a matrix norm consistent with the 2-norm on $\mathbb{R}^n$ and $|\mathbf{g}|_H^2 = |g_1|_H^2 + \cdots + |g_n|_H^2$. Then, the arithmetic term structure model $f(t, \cdot)$ in (5.15) is an element of H for every $t \geq 0$, $\mathbb{P} - a.s.$.*

Proof First, observe that by denoting $\widetilde{\mathbf{L}}(t) := \mathbf{L}(t) - \mathbb{E}[\mathbf{L}(1)]\,t$, $t \mapsto \widetilde{\mathbf{L}}(t)$ is a martingale (due to square integrability of $\mathbf{L}$). Hence, we have

$$f(t,x) = \Lambda(t+x) + \widetilde{\mu}_Q(t+x,t) + \int_0^t \mathbf{g}(t+x,s)^\top \mathbf{\Sigma}(s)\,d\widetilde{\mathbf{L}}(s)\,,$$

with $\widetilde{\mu}_Q(t+x,t) = \mu_Q(t+x,t) + \int_0^t \mathbf{g}(t+x,s)^\top \mathbf{\Sigma}(s)\mathbb{E}[\mathbf{L}(1)]\,ds$. By the triangle inequality we find that

$$|f(t,\cdot)|_H \le |\Lambda(t+\cdot)|_H + |\widetilde{\mu}_Q(t+\cdot,t)|_H + |\int_0^t \mathbf{g}(t+\cdot,s)^\top \mathbf{\Sigma}(s)\,d\mathbf{L}(s)|_H$$

where we have denoted the norm in H by $|\cdot|_H$. By assumption, $|\Lambda(t+\cdot)|_H < \infty$. Furthermore, we find that

$$|\widetilde{\mu}_Q(t+\cdot,t)|_H \le |\mu_Q(t+\cdot,t)|_H + |\int_0^t \mathbf{g}(t+\cdot,s)^\top \mathbf{\Sigma}(s)\mathbb{E}[\mathbf{L}(1)]\,ds|_H\,.$$

By the assumption on μ_Q, we find that $\mathbb{E}[|\mu_Q(t+\cdot,t)|_H] < \infty$, and therefore $|\mu_Q(t+\cdot,t)|_H < \infty$, $\mathbb{P}-a.s.$. Moreover, as the Lebesgue integral $\int_0^t \mathbf{g}(t+x,s)^\top \mathbf{\Sigma}(s)\mathbb{E}[\mathbf{L}(1)]\,ds$ can be viewed as a Bochner integral in H, Bochner's inequality yields

$$|\int_0^t \mathbf{g}(t+\cdot,s)^\top \mathbf{\Sigma}(s)\mathbb{E}[\mathbf{L}(1)]\,ds|_H \le \int_0^t |\mathbf{g}(t+\cdot,s)^\top \mathbf{\Sigma}(s)\mathbb{E}[\mathbf{L}(1)]|_H\,ds\,.$$

But

$$\begin{aligned}|\mathbf{g}(t+\cdot,s)^\top \mathbf{\Sigma}(s)\mathbb{E}[\mathbf{L}(1)]|_H &\le |\mathbf{g}(t+\cdot,s)|_H|\mathbf{\Sigma}(s)\mathbb{E}[\mathbf{L}(1)]| \\ &\le |\mathbf{g}(t+\cdot,s)|_H\|\mathbf{\Sigma}(s)\|\mathbb{E}[|\mathbf{L}(1)|]\,,\end{aligned}$$

where $|\cdot|$ means the canonical Euclidean 2-norm on $\mathbb{R}^n$. Hence, after appealing to Cauchy-Schwarz' inequality twice, we find

$$\int_0^t |\mathbf{g}(t+\cdot,s)|_H\mathbb{E}[\|\mathbf{\Sigma}(s)\|]\,ds \le \sqrt{t}(\int_0^t |\mathbf{g}(t+\cdot,s)|_H^2\mathbb{E}[\|\mathbf{\Sigma}(s)\|^2]\,ds)^{1/2}\,.$$

Therefore, by the assumption, $\mathbb{E}[|\int_0^t \mathbf{g}(t+\cdot,s)^\top \mathbf{\Sigma}(s)\mathbb{E}[\mathbf{L}(1)]\,ds|_H] < \infty$ and $|\int_0^t \mathbf{g}(t+\cdot,s)^\top \mathbf{\Sigma}(s)\mathbb{E}[\mathbf{L}(1)]\,ds|_H < \infty$, $\mathbb{P}-a.s.$

To conclude, we note by the isometry for stochastic integrals in H

$$\mathbb{E}\left[|\int_0^t \mathbf{g}(t+\cdot,s)^\top \mathbf{\Sigma}(s)\,d\widetilde{\mathbf{L}}(s)|_H^2\right] = \mathbb{E}\left[\int_0^t \|\mathbf{g}(t+\cdot,s)^\top \mathbf{\Sigma}(s)Q^{1/2}\|_{\mathrm{HS}}^2\,ds\right]$$

where $\|\cdot\|_{\text{HS}}$ denotes the Hilbert-Schmidt norm for operators from $\mathbb{R}^n$ to H. The matrix Q is the covariance operator of $\widetilde{\mathbf{L}}$, here being $Q = \mathbb{E}[\widetilde{\mathbf{L}}(1)\widetilde{\mathbf{L}}(1)^\top] = \text{Var}(\mathbf{L}(1))$, i.e., the $n \times n$ variance-covariance matrix of the n-dimensional random variable $\mathbf{L}(1)$. If $(e_i)_{i=1}^n$ is an orthonormal basis in $\mathbb{R}^n$, we find by definition of the Hilbert-Schmidt norm that

$$
\begin{aligned}
\|\mathbf{g}(t+\cdot,s)^\top \boldsymbol{\Sigma}(s) Q^{1/2}\|_{\text{HS}}^2 &= \sum_{i=1}^n |\mathbf{g}(t+\cdot,s)^\top \boldsymbol{\Sigma}(s) Q^{1/2}|_H^2 \\
&\leq |\mathbf{g}(t+\cdot,s)|_H^2 \sum_{i=1}^n |\boldsymbol{\Sigma}(s) Q^{1/2} e_i|^2 \\
&\leq |\mathbf{g}(t+\cdot,s)|_H^2 \|\boldsymbol{\Sigma}(s)\|^2 \text{Tr}(Q)
\end{aligned}
$$

This gives, by assumption,

$$
\begin{aligned}
&\mathbb{E}\left[\left| \int_0^t \mathbf{g}(t+\cdot,s)^\top \boldsymbol{\Sigma}(s)\, d\widetilde{\mathbf{L}}(s) \right|_H^2 \right] \\
&\qquad \leq \text{Tr}(\text{Var}(\mathbf{L}(1))) \int_0^t |\mathbf{g}(t+\cdot,s)|_H^2 \mathbb{E}[\|\boldsymbol{\Sigma}(s)\|^2]\, ds < \infty\,.
\end{aligned}
$$

Hence, $|\int_0^t \mathbf{g}(t+\cdot,s)^\top \boldsymbol{\Sigma}(s)\, d\widetilde{\mathbf{L}}(s)|_H < \infty$, $\mathbb{P}-a.s.$, and the proof is complete. □

In the above Proposition, we work with the $\mathbb{P}$-dynamics of the term structure $f(t,\cdot)$. We could equally well have done the analysis with respect to the pricing measure $\mathbb{Q}$, which essentially means that we must substitute the conditions on μ_Q and $\mathbf{g}(t+\cdot,s)^\top \boldsymbol{\Sigma}(s)$ in Prop. 5.3 with $\mu_Q(t+\cdot,t) \in L^1(H;\mathbb{Q})$ and $\int_0^t |\mathbf{g}(t+\cdot,s)|_H^2 \mathbb{E}_{\mathbb{Q}}[\|\boldsymbol{\Sigma}(s)\|^2]\, ds < \infty$ for every $t \geq 0$. In that case, we can argue that $f(t,\cdot) \in H$, $\mathbb{Q}-a.s.$ In any case, our analysis shows that under rather mild conditions on Λ, $\mathbf{g}$, $\boldsymbol{\Sigma}$ and $\mathbf{L}$, the induced forward price term structure dynamics from the spot model belongs to a Hilbert space.

Remark 5.1 The Hilbert space H can be chosen to be H_w, the Filipović space. However, in that case it is not reasonable to have $\Lambda(t+\cdot) \in H_w$. The seasonality function is typically of a trigonometric form with a yearly periodicity, say (see for example Lucia and Schwartz [149] and Benth *et al.* [31]), and therefore its derivative will not tend to zero when x grows. Hence, we study the term structure curve of *deseasonalized* forward prices instead, that is, $f(t,x)-\Lambda(t+x)$, which will belong to H_w under the conditions given for $\mathbf{g}$, Σ and $\mathbf{L}$ in Prop. 5.3.

Considering the forward price for the geometric spot model derived in Prop. 5.2, we find the term structure to be

$$
f(t,x) = \exp\left(\Lambda(t+x) + \mu_Q(t+x,t) + \int_0^t \mathbf{g}(t+x,s)^\top \boldsymbol{\Sigma}(s)\, d\mathbf{L}(s) \right), \quad x \geq 0\,. \tag{5.16}
$$

Rather than a sum of three terms that individually can be analysed, we exponentiate these three terms in the geometric case. A convenient approach to analyse the term structure as a dynamics in a Hilbert space H is to assume that H is a Banach algebra. In that case, if $f \in H$, then $\exp(f) \in H$. Thus, under the conditions of Prop. 5.3, we find $\ln f(t,\cdot) \in H$. Hence, for H being a Banach algebra, we can therefore conclude that $f(t,\cdot) \in H$. Recall from Prop. 3.1 that the Filipovic space H_w is a Banach algebra, at least after appropriate scaling of the norm. We also note that the seasonality function may be periodic and $\Lambda(t+\cdot) \notin H$. In that case, following the discussion in Remark 5.1, we consider the *deseasonalized* term structure of forward prices $\mathbb{R}_+ \ni x \mapsto f(t,x)\exp(-\Lambda(t+x))$.

Recall the example at the end of Section 5.2 in this Chapter, where we derived the futures price dynamics $F(t,T)$ in (5.12) under the pricing measure $\mathbb{Q}$ for a specific two-factor model. In the Musiela parametrization, we have

$$f(t,x) = \exp\left(\Phi(x) + X(t) + e^{-\alpha x} Y(t)\right) ,$$

where we recall X to be a drifted Brownian motion, Y a Lévy-driven Ornstein–Uhlenbeck process independent of X (under the pricing measure $\mathbb{Q}$), and

$$\Phi(x) = (\beta + \frac{1}{2}\eta^2)x + \theta(1-e^{-\alpha x}) + \int_0^x \varphi(e^{-\alpha s})\, ds .$$

Here, φ is the logarithm of the moment generating function under $\mathbb{Q}$ of the Lévy process driving the factor process Y. Thus, for every $t \geq 0$, we see that $x \mapsto f(t,x)$ is continuously differentiable with

$$\partial_x f(t,x) = f(t,x)\left(\beta + \frac{1}{2}\eta^2 + \alpha e^{-\alpha x}(\theta - Y(t)) + \varphi(e^{-\alpha x})\right) .$$

Let us analyse the stochastic process $(f(t,\cdot))_{t\geq 0}$ in light of the Filipović space. To this end, let

$$g(t,x) = \theta(1-e^{-\alpha x}) + X(t) + e^{-\alpha x} Y(t) ,$$

which is a continuously differentiable function with respect to x, for each $t \geq 0$. Using that $g(t,0) = X(t) + Y(t) = \ln S(t)$ and

$$\partial_x g(t,x) = \alpha e^{-\alpha x}(\theta - Y(t)) + \varphi(e^{-\alpha x}) ,$$

it follows after using the elementary inequality $(a+b)^2 \leq 2a^2 + 2b^2$,

$$
\begin{aligned}
|g(t,\cdot)|_w^2 &= |\ln S(t)|^2 + \int_0^\infty w(x)|\alpha e^{-\alpha x}(\theta - Y(t))|^2\,dx \\
&\leq |\ln S(t)|^2 + 2\alpha^2(\theta - Y(t))^2 \int_0^\infty w(x)e^{-2\alpha x}\,dx \\
&\quad + 2\int_0^\infty w(x)\varphi^2(e^{-\alpha x})\,dx.
\end{aligned}
$$

Thus, assuming that

$$
\int_0^\infty w(x)e^{-2\alpha x}\,dx < \infty\,, \quad \text{and} \quad \int_0^\infty w(x)\varphi^2(e^{-\alpha x})\,dx < \infty\,,
$$

we have $g(t,\cdot) \in H_w$ for every $t \geq 0$ since X and Y are right continuous and therefore with bounded paths on each compact almost surely. Hence, since by Assumption 3.1 $w^{-1} \in L^1(\mathbb{R}_+)$ and appealing to the Banach algebra structure of H_w (recall Prop. 3.1), if follows that

$$
\exp(-(\beta + \frac{1}{2}\eta^2)\cdot)f(t,\cdot) = \exp(g(t,\cdot)) \in H_w\,.
$$

By discounting with the linear trend induced from the non-stationary factor X, i.e. $(\beta + \frac{1}{2}\eta^2)x$, we obtain a stochastic process with values in the Filipović space. As elements of H_w must have a (weak) derivative that vanishes at infinity, we cannot have $f(t,\cdot) \in H_w$ due to the drift inherited from the non-stationary factor X. Let us end this example arguing that the the condition $\int_0^\infty w(x)\varphi^2(e^{-\alpha x})\,dx < \infty$ is superfluous. From the assumptions on L,

$$
\varphi(y) = \frac{1}{2}\sigma^2 y^2 + \int_{\mathbb{R}} \{e^{yz} - 1 - yz\}\,\nu(dz)
$$

where ν is the Lévy measure of L. We find from the domintated convergence theorem and L'Hopital's rule that

$$
\lim_{y\downarrow 0} \frac{1}{y^2}\int_{\mathbb{R}} \{e^{yz} - 1 - yz\}\,\nu(dz) = \frac{1}{2}\int_{\mathbb{R}} z^2\,\nu(dz)\,,
$$

which implies that $\varphi(y) \sim y^2$ when y is small. This yields that $\int_0^\infty w(x)\varphi^2(e^{-\alpha x})\,dx < \infty$ whenever $\int_0^\infty w(x)e^{-2\alpha x}\,dx < \infty$.

Remark 5.2 If $X = 0$, that is, we do not have any non-stationary factor in the spot price dynamics, we see that $(f(t,\cdot))_{t\geq 0}$ becomes a H_w-valued stochastic process whenever $w(\cdot)\exp(-2\alpha\cdot) \in L^1(\mathbb{R}_+)$. This can be achieved by appropriately selecting the weight function w, for example, $w(x) = \exp(\rho x)$ for $0 < \rho < 2\alpha$.

Moreover, it is simple to extend the example to the multivariate case of n Ornstein–Uhlenbeck factor processes $Y_1, \ldots, Y_2$, where we can also show that $(f(t, \cdot))_{t \geq 0}$ is a H_w-valued process under appropriate integrability hypotheses on the weight function and the speeds of mean reversion.

Remark 5.3 As discussed in Section 5.1, some classical models of the spot price dynamics S consist of a non-stationary factor X, and for these examples we cannot use the Filipović space as the canonical choice for realizing the forward price term structure curve. In power markets, say, there is empirical evidence suggesting stationary spot prices (see Barndorff-Nielsen *et al.* [13]). Samuelson [174] also made a strong economical argument against non-stationarity (random walk) in commodity spot markets. One may as well speculate whether signs of non-stationarity may in fact be a very slow mean reversion that is difficult to detect statistically. In any case, although non stationary factors may resolve the issue of random forward prices in the long end, this can also be achieved by using infinite dimensional noise in the dynamics, which we will come back to (see discussion at the end of Section 6.3 in Chapter 6).

Observe that in a finite dimensional term structure model, all the uncertainty is spanned by a finite selection of forward contracts. For example, looking back at the two-factor dynamics in (5.12), by selecting two different maturity times T_1 and T_2 we recover the long- and short-term factors X and Y by the following transformations of the forward prices

$$X(t) = \frac{e^{-\alpha(T_1-t)} \ln F(t, T_2) - e^{-\alpha(T_2-t)} \ln F(t, T_1)}{e^{-\alpha(T_1-t)} - e^{-\alpha(T_2-t)}} + \frac{e^{-\alpha(T_2-t)} \Phi(T_1 - t) - e^{-\alpha(T_1-t)} \Phi(T_2 - t)}{e^{-\alpha(T_1-t)} - e^{-\alpha(T_2-t)}}$$

$$Y(t) = \frac{\ln(F(t, T_1)/F(t, T_2))}{e^{-\alpha(T_1-t)} - e^{-\alpha(T_2-t)}} + \frac{\Phi(T_2 - t) - \Phi(T_1 - t)}{e^{-\alpha(T_1-t)} - e^{-\alpha(T_2-t)}}$$

where $t \leq \min(T_1, T_2)$. Thus, a two-factor model states that the market is spanned by two forward contracts only, and remaining forwards can be replicated using these two. In an infinite dimensional term structure model, *each* forward contract across all the maturities has its own idiosyncratic risk. In other words, in such a model, you can never hedge away all the risk by a finite collection of forwards. This might be more realistic if one thinks that trading in different maturities is connected with different risk, since one faces different aspects of future conditions in the market as well as different counter-parties. For example, producers may be interested in long-term hedging, while speculators can have a more short-term view on their trades. Also, various predictions on future market conditions can give rise to different sources of risk, like uncertainty in short and long term weather forecasts for production and demand in power markets and possible outages and power plant maintenance, planned or unplanned (we refer the reader to the concept of *information premium*, introduced in Benth and Meyer-Brandis [39]).

5.4 A Stochastic Partial Differential Equation for the Term Structure Dynamics

In this Section we show that the term structure dynamics is the solution of a first-order stochastic partial differential equation (SPDE), both for the arithmetic and geometric cases. This class of SPDEs will be treated in generality in later Chapters, where we apply the so-called Heath-Jarrow-Morton approach to term structure modelling.

Let us consider the arithmetic term structure dynamics in (5.15) first. As we will see, a stochastic Fubini theorem is needed in deriving the SPDE. Although stochastic Fubini theorems are known for rather general situations (see, e.g., Protter [169]), we include a tailormade version here for the convenience of the reader.

Lemma 5.1 *Suppose* $[0,\tau]^2 \ni (u,v) \mapsto \mathbf{h}(u,v) \in \mathbb{R}^n$ *is a measurable function such that* $u \mapsto \mathbf{h}(s,u)^\top \boldsymbol{\Sigma}(u)$ *and* $u \mapsto \int_0^\tau \mathbf{h}(s,u)^\top ds\, \boldsymbol{\Sigma}(u)$ *are* **L***-integrable on* $[0,s]$ *for all* $s \in [0,\tau]$, *and* $s \mapsto \int_0^\tau \mathbf{h}(s,u)^\top \boldsymbol{\Sigma}(u)\, d\mathbf{L}(u)$ *is Lebesgue integrable on* $[0,\tau]$. *Then,*

$$\int_0^\tau \int_0^\tau \mathbf{h}(s,u)^\top \boldsymbol{\Sigma}(u)\, d\mathbf{L}(u)\, ds = \int_0^\tau \int_0^\tau \mathbf{h}(s,u)^\top ds\, \boldsymbol{\Sigma}(u)\, d\mathbf{L}(u)\,, a.s.$$

Proof First, we observe that the conditions ensure the existence of all the involved integrals. It holds that,

$$\begin{aligned}
&\mathbb{E}\left[\left(\int_0^\tau \int_0^\tau \mathbf{h}(s,u)^\top \boldsymbol{\Sigma}(u)\, d\mathbf{L}(u)\, ds - \int_0^\tau \int_0^\tau \mathbf{h}(s,u)^\top ds\, \boldsymbol{\Sigma}(u)\, d\mathbf{L}(u)\right)^2\right] \\
&= \mathbb{E}\left[\left(\int_0^\tau \int_0^\tau \mathbf{h}(s,u)^\top \boldsymbol{\Sigma}(u)\, d\mathbf{L}(u)\, ds\right)^2\right] \\
&\quad - 2\mathbb{E}\left[\int_0^\tau \int_0^\tau \mathbf{h}(s,u)^\top \boldsymbol{\Sigma}(u)\, d\mathbf{L}(u) \int_0^\tau \int_0^\tau \mathbf{h}(s,u)^\top ds\, \boldsymbol{\Sigma}(u)\, d\mathbf{L}(u)\, ds\right] \\
&\quad + \mathbb{E}\left[\left(\int_0^\tau \int_0^\tau \mathbf{h}(s,u)^\top ds\, \boldsymbol{\Sigma}(u)\, d\mathbf{L}(u)\right)^2\right]. \qquad (5.17)
\end{aligned}$$

From the Itô isometry, the third term above is equal to

$$\begin{aligned}
&\mathbb{E}\left[\left(\int_0^\tau \int_0^\tau \mathbf{h}(s,u)^\top ds\, \boldsymbol{\Sigma}(u)\, d\mathbf{L}(u)\right)^2\right] \\
&= \int_0^\tau \int_0^\tau \mathbf{h}(s,u)^\top ds \mathbb{E}\left[\boldsymbol{\Sigma}(u) \mathrm{Var}(\mathbf{L}(1)) \boldsymbol{\Sigma}(u)^\top\right] \int_0^\tau \mathbf{h}(s,u)\, ds\, du\,.
\end{aligned}$$

Let us consider the second term in (5.17). By the Itô isometry,

$$\mathbb{E}\left[\int_0^\tau \int_0^\tau \mathbf{h}(s,u)^\top \boldsymbol{\Sigma}(u)\, d\mathbf{L}(u)\, ds \int_0^\tau \int_0^\tau \mathbf{h}(s,u)^\top\, ds \boldsymbol{\Sigma}(u)\, d\mathbf{L}(u)\right]$$
$$= \int_0^\tau \mathbb{E}\left[\int_0^\tau \mathbf{h}(z,u)^\top \boldsymbol{\Sigma}(u)\, d\mathbf{L}(u) \int_0^\tau \int_0^\tau \mathbf{h}(s,u)^\top\, ds \boldsymbol{\Sigma}(u)\, d\mathbf{L}(u)\right] dz$$
$$= \int_0^\tau \int_0^\tau \mathbf{h}(z,u)^\top \mathbb{E}[\boldsymbol{\Sigma}(u)\mathrm{Var}(\mathbf{L}(1))\boldsymbol{\Sigma}(u)^\top] \int_0^\tau \mathbf{h}(s,u)\, ds\, du\, dz$$
$$= \int_0^\tau \int_0^\tau \mathbf{h}(s,u)^\top\, ds \mathbb{E}[\boldsymbol{\Sigma}(u)\mathrm{Var}(\mathbf{L}(1))\boldsymbol{\Sigma}(u)^\top] \int_0^\tau \mathbf{h}(s,u)\, ds\, du\,,$$

where the first and last equality hold by the Fubini theorem (see Folland [104]). When computing the first term in (5.17), we note the following: for an integrable function $r : [0,\tau] \to \mathbb{R}$, we find that $(\int_0^\tau r(s)\, ds)^2 = 2\int_0^\tau \int_0^s r(s) r(z)\, dz\, ds$. Hence, by the Fubini theorem and Itô isometry

$$\mathbb{E}\left[\left(\int_0^\tau \int_0^\tau \mathbf{h}(s,u)^\top \boldsymbol{\Sigma}(u)\, d\mathbf{L}(u)\, ds\right)^2\right]$$
$$= 2\mathbb{E}\left[\int_0^\tau \int_0^s \int_0^\tau \mathbf{h}(z,u)^\top \boldsymbol{\Sigma}(u)\, d\mathbf{L}(u) \int_0^\tau \mathbf{h}(s,u)^\top \boldsymbol{\Sigma}(u)\, d\mathbf{L}(u)\, dz\, ds\right]$$
$$= 2\int_0^\tau \int_0^s \mathbb{E}\left[\int_0^\tau \mathbf{h}(z,u)^\top \boldsymbol{\Sigma}(u)\, d\mathbf{L}(u) \int_0^\tau \mathbf{h}(s,u)^\top \boldsymbol{\Sigma}(u)\, d\mathbf{L}(u)\right] dz\, ds$$
$$= 2\int_0^\tau \int_0^s \int_0^\tau \mathbf{h}(z,u)^\top \mathbb{E}[\boldsymbol{\Sigma}(u)\mathrm{Var}(\mathbf{L}(1))\boldsymbol{\Sigma}(u)^\top]\mathbf{h}(s,u)\, du\, dz\, ds$$
$$= 2\int_0^\tau \int_0^\tau \int_0^s \mathbf{h}(z,u)^\top \mathbb{E}[\boldsymbol{\Sigma}(u)\mathrm{Var}(\mathbf{L}(1))\boldsymbol{\Sigma}(u)^\top]\mathbf{h}(s,u)\, dz\, ds\, du\,.$$

A straightforward differentiation yields that

$$\frac{d}{d\tau}\int_0^\tau \mathbf{h}(s,u)^\top\, ds \mathbb{E}[\boldsymbol{\Sigma}(u)\mathrm{Var}(\mathbf{L}(1))\boldsymbol{\Sigma}(u)^\top] \int_0^\tau \mathbf{h}(s,u)\, ds$$
$$= \mathbf{h}(\tau,u)^\top \mathbb{E}[\boldsymbol{\Sigma}(u)\mathrm{Var}(\mathbf{L}(1))\boldsymbol{\Sigma}(u)^\top] \int_0^\tau \mathbf{h}(s,u)\, ds$$
$$+ \int_0^\tau \mathbf{h}(s,u)^\top\, ds \mathbb{E}[\boldsymbol{\Sigma}(u)\mathrm{Var}(\mathbf{L}(1))\boldsymbol{\Sigma}(u)^\top]\mathbf{h}(\tau,u)\,,$$

and therefore

$$\int_0^\tau \mathbf{h}(s,u)^\top ds \mathbb{E}[\boldsymbol{\Sigma}(u)\mathrm{Var}(\mathbf{L}(1))\boldsymbol{\Sigma}(u)^\top] \int_0^\tau \mathbf{h}(s,u)\,ds$$
$$= 2\int_0^\tau \int_0^s \mathbf{h}(z,u)^\top \mathbb{E}[\boldsymbol{\Sigma}(u)\mathrm{Var}(\mathbf{L}(1))\boldsymbol{\Sigma}(u)^\top]\mathbf{h}(s,u)\,dz\,ds\,,$$

by symmetry of the matrix $\mathbb{E}[\boldsymbol{\Sigma}(u)\mathrm{Var}(\mathbf{L}(1))\boldsymbol{\Sigma}(u)^\top]$. Thus

$$\mathbb{E}\left[\left(\int_0^\tau \int_0^\tau \mathbf{h}(s,u)^\top \boldsymbol{\Sigma}(u)\,d\mathbf{L}(u)\,ds\right)^2\right]$$
$$= \int_0^\tau \int_0^\tau \mathbf{h}(s,u)^\top ds \mathbb{E}[\boldsymbol{\Sigma}(u)\mathrm{Var}(\mathbf{L}(1))\boldsymbol{\Sigma}(u)^\top] \int_0^\tau \mathbf{h}(s,u)\,ds\,,$$

and the Lemma follows. □

We next derive the SPDE for the arithmetic term structure dynamics.

Proposition 5.4 *Let the random field $(f(t,x))_{t\geq 0, x\geq 0}$ be given as in (5.15), and assume $\Lambda \in C^1(\mathbb{R}_+)$ and $\mu_Q \in C^1(\mathbb{R}_+^2)$ and $\mathbf{g}(\cdot,t) \in C^1(\mathbb{R}_+)$ for almost every $t \geq 0$. Furthermore, we assume that $s \mapsto \partial_x \mathbf{g}(u,s)^\top \boldsymbol{\Sigma}(s)$ and $s \mapsto \mathbf{g}(0,s)^\top \boldsymbol{\Sigma}(s)$ are integrable with respect to $\mathbf{L}$, and that $u \mapsto \int_0^t \partial_x \mathbf{g}(u,s)^\top \boldsymbol{\Sigma}(s)\,d\mathbf{L}(s) \in L^1(\mathbb{R}_+)$. Then, for almost every $t \geq 0$, $x \mapsto f(t,x) \in C^1(\mathbb{R}_+)$ and the stochastic process $(f(t,x))_{t\geq 0,x\geq 0}$ is a (strong) solution to the stochastic partial differential equation*

$$df(t,x) = \left(\partial_x f(t,x) + \partial_2 \mu_Q(t+x,t)\right) dt + \mathbf{g}(t+x,t)^\top \boldsymbol{\Sigma}(t)\,d\mathbf{L}(t)\,,$$

with initial condition $f(0,x) = \Lambda(x) + \mu_Q(x,0)$. Here, $\partial_2 \mu_Q(t+x,t)$ means the derivative of μ_Q with respect to its second argument and $\partial_x \mathbf{g}$ means the derivative with respect to the first argument of each coordinate in the vector-valued function $\mathbf{g}$.

Proof By the assumption $\mathbf{g}(\cdot,t) \in C^1(\mathbb{R}_+)$, we have for every $t, x \geq 0$ that

$$\mathbf{g}(t+x,s) = \mathbf{g}(0,s) + \int_0^{t+x} \partial_x \mathbf{g}(u,s)\,du\,.$$

By the stochastic Fubini theorem (recall Lemma 5.1 using $\mathbf{h} := \partial_x \mathbf{g}$), it follows

$$\int_0^t \mathbf{g}(t+x,s)^\top \boldsymbol{\Sigma}(s)\,d\mathbf{L}(s) = \int_0^t \mathbf{g}(0,s)^\top \boldsymbol{\Sigma}(s)\,d\mathbf{L}(s)$$
$$+ \int_0^{t+x} \int_0^t \partial_x \mathbf{g}(u,s)^\top \boldsymbol{\Sigma}(s)\,d\mathbf{L}(s)\,du\,.$$

Hence, from the fundamental theorem of calculus for Lebesgue integrals (see Folland [104, Thm. 3.36]), it holds

$$\partial_x \int_0^t \mathbf{g}(t+x,s)^\top \mathbf{\Sigma}(s)\, d\mathbf{L}(s) = \int_0^t \partial_x \mathbf{g}(t+x,s)^\top \mathbf{\Sigma}(s)\, d\mathbf{L}(s)\,.$$

From the assumptions on Λ and μ_Q, we can conclude that $x \mapsto f(t,x)$ is continuously differentiable. Furthermore,

$$\partial_x f(t,x) = \partial_x \Lambda(t+x) + \partial_x \mu_Q(t+x,t) + \int_0^t \partial_x \mathbf{g}(t+x,s)^\top \mathbf{\Sigma}(s)\, d\mathbf{L}(s)\,.$$

Since

$$\frac{d}{dt}\mu_Q(t+x,t) = \partial_x \mu_Q(t+x,t) + \partial_2 \mu_Q(t+x,t)\,,$$

it holds

$$\mu_Q(t+x,t) = \mu_Q(x,0) + \int_0^t \partial_x \mu_Q(s+x,s)\, ds + \int_0^t \partial_2 \mu_Q(s+x,s)\, ds\,.$$

Hence, we find after using the stochastic Fubini theorem (Lemma 5.1)

$$\begin{aligned}
\int_0^t \partial_x f(s,x)\, ds &= \int_0^t \partial_x \Lambda(s+x) + \partial_x \mu_Q(s+x,s)\, ds \\
&\quad + \int_0^t \int_0^s \partial_x \mathbf{g}(s+x,u)^\top \mathbf{\Sigma}(u)\, d\mathbf{L}(u)\, ds \\
&= \Lambda(t+x) - \Lambda(x) + \mu_Q(t+x,t) - \mu_Q(x,0) \\
&\quad - \int_0^t \partial_2 \mu_Q(s+x,s)\, ds \\
&\quad + \int_0^t \int_u^t \partial_x \mathbf{g}(s+x,u)^\top\, ds\, \mathbf{\Sigma}(u)\, d\mathbf{L}(u) \\
&= \Lambda(t+x) - \Lambda(x) + \mu_Q(t+x,t) - \mu_Q(x,0) \\
&\quad - \int_0^t \partial_2 \mu_Q(s+x,s)\, ds \\
&\quad + \int_0^t (\mathbf{g}(t+x,u) - \mathbf{g}(u+x,u))^\top \mathbf{\Sigma}(u)\, d\mathbf{L}(u)
\end{aligned}$$

$$= f(t,x) - \Lambda(x) - \mu_Q(x,0) - \int_0^t \partial_2 \mu_Q(s+x,s)\,ds$$

$$- \int_0^t \mathbf{g}(s+x,s)^\top \boldsymbol{\Sigma}(s)\,d\mathbf{L}(s)\,.$$

As we have $f(0,x) = \Lambda(x) + \mu_Q(x,0)$, the proof is complete. □

The Proposition above shows that f is the solution of a first-order linear SPDE. The dynamics of the term structure depends on the derivative of f along the time to maturity, and the noise term is driven by the Lévy process $\mathbf{L}$, scaled by the stochastic volatility matrix $\boldsymbol{\Sigma}$ and the function $\mathbf{g}$. Such Lévy-driven SPDEs were discussed in more generality in Chapter 4.

Let us now study in more detail the forward dynamics in the Musiela parametrization when the forward price is given as in Prop. 5.1 with a stochastic volatility being the identity matrix, that is, when

$$\mu_Q(T,t) = \mathbb{E}_{\mathbb{Q}}\left[\int_t^T \mathbf{g}(T,s)^\top\,d\mathbf{L}(s)\,|\,\mathcal{F}_t\right].$$

Moreover, we suppose that $\mathbf{L}$ is a Lévy process under both $\mathbb{P}$ and $\mathbb{Q}$ with finite first moment. Then we find,

$$\mu_Q(T,t) = \int_t^T \mathbf{g}(T,s)^\top\,ds\,\mathbb{E}_{\mathbb{Q}}[\mathbf{L}(1)]\,,$$

and thus

$$\partial_2 \mu_Q(T,t) = -g(T,t)^\top \mathbb{E}_{\mathbb{Q}}[\mathbf{L}(1)]\,.$$

This yields a forward dynamics from Prop. 5.4 given by

$$df(t,x) = (\partial_x f(t,x) - \mathbf{g}(t+x,t)^\top \mathbb{E}_{\mathbb{Q}}[\mathbf{L}(1)])\,dt + \mathbf{g}(t+x,t)^\top\,d\mathbf{L}(t)\,.$$

Denoting by $\widetilde{\mathbf{L}}_{\mathbb{Q}}(t) := \mathbf{L}(t) - \mathbb{E}_{\mathbb{Q}}[\mathbf{L}(1)]\,t$, which is a martingale Lévy process, we see that the $\mathbb{Q}$-dynamics of f becomes

$$df(t,x) = \partial_x f(t,x)\,dt + \mathbf{g}(t+x,t)^\top\,d\widetilde{\mathbf{L}}_{\mathbb{Q}}(t)\,.$$

Hence, the dynamics is driven by a martingale process and does not have any drift except $\partial_x f(t,x)$. The $\mathbb{P}$-dynamics, on the other hand, will be

$$df(t,x) = (\partial_x f(t,x) + \mathbf{g}(t+x,t)^\top (\mathbb{E}_{\mathbb{Q}}[\mathbf{L}(1)] - \mathbb{E}[\mathbf{L}(1)]))\,dt + \mathbf{g}(t+x,t)^\top\,d\widetilde{\mathbf{L}}(t)\,,$$

where $\widetilde{\mathbf{L}}(t) = \mathbf{L}(t) - \mathbb{E}[\mathbf{L}(1)]\,t$ is a martingale process under $\mathbb{P}$. The additional drift term is sometimes referred to as the *market price of risk*.

Let us consider the geometric term structure dynamics defined in (5.16). When deriving the dynamics of this, it will be convenient to use the Lévy-Itô decomposition of $\mathbf{L}$. Since $\mathbf{L}$ is square-integrable, we have from Applebaum [7, Sect. 2.4] that

$$\mathbf{L}(t) = \mathbf{c}t + C\mathbf{B}(t) + \int_0^t \int_{\mathbb{R}^n} \mathbf{z}\,\widetilde{N}(dt, d\mathbf{z})\,. \tag{5.18}$$

Here, $\mathbf{c} \in \mathbb{R}^n$, $C \in \mathbb{R}^{n\times n}$, $\mathbf{B} \in \mathbb{R}^n$ is a standard Brownian motion and $\widetilde{N}$ is the compensated Poisson random measure associated with $\mathbf{L}$. With this notation at hand, we find the following:

Proposition 5.5 *Let the random field $(f(t,x))_{t\geq 0, x\geq 0}$ be given as in* (5.16), *and assume the conditions in Prop. 5.4. With the Lévy-Itô decomposition of* $\mathbf{L}$ *in* (5.18), *it holds*

$$\begin{aligned} df(t,x) &= \left(\partial_x f(t,x) + f(t,x)(\partial_2\mu_Q(t+x,t) + \theta(t,x))\right) dt \\ &\quad + f(t,x)\mathbf{g}(t+x,t)^\top \mathbf{\Sigma}(t) C\, d\mathbf{B}(t) \\ &\quad + f(t-,x)\int_{\mathbb{R}^n} (e^{\mathbf{g}(t+x,t)^\top \mathbf{\Sigma}(t)\mathbf{z}} - 1)\,\widetilde{N}(dt, d\mathbf{z})\,, \end{aligned}$$

where

$$\begin{aligned} \theta(t,x) &= \mathbf{g}(t+x,t)^\top \mathbf{\Sigma}(t)\mathbf{c} + \frac{1}{2}\mathbf{g}(t+x,t)^\top \mathbf{\Sigma}(t) C C^\top \mathbf{\Sigma}(t)^\top \mathbf{g}(t+x,t) \\ &\quad + \int_{\mathbb{R}^n} (e^{\mathbf{g}(t+x,t)^\top \mathbf{\Sigma}(t)\mathbf{z}} - 1 - \mathbf{g}(t+x,t)^\top \mathbf{\Sigma}(t)\mathbf{z})\,\nu(d\mathbf{z}) \end{aligned}$$

and $\nu(d\mathbf{z})$ is the Lévy measure of $\mathbf{L}$

Proof We apply the Itô Formula with respect to a multivariate Lévy process (see, e.g., Applebaum [7, Sect. 4.4]) to show the claimed dynamics: To this end, introduce the process

$$\widetilde{f}(t,x) := \ln f(t,x) = \Lambda(t+x) + \mu_Q(t+x,t) + \int_0^t \mathbf{g}(t+x,s)^\top \mathbf{\Sigma}(s)\, d\mathbf{L}(s)\,.$$

From Prop. 5.4, we find that the dynamics of $\widetilde{f}$ is

$$d\widetilde{f}(t,x) = \left(\partial_x \widetilde{f}(t,x) + \partial_2\mu_Q(t+x,t)\right) dt + \mathbf{g}(t+x,t)^\top \mathbf{\Sigma}(t)\, d\mathbf{L}(t)\,.$$

Introducing the Lévy-Itô decomposition of $\mathbf{L}$, we find,

$$d\widetilde{f}(t,x) = \left(\partial_x \widetilde{f}(t,x) + \partial_2\mu_Q(t+x,t) + \mathbf{g}(t+x,t)^\top \boldsymbol{\Sigma}(t)\mathbf{c}\right) dt$$
$$+ \mathbf{g}(t+x,t)^\top \boldsymbol{\Sigma}(t) C\, d\mathbf{B}(t) + \int_{\mathbb{R}^n} \mathbf{g}(t+x,t)^\top \boldsymbol{\Sigma}(t)\mathbf{z}\widetilde{N}(dt, d\mathbf{z}).$$

It then follows from Itô's Formula that, (with $\widetilde{f}^c(t,x)$ denoting the continuous part of $\widetilde{f}(t,x)$)

$$df(t,x) = e^{\widetilde{f}(t,x)}\, d\widetilde{f}^c(t,x) + \frac{1}{2}e^{\widetilde{f}(t,x)}\mathbf{g}(t+x,t)^\top \boldsymbol{\Sigma}(t) C C^\top \boldsymbol{\Sigma}(t)^\top \mathbf{g}(t+x,t)\, dt$$
$$+ \int_{\mathbb{R}^n} e^{\widetilde{f}(t-,x)+\mathbf{g}(t+x,t)^\top \boldsymbol{\Sigma}(t)\mathbf{z}} - e^{\widetilde{f}(t-,x)}\, \widetilde{N}(dt, d\mathbf{z})$$
$$+ \int_{\mathbb{R}^n} e^{\widetilde{f}(t-,x)+\mathbf{g}(t+x,t)^\top \boldsymbol{\Sigma}(t)\mathbf{z}}$$
$$- e^{\widetilde{f}(t-,x)} - e^{\widetilde{f}(t-,x)}\mathbf{g}(t+x,t)^\top \boldsymbol{\Sigma}(t)\mathbf{z}\, \nu(d\mathbf{z})\, dt.$$

Noting that $\partial_x f(t,x) = f(t,x)\partial_x \widetilde{f}(t,x)$ and collecting the dt-terms yield the result. □

Notice above that we have not been specifying whether we work with the $\mathbb{Q}$- or $\mathbb{P}$-dynamics. This goes back to the Lévy-Kintchine representation of $\mathbf{L}$ that we have started out with in (5.18), which can be interpreted as either the risk-neutral dynamics of the noise, or the noise under the market probability. For example, if we let $\mathbf{L}$ be under $\mathbb{Q}$, then the Lévy measure ℓ is also under $\mathbb{Q}$, of course.

Recalling the example at the end of Section 5.2 in this Chapter, we have a geometric term structure model under $\mathbb{Q}$ with a two-dimensional Lévy process $\mathbf{L}(t) = (\eta W(t), L(t))^\top$ with L being a real-valued, square integrable Lévy process with zero mean. In this case, the matrix C will consist of zeros except the first element on the first row, which is equal to η. The compensated Poisson process $\widetilde{N}$ for $\mathbf{L}$ will only charge on the second variable where it is equal to the compensated Poisson process of L, denoted $\widetilde{N}(dt, dz)$. Since in this example

$$\mu_Q(T,t) = \frac{1}{2}\eta^2(T-t) + \int_0^{T-t} \varphi(e^{-\alpha s})\, ds$$

for φ being the log-moment generating function of L, it holds that

$$\partial_2\mu_Q(T,t) = -\frac{1}{2}\eta^2 - \varphi(e^{-\alpha(T-t)}).$$

Hence, since $\Sigma(t) = I_2$ and $\mathbf{g}(T,t)^\top = (1, \exp(-\alpha(T-t)))$, we find

$$\theta(t,x) = \frac{1}{2}\eta^2 + \int_{\mathbb{R}} (e^{ze^{-\alpha x}} - 1 - ze^{-\alpha x})\,\nu(dz) = \frac{1}{2}\eta^2 + \varphi(e^{-\alpha x})\,.$$

We obtain from Prop. 5.5 that the $\mathbb{Q}$-dynamics becomes,

$$df(t,x) = \partial_x f(t,x)\,dt + \eta f(t,x)\,dW(t) + f(t-,x)\int_{\mathbb{R}} (e^{ze^{-\alpha x}} - 1)\,\widetilde{N}(dt,dz)\,.$$

Applying exponential jump models in the spot lead to a term structure dynamics where we do not recover the Lévy process explicitly in the jump term, as we see above. However, we observe that for small z, $\exp(ze^{-\alpha x}) - 1 \approx ze^{-\alpha x}$, and therefore

$$\int_{|z|<\epsilon} (e^{ze^{-\alpha x}} - 1)\,\widetilde{N}(dt,dz) \approx e^{-\alpha x}\int_{|z|<\epsilon} z\widetilde{N}(dt,dz)\,.$$

Thus, the small jumps of L scaled by $f(t-,x)\exp(-\alpha x)$ coincides approximately with the small jump part of the the term structure dynamics. We see in particular the Samuelson effect as a result of the mean reversion property of the jump part of the spot price, given by the factor process Y.

Recalling the wind index forward contract mentioned in Section 5.1 of this Chapter, the underlying dynamics is the exponential of the *negative* of an Ornstein–Uhlenbeck process driven by an integrable subordinator. Thus, in this pure-jump one-dimensional model under the market probability $\mathbb{P}$, we have $g(T,t) = -\exp(-\alpha(T-t))$ and $L(t) = \mathbb{E}[L(1)]t + \int_0^t \int_0^\infty z\widetilde{N}(ds,dz)$. Hence, $c = \mathbb{E}[L(1)]$. Moreover,

$$\mu_Q(T,t) = \ln \mathbb{E}_{\mathbb{Q}}[\exp(-\int_t^T e^{-\alpha(T-s)}dL(s))\,|\,\mathcal{F}_t]\,.$$

If we assume L to be integrable under $\mathbb{Q}$, we find

$$\partial_2\mu_Q(T,t) = -\varphi_{\mathbb{Q}}(-e^{-\alpha(T-t)})\,,$$

and

$$\theta(t,x) = -e^{-\alpha x}\mathbb{E}[L(1)] + \varphi_{\mathbb{P}}(-e^{-\alpha x})\,,$$

for $\varphi_{\mathbb{Q}}$ and $\varphi_{\mathbb{P}}$ being the log-moment generating functions with respect to $\mathbb{Q}$ and $\mathbb{P}$, respectively. Therefore,

$$df(t,x) = (\partial_x f(t,x) + f(t,x)(\varphi_{\mathbb{P}}(-e^{-\alpha x}) - \varphi_{\mathbb{Q}}(-e^{-\alpha x}) - e^{-\alpha x}\mathbb{E}[L(1)]))\,dt + f(t-,x)\int_0^\infty (e^{-ze^{-\alpha x}} - 1)\,\widetilde{N}(dt,dz)\,.$$

This gives the $\mathbb{P}$-dynamics of the wind power forwards considered in Benth and Pircalabu [43].

In the derivation of an SPDE for both the arithmetic and geometric forward price models, we assumed in Propositions 5.1 and 5.2 rather strong regularity of μ_Q, e.g., we supposed that μ_Q is continuously differentiable in both variables. Most stochastic volatility models will lead to a μ_Q which will not be differentiable in the second variable. Let us focus our considerations on the arithmetic spot price model with a stochastic volatility, recalling the forward price dynamics in Proposition 5.1. If $\mathbf{\Sigma}$, the stochastic volatility process of the spot price dynamics, is independent of the driving Lévy process $\mathbf{L}$, then we find that

$$\mu_Q(T,t) = \int_t^T \mathbf{g}(T,s)^\top \mathbb{E}_{\mathbb{Q}}[\mathbf{\Sigma}(s)\,|\,\mathcal{F}_t]\,ds\,\mathbb{E}_{\mathbb{Q}}[\mathbf{L}(1)]$$

after appealing to the tower law of conditional expectation and the independent increment property of Lévy processes. Typically, the stochastic volatility process $s \mapsto \mathbf{\Sigma}(s)$ is a semimartingale (with respect to $\mathbb{Q}$), and therefore we get an explicit dependency of $\mathbf{\Sigma}(t)$ in the expression of μ_Q. Hence, any differentiation with respect to the second argument of μ_Q will lead to a differentiation of the path of $t \mapsto \mathbf{\Sigma}(t)$, which is not possible in general. As an example, consider a variant of the BNS stochastic volatility model proposed by Veraart [185] (see Barndorff-Nielsen and Shephard [18] for the BNS stochastic volatility model) in the univariate case, defined by

$$d\Sigma(t) = -\lambda\Sigma(t)\,dt + d\Theta(t), \tag{5.19}$$

with Θ being an integrable subordinator on $\mathbb{R}_+$ and $\lambda > 0$ a constant. Since for $s \geq t \geq 0$,

$$\Sigma(s) = \Sigma(t)e^{-\lambda(s-t)} + \int_t^s e^{-\lambda(s-u)}\,d\Theta(u)$$

we get

$$\mathbb{E}_{\mathbb{Q}}[\Sigma(s)\,|\,\mathcal{F}_t] = \Sigma(t)e^{-\lambda(s-t)} + \lambda^{-1}\mathbb{E}_{\mathbb{Q}}[\Theta(1)]\left(1 - e^{-\lambda(s-t)}\right).$$

Therefore,

$$\mu_Q(T,t) = \int_t^T g(T,s)\left(\Sigma(t)e^{-\lambda(s-t)} + \lambda^{-1}\mathbb{E}_{\mathbb{Q}}[\Theta(1)](1-e^{-\lambda(s-t)})\right) ds$$
$$\times\mathbb{E}_{\mathbb{Q}}[L(1)],$$

which implies that μ_Q is not even continuous in its second argument as Θ is only RCLL. However, we observe that the forward price $f(t,x)$ becomes dependent on $\Sigma(t)$ explicitly, and we can use the Itô Formula to analyse its dynamics.

Let us take a more general perspective, going back to an arithmetic random field $(f(t,x))_{t\geq 0, x\geq 0}$ given as in (5.15), where we do not require continuous differentiability of μ_Q with respect to the second variable. We have the following result on its dynamics.

Proposition 5.6 *Let the random field $(f(t,x))_{t\geq 0,x\geq 0}$ be given as in (5.15), assume $\Lambda \in C^1(\mathbb{R}_+)$ and $\mu_Q(\cdot,t), \mathbf{g}(\cdot,t) \in C^1(\mathbb{R}_+)$ for almost every $t \geq 0$. Furthemore, we assume that $s \mapsto \partial_x\mathbf{g}(u,s)^\top\mathbf{\Sigma}(s)$ and $s \mapsto \mathbf{g}(0,s)^\top\mathbf{\Sigma}(s)$ are integrable with respect to $\mathbf{L}$, and that $u \mapsto \int_0^t \partial_x\mathbf{g}(u,s)^\top\mathbf{\Sigma}(s)\,d\mathbf{L}(s) \in L^1(\mathbb{R}_+)$. Then, for almost every $t \geq 0$, $x \mapsto f(t,x) \in C^1(\mathbb{R}_+)$ and the stochastic process $(f(t,x))_{t\geq 0,x\geq 0}$ is a (strong) solution to the stochastic partial differential equation*

$$df(t,x) = \left(\partial_x f(t,x) + \partial_x\mu_Q(t+x,t)\right) dt$$
$$+d\mu_Q(t+x,t) + \mathbf{g}(t+x,t)^\top\mathbf{\Sigma}(t)\,d\mathbf{L}(t)\,,$$

with initial condition $f(0,x) = \Lambda(x) + \mu_Q(x,0)$. Here, $\partial_x\mathbf{g}$ means the derivative with respect to the first argument of each coordinate in the vector-valued function $\mathbf{g}$.

Proof We recall that

$$f(t,x) = \Lambda(t+x) + \mu_Q(t+x,t) + \int_0^t \mathbf{g}(t+x,u)^\top\mathbf{\Sigma}(u)\mathbf{L}(u).$$

We argue as in the proof of Prop. 5.4 to show that $x \mapsto f(t,x)$ is continuously differentiable. Thus, by Lemma 5.1

$$\int_0^t \partial_x f(s,x)\,ds = \int_0^t \partial_x\Lambda(s+x)\,ds + \int_0^t \partial_x\mu_Q(s+x,s)\,ds$$
$$+\int_0^t\int_0^s \partial_x\mathbf{g}(s+x,u)^\top\mathbf{\Sigma}(u)\,d\mathbf{L}(u)\,ds$$
$$= \Lambda(t+x) - \Lambda(x) + \int_0^t \partial_x\mu_Q(s+x,s)\,ds$$
$$+\int_0^t \mathbf{g}(t+x,u)^\top\mathbf{\Sigma}(u)\,d\mathbf{L}(u)$$

$$
\begin{aligned}
&-\int_0^t \mathbf{g}(u+x,u)^\top \mathbf{\Sigma}(u)\,d\mathbf{L}(u) \\
&= f(t,x) - \Lambda(x) - \mu_Q(t+x,t) + \int_0^t \partial_x \mu_Q(s+x,s)\,ds \\
&-\int_0^t \mathbf{g}(u+x,u)^\top \mathbf{\Sigma}(u)\,d\mathbf{L}(u).
\end{aligned}
$$

The result follows. □

We can do similar considerations for the geometric case. We notice that the differential of $\mu_Q(t+x,t)$ will yield both a drift term and a stochastic forcing. Nevertheless, we obtain an SPDE for $f(t,x)$ which has the derivative operator as generator. In the next chapter, we will consider general SPDEs as models for the forward price, taking the dynamics we have obtained here as the starting point.

5.5 Pricing Measures, Storage Costs and Convenience Yields for Commodities

To end this chapter on spot price models and forward pricing, we make some considerations around pricing measures $\mathbb{Q}$ and their relation to classical concepts as storage costs and convenience yield in commodity pricing. Indeed, what we shall see, is that convenience yields and storage costs can be conveniently modelled by specifying the pricing measure $\mathbb{Q}$. This puts the classical theory of commodity pricing into a general and mathematically attractive framework.

To facilitate the discussion, we assume a fairly general stochastic dynamics for the spot price, deviating from the previous analysis in this Chapter. Under probability $\mathbb{P}$, suppose the spot price of a commodity has the dynamics

$$
\frac{dS(t)}{S(t)} = \mu(t)dt + \sigma(t)dB(t), \tag{5.20}
$$

where B is a Brownian motion and μ and σ are adapted stochastic processes with μ being integrable and σ square-integrable on compacts.

By Itô's Formula we have the representation for $T \geq t \geq 0$,

$$
S(T) = S(t)\exp\left(\int_t^T \mu(s) - \frac{1}{2}\sigma^2(s)ds + \int_t^T \sigma(s)dB(s)\right).
$$

Now, introduce the stochastic process W by

$$
dW(t) = dB(t) - \theta(t)dt. \tag{5.21}
$$

Under appropriate conditions on the stochastic process θ, Girsanov's Theorem assures the existence of $\mathbb{Q} \sim \mathbb{P}$ such that W is a $\mathbb{Q}$-Brownian motion on $[0, T]$. But then we have

$$S(T) = S(t)\exp\left(\int_t^T \mu(s) + \sigma(s)\theta(s) - \frac{1}{2}\sigma^2(s)ds + \int_t^T \sigma(s)dW(s)\right). \tag{5.22}$$

Suppose there is a forward contract written on the spot commodity, with delivery at time T. The forward price at time $t \leq T$ is as usual denoted by $F(t, T)$, and we have from no-arbitrage arguments that $F(T, T) = S(T)$. By the arbitrage pricing theory of mathematical finance (see, e.g., Björk [58] and Duffie [88]), we know that the forward price dynamics $t \mapsto F(t, T)$ must be a (local) martingale with respect to some measure $\mathbb{Q} \sim \mathbb{P}$. The above Girsanov transformation (5.21) provides us with a huge class of such probabilities. Hence, under the assumption that S is $\mathbb{Q}$-integrable,

$$F(t, T) = \mathbb{E}_{\mathbb{Q}}[F(T, T) \,|\, \mathcal{F}_t] = \mathbb{E}_{\mathbb{Q}}[S(T) \,|\, \mathcal{F}_t]. \tag{5.23}$$

If the spot can be traded without frictions, then the $\mathbb{Q}$ is the so-called equivalent martingale measure, i.e., the probability $\mathbb{Q} \sim \mathbb{P}$ such that $t \mapsto \exp(-rt)S(t)$ is a $\mathbb{Q}$-(local) martingale. Here, the constant r is the risk-free interest rate. Moreover, it is easily seen that

$$F(t, T) = \mathbb{E}_{\mathbb{Q}}[S(T) \,|\, \mathcal{F}_t] = S(t)e^{r(T-t)}, \tag{5.24}$$

by appealing to the martingale property of the discounted spot price. We recognize this as the derived forward price from a buy-and-hold hedging argument. As we have mentioned several times throughout this book, a commodity can in practice not be traded liquidly. There are frictions like storage and transportation costs (for example for metals, oil and agriculture), or not possible to store at all (like for example power).

Given the $\mathbb{Q}$-dynamics of S in (5.22) above, we obtain

$$F(t, T) = S(t)\mathbb{E}_{\mathbb{Q}}\left[\exp\left(\int_t^T \mu(s) + \sigma(s)\theta(s) - \frac{1}{2}\sigma^2(s)ds + \int_t^T \sigma(s)dW(s)\right) \,|\, \mathcal{F}_t\right].$$

To this end, define for $t \leq T$ the $\mathbb{Q}$ (local) martingale process

$$M(t) := \exp\left(\int_0^t \sigma(s)dW(s) - \frac{1}{2}\int_0^t \sigma^2(s)ds\right).$$

Under appropriate conditions on σ, M becomes a true martingale with expectation 1, and hence by Girsanov's theorem there exists $\widetilde{\mathbb{Q}} \sim \mathbb{Q}$ with density $\frac{d\widetilde{\mathbb{Q}}}{d\mathbb{Q}}|_{\mathcal{F}_t} = M(t)$. From this it follows by the conditional Bayes formula (see, e.g., Karatzas and Shreve [133, p. 193]),

$$\mathbb{E}_{\widetilde{\mathbb{Q}}}\left[e^{\int_t^T \mu(s)+\sigma(s)\theta(s)ds} \mid \mathcal{F}_t\right] = \mathbb{E}_{\mathbb{Q}}\left[e^{\int_t^T \mu(s)+\sigma(s)\theta(s)ds}\frac{M(T)}{M(t)} \mid \mathcal{F}_t\right].$$

Therefore,

$$F(t,T) = S(t)\mathbb{E}_{\widetilde{\mathbb{Q}}}\left[\exp\left(\int_t^T \mu(s) + \theta(s)\sigma(s)ds\right) \mid \mathcal{F}_t\right]. \tag{5.25}$$

Thus, the forward price becomes the product of the current spot price and a positive random variable depending on t and T. We investigate the expression in (5.25) in two particular cases:

5.5.1 *Constant Storage Costs and Convenience Yield*

Let μ and σ in the spot dynamics (5.20) be deterministic functions, and choose θ in the Girsanov transform (5.21) so that

$$\theta(s) := \frac{r + c - \delta - \mu(s)}{\sigma(s)}$$

for two constants δ and c. Assuming in addition that σ is positive and strictly bounded away from zero, we can show that all the steps above are validated and

$$F(t,T) = S(t)\exp\left(r(T-t) + c(T-t) - \delta(T-t)\right). \tag{5.26}$$

From Geman [111, p. 37], we recognize this relationship between the forward and spot price from the classical theory of storage and convenience yield. Indeed, assuming constant rate of storage costs given by c, one adds to the cost of borrowing r the storage costs c in the buy-and-hold hedging strategy. But there is a convenience of actually holding the commodity over owning it through a forward contract, which is coined the convenience yield. Here we have included this as a constant rate δ.

Worth observing is that we can recover the classical forward-spot relationship with constant storage costs and convenience yield via a specific choice of a pricing measure $\mathbb{Q}$. This measure is described via our choice of θ above.

5.5.2 A Stochastic Convenience Yield

Again assume that μ and σ are deterministic functions with σ being positive and strictly bounded away from zero. Define the Girsanov transform

$$\theta(t) := \frac{r - \mu(t) - Y(t)}{\sigma(t)}$$

for an adapted stochastic process Y, where the Novikov condition (see, e.g., Karatzas and Shreve [133, p. 198]) is supposed to hold

$$\mathbb{E}_{\mathbb{Q}}\left[\exp\left(\frac{1}{2}\int_0^T \frac{Y^2(s)}{\sigma^2(s)}ds\right)\right] < \infty.$$

Again all the steps above are validated and hence,

$$\int_t^T \mu(s) + \sigma(s)\theta(s)ds = r(T-t) - \int_t^T Y(s)ds.$$

From this it follows that

$$F(t,T) = S(t)e^{r(T-t)}\mathbb{E}_{\widetilde{\mathbb{Q}}}\left[\exp\left(-\int_t^T Y(s)ds\right) | \mathcal{F}_t\right]. \tag{5.27}$$

Suppose we model Y as an affine stochastic process under $\widetilde{\mathbb{Q}}$, which means that there exist deterministic functions $\alpha(t,T)$ and $\beta(t,T)$ such that it holds

$$F(t,T) = S(t)\exp\left(r(T-t) - \alpha(t,T)Y(t) + \beta(t,T)\right). \tag{5.28}$$

This dynamics generalises Eq. (6.17) on page 98 in Clewlow and Strickland [72]. Their dynamics can be recovered by choosing Y to be an Ornstein-Uhlenbeck process. In their context, Y is interpreted as a *stochastic convenience yield.* Again we observe how we can encode a classical notion from commodity markets into a pricing measure.

5.6 Bibliographical Notes

There is an extensive literature on spot price models for commodities in general, and power in particular. The classical Schwartz one- and two-factor models with mean reversion (see Lucia and Schwartz [150]) have been extended in various ways, and prominent contributions are discussed in the survey paper by Deschatre, Féron and Gruet [82]. Seasonality is salient in power spot prices and addressed in multi-factor models assuming mean reversion towards a deterministic seasonal level (see Weron [188]). Power prices tend to spike in periods with sudden imbalances in supply and demand, and many models have explained this by Lévy processes combined with a fast speed of mean reversion. Alternatives are regime switch models or jump processes with state dependent intensities, like for example proposed in Geman and Roncoroni [113], Janczura and Weron [131], Kholodnyi [134] and Paraschiv, Fleten and Schürle [161].

Polynomial processes, as introduced by Cuchiero [79], have been applied in energy market modeling as well. Ware [187] uses polynomial processes to model Alberta power spot prices in Canada, while Kleisinger-Yu *et al.* [140] analyse long-term forwards in the EEX market in this framework. As discussed in Benth [24], a large class of popular multi-factor models for the spot dynamics in commodity and energy markets are encompassed by the class of polynomial processes.

Borovkova, Permana and Pavlyukevich [62] propose a potential Lévy diffusion model to explain the spike behaviour of power prices. Rather than considering an Ornstein–Uhlenbeck process with fast mean reversion, they suggest a potential function in the drift to allow for a continuum of state-dependent reversion rates. A stochastic time-change model is proposed by Borovkova and Schmeck [63]. Self-exciting jump processes have been combined with mean reversion processes to model power prices in the UK market by Eyjolfsson and Tjøstheim [98]. Jiao *et al.* [132] also consider a class of stochastic dynamics inspired by Hawkes processes to explain the behaviour of power prices.

Stochastic volatility is another extension of the classical commodity spot price models. Hikspoors and Jaimungal [123] suggest multi-factor stochastic processes with an Ornstein–Uhlenbeck process modeling the stochastic volatility dynamics and analyse forward and option prices on these. Benth [21] applies the Barndorff-Nielsen and Shephard stochastic volatility dynamics (see Barndorff-Nielsen and Shephard [18]) to model UK gas prices. Benth [22] and Benth and Vos [54] extend this approach to cross-commodity markets. Classical stochastic volatility dynamics in modeling prices in commodity markets are presented and discussed in Eydeland and Wolyniec [97, pages 175–181].

The spot dynamics we have presented and analysed in this chapter belong to so-called *reduced-form models*. The supply and demand in commodity and power markets are obviously driven by many physical factors, for example wind and temperature in electricity markets or precipitation in agriculture. Storage capacity and transportation availability are other factors. Hence, in determining the spot price evolution, one may model the dynamics of the important factors driving supply and demand. Such models may become very complex and analytically challenging for risk management purposes. Reduced-form models, on the other hand, are comparably simple, but fail in explaining the impact of fundamental variables on the price dynamics. A compromise is found in *structural models*, as suggested by Aïd *et al.* [2], Alasseur and Féron [3], Carmona, Coulon and Schwarz [68] and Kiesel and Kustermann [138]. We refer to Pirrong [167] for a monograph on this approach in general commodity markets focusing on storage problems. Reduced-form models with wind as explanatory factor are proposed for energy market pricing by Deschatre and Veraart [83] and Rowinska, Veraart and Gruet [171].

The spot-forward relationship is challenging to model in commodities, and even more so in power. In Section 5.5 we have provided a small window into this vast strand of literature. The basic relationship is established using the so-called buy-and-hold hedging strategy, which requires a perfectly liquid spot market (see Duffie [88]). Geman [111] and Eydeland and Wolyniec [97] present the economic theory on storage, inventory and convenience yield, which extends the risk-neutral pricing of forwards to commodities. In power, storage is very restricted, and the risk premium may be much more complex. In this chapter have encoded it in a pricing probability $\mathbb{Q}$.

Chapter 6
Heath-Jarrow-Morton Type Models

Abstract In this main chapter of the book, infinite-dimensional stochastic processes are defined for the forward dynamics using a Hilbert space as state space for the term structures. Arithmetic and geometric models are introduced, where the noise driver is a Wiener process or a Lévy process and the context is cross-commodity markets. Moreover, we also allow for a class of stochastic volatility models in the forward dynamics. Drift conditions are derived ensuring a risk-neutral dynamics, i.e., a no-arbitrage dynamics under a pricing measure. We furthermore study swap prices (forward with delivery period) and finite factor models in this HJM-framework. To include seasonality and modeling under the market probability require a study of measure change, where one may apply the Girsanov and Esscher transform in our context. To have available data and the initial forward curve, a smoothing approach based on a combination of parametric curves (i.e., the Nelson-Siegel model) the the interpolation technique *kriging* is proposed and applied in an empirical example.

Rather than *deriving* the forward price dynamics based on a spot model, we will in this Chapter consider the so-called Heath-Jarrow-Morton (HJM) approach where the idea is to define the forward price dynamics directly. The HJM modelling paradigm was first introduced by Heath, Jarrow and Morton [121] in fixed-income theory, where the authors proposed to model the forward rates instead of the short rates. The analogy in commodity markets is to model the forward prices instead of the spot, thereby circumvent the possibly complex link between spot and forward prices in such markets. As commodity markets in general, and energy markets in particular, are incomplete due to frictions in trading the spot, it may be difficult to identify reasonable models for the risk premium (or, the market price of risk) encoded in the choice of the pricing measure $\mathbb{Q}$. If the purpose is have available the forward priced dynamics for production and portfolio management purposes, or to price options on forwards, then it may be convenient to model the forward prices directly. Forward contracts in commodity markets are in many circumstances liquidly traded, at least the front contracts. Also, in power markets, the forwards are financial products that

F. E. Benth, P. Krühner, *Stochastic Models for Prices Dynamics in Energy and Commodity Markets*, Springer Finance,
https://doi.org/10.1007/978-3-031-40367-5_6

do not require any physical capacity, enabling banks and speculators access to the market.

Early accounts on the HJM modelling of energy forwards can be found in the works of Clewlow and Strickland [72], Kiesel, Börger and Schindlmayer [137] and Benth and Koekebakker [32] (see also Benth, Šaltytė Benth and Koekebakker [25] for a thorough discussion and analysis in electricity). The infinite dimensional viewpoint has been studied in Benth and Krühner [37].

In commodity markets, most forwards are traded with a given date of delivery, while in power and gas and weather markets, the forwards usually deliver over a contractual period of time. We will first consider fixed-delivery forwards, and later analyse the dynamics of forwards with a delivery period. We also set our dynamical models in the Musiela parametrisation, meaning that we model forward price with respect to *time to maturity* rather than *time at maturity*.

6.1 Musiela-type Forward Dynamics

We want to model the dynamics of $d \in \mathbb{N}$ forward prices $\mathbf{f} = (f_1, \dots, f_d)^\top$, i.e., a cross-commodity dynamics in a d-dimensional market of forward contracts. We will restrict our attention to the *risk-neutral* dynamics, that is, we consider the model with respect to the pricing measure $\mathbb{Q}$. Let $H = H_1 \times H_2 \times \cdots \times H_d$ be a product Hilbert space of the d separable Hilbert spaces H_j, $j = 1, \dots, d$. We assume that each H_j is a Hilbert space of real-valued functions on $\mathbb{R}_+$, where we suppose that the evaluation map $\delta_x(g) = g(x)$ for $g \in H_j$ is a linear functional for each $x \geq 0$, i.e., $\delta_x \in H_j^*$ for $x \geq 0$. This implies that the evaluation map is also well-defined as an $\mathbb{R}^d$-valued functional on H, being $\delta_x(\mathbf{g}) = (\delta_x(g_1), \dots, \delta_x(g_d))'$ (here we have slightly misused the notation of δ_x, but trusting that the reader knows which one is used given the context). Furthermore, we assume that the shift operator $(\mathcal{S}_t^j)_{t\geq 0}$ where $\mathcal{S}_t^j(g) = g(\cdot + t)$ for $g \in H_j$ defines a quasi-contractive C_0-semigroup on H_j with a densely defined generator being ∂_x. We recall that the Filipović space discussed in Chapter 3 satisfies these conditions and thus provides us with a possible specification of H_j. Notice that H is separable as each H_j is assumed to be separable. Finally, we let $(\mathbf{L}(t))_{t\geq 0}$ be a square-integrable H-valued Lévy process with zero mean (i.e., a square-integrable martingale process in H). Its covariance operator is denoted by $\mathcal{Q}$.

6.1.1 Simple HJM Models

Let $(\boldsymbol{\mu}(t))_{t\geq 0}$ be a given H-valued predictable stochastic process which is square-integrable and its second moment is locally integrable on $\mathbb{R}_+$, i.e.,

$$\mathbb{E}_{\mathbb{Q}}[\int_0^T |\boldsymbol{\mu}(t)|_H^2 dt] < \infty \tag{6.1}$$

for all $T < \infty$. Recalling Definition 4.1, we furthermore let $(\mathbf{\Sigma}(t))_{t\geq 0}$ be a predictable stochastic process with values in $L(H)$, satisfying

$$\mathbb{E}_{\mathbb{Q}}\left[\int_0^T \|\mathbf{\Sigma}(t)Q^{1/2}\|_{\mathrm{HS}}^2 dt\right] < \infty \tag{6.2}$$

for all $T < \infty$. Consider the SPDE

$$d\mathbf{f}(t) = \nabla_x \mathbf{f}(t)\,dt + \boldsymbol{\mu}(t)\,dt + \mathbf{\Sigma}(t)\,d\mathbf{L}(t), \quad t \geq 0 \tag{6.3}$$

with initial condition $\mathbf{f}(0) = \mathbf{f}_0 \in H$. We use (6.3) as the model for the dynamics of the forward prices in the d-dimensional market. With ∇_x, we mean to the (unbounded, densely defined) operator

$$\nabla_x := \begin{bmatrix} \partial_x & 0 & \cdots & 0 \\ 0 & \partial_x & \cdots & 0 \\ \cdot & \cdot & \cdots & \cdot \\ 0 & 0 & \cdots & \partial_x \end{bmatrix}, \tag{6.4}$$

thus, $\nabla_x \mathbf{f}(t) = (\partial_x f_1(t), \ldots, \partial_x f_d(t))^\top$. Introduce now the shift semigroup on H, denoted by $(\mathcal{S}_t)_{t\geq 0}$, as the matrix-valued operator,

$$\mathcal{S}_t := \begin{bmatrix} \mathcal{S}_t^1 & 0 & \cdots & 0 \\ 0 & \mathcal{S}_t^2 & \cdots & 0 \\ \cdot & \cdot & \cdots & \cdot \\ 0 & 0 & \cdots & \mathcal{S}_t^d \end{bmatrix}. \tag{6.5}$$

It is simple to see that $(\mathcal{S}_t)_{t\geq 0}$ defines a C_0-semigroup on H with generator ∇_x. We have the following

Proposition 6.1 *The H-valued, square-integrable and predictable stochastic process*

$$\mathbf{f}(t) = \mathcal{S}_t \mathbf{f}_0 + \int_0^t \mathcal{S}_{t-s}\boldsymbol{\mu}(s)\,ds + \int_0^t \mathcal{S}_{t-s}\mathbf{\Sigma}(s)\,d\mathbf{L}(s)$$

is a mild solution of (6.3) *on $t \leq T$ for any $T < \infty$.*

Proof As $\mathcal{S}_t$ is a C_0-semigroup, it holds that $[0, t] \ni s \mapsto \mathcal{S}_{t-s}\boldsymbol{\mu}(s)$ is H-valued and integrable on $[0, t]$ in the sense of Bochner. Additionally, by Cauchy-Schwarz' inequality, $\int_0^t \mathcal{S}_{t-s}\mu(s)ds$ will have a finite second moment uniformly bounded in $t \in [0, T]$ for any $T < \infty$. Also, by the semigroup property of $\mathcal{S}_t$, $[0, t] \ni s \mapsto \mathcal{S}_{t-s}\mathbf{\Sigma}(s)$ is $L(H)$-valued, predictable and integrable with respect to $\mathbf{L}$ (due to the exponential bound on the operator norm of $\mathcal{S}_t$) according to Definition 4.1. Thus, the integral

$$\int_0^t \mathcal{S}_{t-s}\mathbf{\Sigma}(s)\, d\mathbf{L}(s)$$

is well-defined and defines an adapted stochastic process with values in H. Moreover, it is square-integrable uniformly in $t \in [0, T]$ for any $T < \infty$.

It is left to show that the solution has RCLL (i.e., cádlág) paths. By representing the operator $\mathbf{\Sigma}(s)$ as a matrix of operators, with elements $\Sigma_{i,j}(s)$, $i, j = 1, \ldots, d$, we find that the ith component of the stochastic integral is a sum of H_i-valued stochastic convolutions of the form

$$\int_0^t \mathcal{S}^i_{t-s}\Sigma_{i,j}(s)\, dL_i(s)$$

for $j = 1, \ldots, d$. From Lemma 3.7 we recall that the shift semigroup is quasi-contractive on the Filipović space, and hence, according to Peszat and Zabczyk [165, Thm. 9.24] the stochastic convolution $\int_0^t \mathcal{S}^i_{t-s}\Sigma_{i,j}(s)dL_j(s)$ has a RCLL (cádlág) modification. As this holds for all i, j, we conclude that $\mathbf{f}(t)$ is a mild solution in H for $t \leq T < \infty$ according to Definition 4.3. □

We remark that we may define a mild solution of (6.3) with only demanding predictability of the solution process. In our Definition 4.3, we have added the assumption of adapted RCLL paths to be a mild solution. By weakening to predictable processes only, we may dispense with the quasi-contractivity of the shift semigroups $\mathcal{S}^i$, which is only required for establishing the RCLL property of the stochastic convolutions. Further, we remark that if we do not require square-integrability from our mild solution, we may allow for less restrictive conditions on μ, say. For example, we may assume only that $t \mapsto |\mu(t)|_H$ are locally integrable, $a.s.$. Similarly, we could extend the definition of stochastic integration by localization to allow for a broader class of $\mathbf{\Sigma}$.

From general no-arbitrage theory, we know that forward contracts are tradable assets which must be (local) martingales under the pricing measure $\mathbb{Q}$. One means here the forwards with given delivery time, e.g., $[0, T] \ni t \mapsto \mathbf{F}(t, T) := \delta_{T-t}(\mathbf{f}(t))$ must be a (local) $\mathbb{Q}$-martingale. We have the following martingale condition for our model in (6.3).

Proposition 6.2 *The process* $[0, T] \ni t \mapsto \mathbf{F}(t, T) := \delta_{T-t}(\mathbf{f}(t))$ *is a (true)* $\mathbb{R}^d$*-valued* $\mathbb{Q}$*-martingale if and only if* $t \mapsto \boldsymbol{\delta}_{T-t}(\boldsymbol{\mu}(t)) = 0$ *a.s. for a.e.* $T \geq t \geq 0$.

Proof Since $\delta_t \mathcal{S}_s = \delta_{t+s}$ for any $s, t \geq 0$, and $\delta_t \in H_1^* \times \cdots \times H_d^*$, we have that

$$\begin{aligned}\delta_{T-t}\mathbf{f}(t) &= \delta_{T-t}\mathcal{S}_t\mathbf{f}_0 + \int_0^t \delta_{T-t}\mathcal{S}_{t-s}\boldsymbol{\mu}(s)\, ds + \int_0^t \delta_{T-t}\mathcal{S}_{t-s}\mathbf{\Sigma}(s)\, d\mathbf{L}(s) \\ &= \delta_T\mathbf{f}_0 + \int_0^t \delta_{T-s}\boldsymbol{\mu}(s)\, ds + \int_0^t \delta_{T-s}\mathbf{\Sigma}(s)\, d\mathbf{L}(s).\end{aligned}$$

Consider the process $[0, T] \ni t \mapsto \int_0^t \delta_{T-s}\boldsymbol{\Sigma}(s)\,d\mathbf{L}(s)$, which is $\mathbb{R}^d$-valued. By assumptions on $\boldsymbol{\Sigma}$ and δ_{T-s} being bounded, we find that this process is square integrable. Moreover, by the independent increment property of $\mathbf{L}$ and its zero mean, it follows that it is a square-integrable martingale. Thus, since for $T \geq t \geq s \geq 0$,

$$\mathbb{E}_{\mathbb{Q}}[\mathbf{F}(t,T)\,|\,\mathcal{F}_s] = \mathbf{F}(s,T) + \int_s^t \mathbb{E}_{\mathbb{Q}}[\delta_{T-u}(\mu(u))\,|\,\mathcal{F}_s]\,du$$

we find that $t \mapsto \mathbf{F}(t,T)$ is a $\mathbb{Q}$-martingale if and only if

$$\int_s^t \mathbb{E}_{\mathbb{Q}}[\delta_{T-u}(\boldsymbol{\mu}(u))\,|\,\mathcal{F}_s]\,du = 0\,a.s.,$$

for every $T \geq t \geq s \geq 0$. Notice that the conditional expectation is well-defined by the integrability assumptions on μ. Obviously, if $\delta_{T-t}(\boldsymbol{\mu}(t)) = 0$ $a.s.$, for $a.e.$ $T \geq t \geq 0$, then this is true and the martingale property follows. Introduce the short-hand notation $G(t,s) = \int_s^t \mathbb{E}_{\mathbb{Q}}[\delta_{T-u}(\boldsymbol{\mu}(u))\,|\,\mathcal{F}_s]\,du$ for $T \geq t \geq s \geq 0$. Since $G(t,s) = 0$ for every $T \geq t \geq s \geq 0$, we find that it is everywhere differentiable with derivative equal to zero, and therefore $\mathbb{E}_{\mathbb{Q}}[\delta_{T-t}(\boldsymbol{\mu}(t))\,|\,\mathcal{F}_s] = 0$ for every $T \geq t \geq s \geq 0$. Choosing $t = s$, it follows by $\mathcal{F}_s$-measurability that $\delta_{T-t}(\boldsymbol{\mu}(t)) = 0$ for every $T \geq t \geq 0$. The result follows. □

In conclusion, we pose the simple HJM dynamics of the forward price model as the mild solution to the SPDE

$$d\mathbf{f}(t) = \nabla_x \mathbf{f}(t)\,dt + \boldsymbol{\Sigma}(t)\,d\mathbf{L}(t), \tag{6.6}$$

with initial condition $\mathbf{f}(0) = \mathbf{f}_0$, that is,

$$\mathbf{f}(t) = \mathcal{S}_t\mathbf{f}_0 + \int_0^t \mathcal{S}_{t-s}\boldsymbol{\Sigma}(s)\,d\mathbf{L}(s). \tag{6.7}$$

In practice, we must model the volatility $\boldsymbol{\Sigma}$ as well as using current data to assess the initial curve $\mathbf{f}_0$.

As an illustration of such a simple HJM-model in the univariate case, we have shown the time dynamics of the term structure in Figure 6.1. The simulation is the daily forward curve over 1 year (with 250 trading days), with time to maturity ranging up to three years. The simulation is based on the HJM-model for the power price

$$f(t) = \mathcal{S}_t f_0 + \int_0^t \mathcal{S}_{t-s}\mathcal{M}dW(s) \tag{6.8}$$

with a fixed initial level $f_0 = 80$ Euro/MWh. Furthermore, $\mathcal{M}$ is a multiplication operator as defined in Definition 3.2 using the function $\exp(-\eta x)$, which introduces

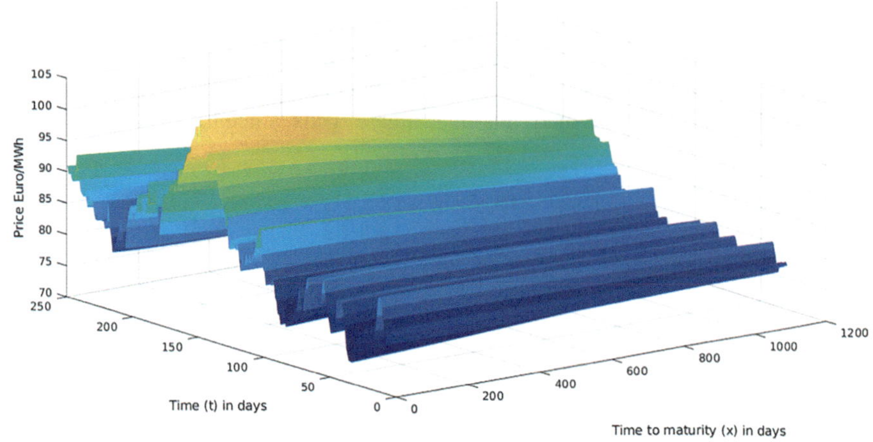

Fig. 6.1 Simulated dynamics of the simple HJM term structure model in (6.8)

a Samuelson effect in the term structure of volatility. We choose to model in the Filipovic space with exponential weight function $w(x) = \exp(\alpha x)$, where $\alpha < 2\eta$. The covariance operator for W is the one discussed in (3.13). We simulated the dynamics in (6.8) iteratively over a time step Δ (which is $1/250$ in the simulated example) by the recursion

$$f(t+\Delta) = \mathcal{S}_\Delta f(t) + \int_t^{t+\Delta} \mathcal{S}_{t+\Delta-s} \mathcal{M} dW(s),$$

where the stochastic integral is a mean zero Gaussian random variable being independent of $f(t)$. A rather crude approximation yields a covariance structure of this Gaussian field between x and y on $\mathbb{R}_+$,

$$\begin{aligned}
\mathbb{E}[\delta_x \int_t^{t+\Delta} \mathcal{S}_{t+\Delta-s} \mathcal{M} dW(s) \cdot \delta_y \int_t^{t+\Delta} \mathcal{S}_{t+\Delta-s} \mathcal{M} dW(s)] \\
= \delta_y \int_t^{t+\Delta} \mathcal{S}_{t+\Delta-s} \mathcal{M} \mathcal{Q} \mathcal{M}^* \mathcal{S}^*_{t+\Delta-s} ds \delta_x^* 1 \\
\approx \delta_y \mathcal{S}_\Delta \mathcal{M} \mathcal{Q} \mathcal{M}^* \mathcal{S}^*_\Delta \delta_x^* 1.
\end{aligned}$$

We know that $\delta_x^* 1 = h_x \in H_w$, and moreover one can derive $\mathcal{S}^*_\Delta h_x = h_{x+\Delta}$. This is readily seen by

$$\mathcal{S}^*_\Delta h_x(y) = \delta_y \mathcal{S}^*_\Delta h_x = \langle h_x, \mathcal{S}_\Delta h_y \rangle = h_y(x+\Delta) = h_{x+\Delta}(y).$$

We also know from Proposition 3.5 that

$$\mathcal{M}^* g = \langle g, \mathcal{M}h.\rangle$$

which implies

$$\mathcal{M}^* \mathcal{S}_\Delta^* \delta_x^* 1 = \delta_x(\mathcal{S}_\Delta \mathcal{M}h.) = e^{-\eta(x+\Delta)} h.(x+\Delta) = e^{-\eta(x+\Delta)} h_{x+\Delta}(\cdot).$$

Thus, $\mathcal{Q}\mathcal{M}^*\mathcal{S}_\Delta^*\delta_x^* 1 = \exp(-\eta(x+\Delta))\mathcal{Q}h_{x+\Delta}$. But then

$$\begin{aligned}\mathbb{E}[\delta_x \int_t^{t+\Delta} \mathcal{S}_{t+\Delta-s}\mathcal{M}dW(s) \cdot \delta_y \int_t^{t+\Delta} \mathcal{S}_{t+\Delta-s}\mathcal{M}dW(s)]\\ \approx e^{-2\eta\Delta} e^{-\eta(x+y)} (\mathcal{Q}h_{x+\Delta})(y+\Delta).\end{aligned} \tag{6.9}$$

We notice that $\mathrm{Cov}(W(1,x)W(1,y)) = \mathcal{Q}h_x(y)$. An easy way to simulate the Gaussian integrals in (6.8) is to draw independent vectors (of suitable size according to the maturities of interest) of mean zero Gaussian random variables, where each vector has a variance-covariance matrix defined by sampling (6.9). This matrix may be ill-conditioned, and further approximations have to be made in simulations when deriving numerically its square-root. In the numerical illustration in Figure 6.1, we chose $\alpha = 0.5$ and $\eta = 0.51$, while the two other parameters in the specification of $\mathcal{Q}$ were chosen to be as in the case study presented in Figure 3.2 of Chapter 3, i.e., $\beta = 0.5$ and $c = 1$.

As a cross-commodity, multivariate example, consider the market for gas and power. Let $d = 2$, and with $h \in L(H_2, H_1)$ being the so-called *heat rate* operator, converting the energy content of gas into power, the spark spread is defined as $p(t) := f_1(t) - hf_2(t)$. Here, $(f_1(t))_{t\geq 0}$ is the forward price curve of power, while $(f_2(t))_{t\geq 0}$ is the forward curve of gas. The spark spread $(p(t))_{t\geq 0}$ becomes an H_1-valued stochastic process. In practice, the heat rate is a constant conversion factor related to the efficiency of gas-fired power plants. Since we consider the spark spread as the difference between power and gas prices along all delivery times, and realise the gas forward prices in another space than H_2, we must define the heat rate as a (linear) operator. If $H_1 = H_2$, which is a natural assumption, we can indeed choose h as a constant. Otherwise, we may imagine a situation where there are different players in the market for different delivery times, yielding a heat rate depending on delivery, In this case, it is natural to use a linear operator representing the heat rate.

So far, we have only focussed on the forward price dynamics in the Musiela parametrization. In the marketplace, we have forward contracts with given maturity dates. I.e., in a d-dimensional setting, we have that the forward price at time $t \geq 0$ of contracts in the d markets maturing at time $T < \infty$, where $t \leq T$, is given by

$$\mathbf{F}(t,T) = \mathbf{f}(t, T-t) = \delta_{T-t}\mathbf{f}(t).$$

Considering the market of gas and power, say, where $d = 2$, where have two forward contracts maturing at the time T with price dynamics $F_1(t, T)$ and $F_2(t, T)$, respectively. In most situations, it is of interest to have available the forward dynamics of many contracts maturing at different times. A particular case being for instance two forward contracts maturing at dates T_1 and T_2 in the same market (meaning $d = 1$). This would mean that we are focussing on a dynamics in $\mathbb{R}^2$ as

$$t \mapsto (\delta_{T_1-t} f(t), \delta_{T_2-t} f(t))$$

with $t \leq T_1 \wedge T_2$, and for $T_1 \wedge T_2 < t \leq T_1 \vee T_2$ we have only a one-dimensional dynamics of the remaining existing contract. A general form is the kd-dimensional dynamics

$$t \mapsto (\delta_{T_1-t}\mathbf{f}(t), \dots, \delta_{T_k-t}\mathbf{f}(t))$$

for $t \leq T_1 \wedge \cdots \wedge T_k$. As we see, the dynamics of the contracts in the marketplace are given as application of the evaluation functional on the forward curve $\mathbf{f}$.

Recalling the finite-dimensional representation theory presented in Section 4.2 of Chapter 4, we may state this dynamics in terms of finite-dimensional stochastic processes. Given the dynamics of $\mathbf{f}$ in (6.7) where $\mathbf{L}(t) = \mathbf{W}(\Theta(t))$ for $\mathbf{W}$ a Wiener process in H and Θ an integrable subordinator on $\mathbb{R}_+$. Then it follows from Thm. 4.2 that

$$\begin{aligned}
\mathbf{F}(t, T) &= \delta_{T-t}\mathbf{f}(t) \\
&= \delta_{T-t}\mathcal{S}_t\mathbf{f}_0 + \delta_{T-t}\int_0^t \mathcal{S}_{t-s}\mathbf{\Sigma}(s)d\mathbf{W}(\Theta(s)) \\
&= \mathbf{f}_0(T) + \int_0^t \boldsymbol{\sigma}(T, s)d\mathbf{N}(s)
\end{aligned}$$

where $t \leq T$ and we have understood δ_{T-t} as an element in $L(H, \mathbb{R}^d)$. Furthermore,

$$\boldsymbol{\sigma}(T, s) := (\delta_{T-t}\mathcal{S}_{t-s}\mathbf{\Sigma}(s)Q_L\mathbf{\Sigma}(s)^*\mathcal{S}_{t-s}^*\delta_{T-t}^*)^{1/2} = (\delta_{T-s}\mathbf{\Sigma}(s)Q_L\mathbf{\Sigma}(s)^*\delta_{T-s}^*)^{1/2}$$

with Q_L being the martingale covariance of $\mathbf{L}$ and $\mathbf{N}$ being a subordinated Brownian motion on $\mathbb{R}^d$. What we read from this representation is that the forward price becomes a finite-dimensional dynamics in terms of a random driver being a d-dimensional subordinated Brownian motion and a time-dependent volatility matrix constructed from the infinite dimensional volatility $\mathbf{\Sigma}(t)$ and the covariance structure of $\mathbf{L}$. The covariance structure of $\mathbf{L}$ is modelled via the covariance operator of $\mathbf{W}$. The covariance operator of $\mathbf{W}$ models the dependency structure among the d commodities for different times to maturity.

In the above case, we can choose Θ to be an inverse Gaussian subordinator. Then, recalling from Subsection 2.4.1 in Chapter 2, $\mathbf{W}(\Theta)$ is an H-valued normal inverse

Gaussian (NIG) Lévy process. But as $\mathbf{N}$ is a subordinated Brownian motion on $\mathbb{R}^d$, with the same subordinator Θ, $\mathbf{N}$ becomes a d-dimensional NIG Lévy process. Andersen, Koekebakker and Westgaard [5] propose models of electricity forward prices based on the multivariate NIG-Lévy process. Further, the studies of Frestad, Benth and Koekebakker [107] and Benth and Paraschiv [42] show the relevance of the NIG Lévy process in power markets, while Benth and Šaltytė-Benth [46] propose NIG-Lévy processes to model gas and oil prices.

As we shall see in Section 6.5 of this chapter, contracts with delivery period can be represented as certain linear functionals on $\mathbf{f}$ as well. Thus, we may extend the above discussion to such contracts.

An infinite dimensional term structure model as we have introduced in this section, assigns idiosyncratic risk to every maturity along the forward curve. In this way, we obtain a consistent model explaining all risk in the term structure dynamics, which is attractive for studying the management of portfolios, say. As a simple case, imagine a situation where you have a portfolio of k forwards, and would like to add another to the portfolio. In an infinite dimensional term structure model we can then ask questions like which maturity will best provide the desired risk exposure of the total portfolio of $k + 1$ forwards. In a finite dimensional model, we could face the situation that the risk factors are already spanned in the portfolio, and thus a totally new $k + 1$-dimensional forward dynamics must be constructed. This would include up to k new correlation parameters. In an infinite dimensional model we have potentially a very low number of parameters describing the complete correlation structure across *all* maturities. We refer the interested reader to Benth and Lempa [38] for an analysis of optimal forward portfolios for finite dimensional term structure models and Carmona [67] for trading strategies using different forward contract maturities.

6.2 Simple Geometric HJM Models

Let us analyse simple geometric HJM-models, where, to this end, we assume that

$$\mathbf{f}(t) := \exp(\mathbf{g}(t)) \tag{6.10}$$

and $\mathbf{g}(t) = (g_1(t), \ldots, g_d(t))^\top$ follows a similar dynamics as in (6.3)

$$d\mathbf{g}(t) = \nabla_x \mathbf{g}(t)dt + \boldsymbol{\mu}(t)dt + \boldsymbol{\Sigma}(t)\,d\mathbf{L}(t) \tag{6.11}$$

with initial condition $\mathbf{g}(0) = \mathbf{g}_0 \in H$. Here, $\exp(\mathbf{g}(t)) = (\exp(g_1(t)), \ldots, \exp(g_d(t))^\top$, $(\boldsymbol{\mu}(t))_{t\geq 0}$ is an H-valued stochastic process and $\boldsymbol{\Sigma}(t))_{t\geq 0}$ is a stochastic process with values in $L(H)$, both satisfying the same conditions as in the previous Subsection 6.1.1. We recall ∇_x from (6.4). Obviously, to have $\mathbf{f}$ well-defined, we must have a Hilbert space H that is an algebra under pointwise multiplication. The Filipović spaces are such, as we recall from Proposition 3.1 in Chapter 3. Moreover, for two elements $\mathbf{g}, \mathbf{h} \in H$, we understand the product $\mathbf{gh} \in H$ in a coordinate-wise sense, i.e.,

$$\mathbf{gh} = (g_1 h_1, \ldots, g_d h_d)^\top.$$

Thus, if each Hilbert space H_i is a Banach algebra, then H will be so as well. This coordinate-wise definition of the product in H is consistent with our definition of the exponential in H above.

With $\mathcal{S}_t$ as in (6.5) and following Proposition 6.1, we find the mild solution of (6.11) to be

$$\mathbf{g}(t) = \mathcal{S}_t \mathbf{g}_0 + \int_0^t \mathcal{S}_{t-s} \boldsymbol{\mu}(s)\, ds + \int_0^t \mathcal{S}_{t-s} \boldsymbol{\Sigma}(s)\, d\mathbf{L}(s). \tag{6.12}$$

Let $(\mathbf{u}_i)_{i=1}^d$ be the canonical basis vectors in $\mathbb{R}^d$. Then we find that $\mathbf{u}_i^\top \mathbf{g}(t) = g_i(t)$. Moreover,

$$dg_i(t) = \partial_x g_i(t)\, dt + \mu_i(t)\, dt + \Sigma_i(t)\, d\mathbf{L}(t) \tag{6.13}$$

where $\mu_i(t) := \mathbf{u}_i^\top \boldsymbol{\mu}(t)$ and $\Sigma_i(t) := \mathbf{u}_i^\top \boldsymbol{\Sigma}(t) \in L(H, H_i)$.

To have a no-arbitrage dynamics of the forward price $t \mapsto \mathbf{f}(t)$, we must impose a drift condition on $\boldsymbol{\mu}$ to ensure that $t \mapsto \mathbf{F}(t, T) := \delta_{T-t}\mathbf{f}(t)$ is a (local) $\mathbb{Q}$-martingale for $t \leq T$. The next proposition provides us with a condition for this.

Proposition 6.3 *Let $h_x^i := \delta_x^* 1 \in H_i$ where $\delta_x, x \geq 0$ is the evaluation operator in H_i, and denote by $\mathbf{h}_x := (h_x^1, \ldots, h_x^d)^\top$. Furthermore, denote by Q_W the covariance operator on H of the continuous part of $\mathbf{L}$ and ν the Lévy measure of $\mathbf{L}$ on H. If $\boldsymbol{\mu}(t)$ is an H-valued process for which*

$$\boldsymbol{\mu}(t) = -\frac{1}{2}\|Q_W^{1/2}\boldsymbol{\Sigma}^*(t)\mathbf{h}_\cdot\|^2 - \int_H \{\exp(\boldsymbol{\Sigma}(t)\mathbf{z}) - 1 - \boldsymbol{\Sigma}(t)\mathbf{z}\}\, \nu(d\mathbf{z})$$

for all $t \geq 0$, then $t \mapsto \mathbf{F}(t, T), t \leq T$ is a (local) $\mathbb{Q}$-martingale. We have used the notations $1 := (1, \ldots, 1)^\top \in H$ and

$$\|Q_W^{1/2}\boldsymbol{\Sigma}^*(t)\mathbf{h}_\cdot\|^2 := (|Q_W^{1/2}\Sigma_1^*(t)h_\cdot^1|_1^2, \ldots, |Q_W^{1/2}\Sigma_d^*(t)h_\cdot^d|_d^2)^\top \in H$$

in the drift condition.

Proof The proof applies the Itô Formula for jump processes in Hilbert spaces, see Peszat and Zabczyk [165, Thm. D.2].

First, we notice that, as in the proof of Prop. 6.1

$$\delta_{T-t}\mathbf{g}(t) = \delta_T \mathbf{g}_0 + \int_0^t \delta_{T-s}\boldsymbol{\mu}(s)\, ds + \int_0^t \delta_{T-s}\boldsymbol{\Sigma}(s)\, d\mathbf{L}(s)$$

where we used that $\delta_t \mathcal{S}_s = \delta_{t+s}$. In particular, for $i = 1, \ldots, d$, using that $\delta_t(a \cdot b) = \delta_t a \cdot \delta_t b$ for $a, b \in H_i$, we have

$$F_i(t, T) := \mathbf{u}_i^\top \delta_{T-t}\mathbf{f}(t) = \exp(\delta_{T-t} g_i(t))$$

and

$$\delta_{T-t} g_i(t) = \delta_T g_{0,i} + \int_0^t \delta_{T-s}\mu_i(s)\, ds + \int_0^t \delta_{T-s}\Sigma_i(s)\, d\mathbf{L}(s).$$

Alternatively, we can express $\delta_{T-t} g_i(t)$ in a more attractive way for applying Itô's Formula,

$$\delta_{T-t} g_i(t) = \delta_0 \mathcal{S}_T^i g_{0,i} + \delta_0 \int_0^t \mathcal{S}_{T-u}^i \mu_i(u)\, du + \delta_0 \int_0^t \mathcal{S}_{T-u}^i \Sigma_i(u)\, d\mathbf{L}(u).$$

Hence, $F_i(t, T) = \exp(\delta_0 \mathcal{S}_T^i g_{0,i} + \delta_0 X(t))$, with X being the H_i-valued process

$$X(t) = \int_0^t \mathcal{S}_{T-u}^i \mu_i(u)\, du + \int_0^t \mathcal{S}_{T-u}^i \Sigma_i(u)\, d\mathbf{L}(u)$$

for $t \leq T$, where the first term in the expression of X is of bounded variation and the second is a martingale. We denote this term by $M(t) := \int_0^t \mathcal{S}_{T-u}^i \Sigma_i(u)\, d\mathbf{L}(u)$. We apply Itô's Formula to $Y(t) := \exp(\delta_0 X(t))$.

First, we observe that $Y(t) = \psi(\delta_0 X(t))$ for the function $\psi : H_i \to \mathbb{R}$ given by $\psi(f) = \exp(\delta_0 f)$. Since for any $h \in H_i$,

$$\psi(f + h) - \psi(f) - \psi(f)\delta_0 h = \psi(f)(\exp(\delta_0 h) - 1 - \delta_0 h) = \psi(f) \sum_{n=2}^{\infty} \frac{(\delta_0 h)^n}{n!}$$

it follows that the Frechet derivative of $\psi(f)$ is $D\psi(f) = \psi(f)\delta_0$. This is so because

$$\begin{aligned}
\frac{|\psi(f + h) - \psi(f) - \psi(f)\delta_0 h|}{|h|_i} &\leq |h|_i^{-1} \sum_{n=2}^{\infty} \frac{\|\delta_0\|_{\mathrm{op}}^n |h|_i^n}{n!} \\
&\leq |h|_i \|\delta_0\|_{\mathrm{op}}^2 \sum_{n=0}^{\infty} \frac{\|\delta_0\|_{\mathrm{op}}^n |h|_i^n}{(n+2)!} \to 0
\end{aligned}$$

when $|h|_i \to 0$, where we recall that $|\cdot|_i$ denotes the norm in H_i. Moreover, $D\psi(f)$ is a linear and bounded operator from H_i to $\mathbb{R}$ since δ_0 is. As $D\psi : H_i \to L(H_i, \mathbb{R})$, we find that $D^2\psi : H_i \to L(H_i, L(H_i, \mathbb{R}))$. Moreover,

$$\frac{\|D\psi(f+h) - D\psi(f) - \psi(f)(\delta_0 h)\delta_0\|_{\text{op}}}{|h|_i} \leq |\psi(f)| \frac{\|(e^{\delta_0 h} - 1 - \delta_0 h)\delta_0\|_{\text{op}}}{|h|_i}$$
$$\leq |\psi(f)| \|\delta_0\|_{\text{op}}^3 |h|_i$$
$$\times \sum_{n=0}^{\infty} \frac{\|\delta_0\|_{\text{op}}^n |h|_i^n}{(n+2)!} \to 0$$

when $|h|_i \to 0$. Therefore $D^2\psi(f)h = \psi(f)(\delta_0 h)\delta_0$, and we can write $D^2\psi(f) = \psi(f)(\delta_0 \otimes \delta_0)$ with the obvious interpretation of the tensor product squared of the operator δ_0. Indeed, as H_i is a separable Hilbert space and $\delta_0 \in H_i^*$ by assumption, there exists $h_0^i \in H_i$ such that $\delta_0 = \langle \cdot, h_0^i \rangle_i$, with $\langle \cdot, \cdot \rangle_i$ being the inner product in H_i. Thus, we can write $D^2\psi(f)(h_0^i \otimes h_0^i)$, with the usual tensor product between two elements in the Hilbert space defining an Hilbert-Schmidt operator, $h_0^i \otimes h_0^i \in L_{\text{HS}}(H_i)$.

Appealing to the Itô Formula on page 393 in Peszat and Zabczyk [165, Thm. D.2] and recalling that $X(0) = 0$, we find,

$$\begin{aligned}
\psi(X(t)) &= \psi(X(0)) + \int_0^t (D\psi(X(s-)), dX(s))_i \\
&\quad + \frac{1}{2} \int_0^t D^2\psi(X(s-)) d[M, M]^c(s) \\
&\quad + \sum_{s \leq t} \Delta\psi(X(s)) - (D\psi(X(s-)), \Delta X(s))_i \\
&= 1 + \int_0^t \psi(X(s))\delta_0 \mathcal{S}_{T-s}^i \mu_i(s)\, ds \\
&\quad + \int_0^t \psi(X(s-))\delta_0 \mathcal{S}_{T-s}^i \Sigma_i(s)\, d\mathbf{L}(s) \\
&\quad + \frac{1}{2} \int_0^t \psi(X(s))(\delta_0 \otimes \delta_0)\, d[M, M]^c(s) \\
&\quad + \sum_{s \leq t} \psi(X(s-))(\exp(\delta_0 \Delta X(s)) - 1) - \psi(X(s-))\delta_0 \Delta X(s).
\end{aligned}$$

Now, suppose that $\mathbf{W}$ is the Wiener process on H being the continuous martingale part of $\mathbf{L}$. Recalling the (operator) angle bracket notations in (4.4) and (4.5) of Chapter 4 (see also p. 392 in Appendix D of Peszat and Zabczyk [165]),

$$[M, M]^c(t) = \sum_{n,m} (e_n \otimes e_m)\langle\langle M_n^c, M_m^c\rangle\rangle^c(t)$$

where $(e_n)_{n\in\mathbb{N}}$ is an ONB of H_i and M_n^c is

$$M_n^c(t) := \langle \int_0^t \mathcal{S}_{T-s}^i \Sigma_i(s)\, d\mathbf{W}(s), e_n\rangle_i$$

(i.e., the continuous part of M projected down to the nth coordinate). Introduce the linear functional $\mathcal{L}_n := \langle\cdot, e_n\rangle_i$, and by (4.4) (see also Corollary 8.17 in Peszat and Zabczyk [165]) it follows that

$$\begin{aligned}
\langle\langle M_n^c, M_m^c\rangle\rangle^c(t) &= \int_0^t \langle \mathcal{L}_n \mathcal{S}_{T-s}^i \Sigma_i(s) Q_W^{1/2}, \mathcal{L}_m \mathcal{S}_{T-s}^i \Sigma_i(s) Q_W^{1/2}\rangle_{\mathrm{HS}}\, ds \\
&= \int_0^t \sum_{k=1}^{\infty} \langle \mathcal{S}_{T-s}^i \Sigma_i(s) Q_W^{1/2} e_k, e_n\rangle_i \langle \mathcal{S}_{T-s}^i \Sigma_i(s) Q_W^{1/2} e_k, e_m\rangle_i\, ds \\
&= \int_0^t \sum_{k=1}^{\infty} \langle e_k, Q_W^{1/2} \Sigma_i^*(s) \mathcal{S}_{T-s}^{i,*} e_n\rangle_i \langle e_k, Q_W^{1/2} \Sigma_i^*(s) \mathcal{S}_{T-s}^{i,*} e_m\rangle_i\, ds \\
&= \int_0^t \langle Q_W^{1/2} \Sigma_i^*(s) \mathcal{S}_{T-s}^{i,*} e_n, Q_W^{1/2} \Sigma_i^*(s) \mathcal{S}_{T-s}^{i,*} e_m\rangle_i\, ds
\end{aligned}$$

where Q_W is the covariance operator of $\mathbf{W}$ and the last equality followed from Parseval's identity. Observe that for a bounded linear operator $\mathcal{M} \in L(H_i)$, we have using $\delta_0^* 1 = h_0$

$$\begin{aligned}
\delta_0 \mathcal{M}^* \mathcal{M} \delta_0^* 1 &= \delta_0 \mathcal{M}^* \mathcal{M} h_0 \\
&= \langle h_0, \mathcal{M}^* \mathcal{M} h_0\rangle_i \\
&= \sum_{n=1}^{\infty} \langle \langle h_0, e_n\rangle_i e_n, \mathcal{M}^* \mathcal{M} \sum_{m=1}^{\infty} \langle h_0, e_m\rangle_i e_m\rangle_i \\
&= \sum_{n,m=1}^{\infty} \langle \mathcal{M} e_n, \mathcal{M} e_m\rangle_i e_n(0) e_m(0).
\end{aligned}$$

Using $\mathcal{M} = Q_W^{1/2}\Sigma_i^*(s)\mathcal{S}_{T-s}^{i,*}$ it follows

$$\begin{aligned}
&\int_0^t \psi(X(s))(\delta_0 \otimes \delta_0) d[M, M]^c(s) \\
&\quad = \int_0^t \psi(X(s)) \sum_{n,m=1}^{\infty} \langle \mathcal{M}e_n, \mathcal{M}e_m \rangle_i e_n(0) e_m(0)\, ds \\
&\quad = \int_0^t \psi(X(s)) \delta_0 \mathcal{S}_{T-s}^i \Sigma_i(s) Q_W \Sigma_i^*(s) \mathcal{S}_{T-s}^{i,*} \delta_0^* 1\, ds.
\end{aligned}$$

Hence, from $\delta_0 \mathcal{S}_{T-s}^i = \delta_{T-s}$, we have

$$\begin{aligned}
\psi(X(t)) &= 1 + \int_0^t \psi(X(s))(\delta_{T-s}\mu_i(s) + \frac{1}{2}\delta_{T-s}(\Sigma_i(s)Q_W\Sigma_i^*(s))\delta_{T-s}^* 1)\, ds \\
&\quad + \sum_{s \le t} \psi(X(s-))(\exp(\Delta\delta_0 X(t)) - 1 - \Delta\delta_0 X(s)) \\
&\quad + \int_0^t \psi(X(s-))\delta_{T-s}\Sigma_i(s)\, d\mathbf{L}(s) \\
&= 1 + \int_0^t \psi(X(s))(\delta_{T-s}\mu_i(s) + \frac{1}{2}\delta_{T-s}(\Sigma_i(s)Q_W\Sigma_i^*(s))\delta_{T-s}^* 1)\, ds \\
&\quad + \int_0^t \int_H \psi(X(s))(\exp(\delta_{T-s}\Sigma_i(s)\mathbf{z}) - 1 - \delta_{T-s}\Sigma_i(s)\mathbf{z})\, \nu(d\mathbf{z})\, ds \\
&\quad + \int_0^t \int_H \psi(X(s-))(\exp(\delta_{T-s}\Sigma_i(s)\mathbf{z}) - 1 - \delta_{T-s}\Sigma_i(s)\mathbf{z})\, \widetilde{\mathbf{N}}(d\mathbf{z}, ds) \\
&\quad + \int_0^t \psi(X(s-))\delta_{T-s}\Sigma_i(s)\, d\mathbf{L}(s),
\end{aligned}$$

where ν is the Lévy measure on H of $\mathbf{L}$ and $\widetilde{\mathbf{N}}$ is the compensated Poisson random measure of $\mathbf{L}$. Notice that

$$\delta_x \Sigma_i(t) Q_W \Sigma_i^*(t) \delta_x^* 1 = \langle h_x^i, \Sigma_i(t) Q_W \Sigma_i^*(t) h_x^i \rangle_i = |Q_W^{1/2} \Sigma_i^*(t) h_x^i|_i^2.$$

The Proposition follows after setting the ds-terms equal to zero. □

In the case of a simple arithmetic HJM dynamics, as considered in the previous section, we recall that we obtained an arbitrage-free dynamics when the drift $\boldsymbol{\mu}(t)$ was identically zero. In the geometric case, we get a non-trivial *drift condition* ensuring an arbitrage-free forward model.

Let us focus on some special cases, and investigate the no-arbitrage drift condition in further detail: Consider the one-dimensional case ($d = 1$) and assume that $H = H_w$, the Filipović space which we recall from Chapter 3. In the case of no jumps, the drift condition becomes

$$\mu(t,x) = -\frac{1}{2}|Q_W^{1/2}\Sigma^*(t)h_x|_w^2 \tag{6.14}$$

according to Proposition 6.3. The function $h_x(\cdot)$ is explicitly given in (3.4) in Chapter 3 for the Filipović space. To have a valid drift condition, we must check if $\mathbb{R}_+ \ni x \mapsto \mu(t,x) \in H_w$, *a.s.* To simplify the exposition, denote by $\mathcal{M} := Q_W^{1/2}\Sigma^*(t) \in L(H_w)$ and introduce, for fixed $\omega \in \Omega$, the function $\varphi : \mathbb{R}_+ \to \mathbb{R}_+$

$$\varphi(x) := |\mathcal{M}h_x|_w^2. \tag{6.15}$$

It is straightforward to check that $\mathcal{M}^*\mathcal{M}$ is a symmetric and positive definite operator on H_w. Furthermore, it is trace class, since

$$\begin{aligned}
\mathrm{Tr}(\mathcal{M}^*\mathcal{M}) &= \sum_{k=1}^{\infty} |Q_W^{1/2}\Sigma^*(t)e_k|_w^2 \\
&= \|Q_W^{1/2}\Sigma^*(t)\|_{\mathrm{HS}}^2 \\
&= \|\Sigma(t)Q_W^{1/2}\|_{\mathrm{HS}}^2 \\
&\leq \|\Sigma(t)\|_{\mathrm{op}}^2 \sum_{k=1}^{\infty} |Q_W^{1/2}e_k|_w^2 \\
&= \|\Sigma(t)\|_{\mathrm{op}}^2 \mathrm{Tr}(Q_W) < \infty.
\end{aligned}$$

Thus, it follows that

$$\varphi(x) = (h_x, \mathcal{M}^*\mathcal{M}h_x)_w = (\mathcal{M}^*\mathcal{M}h_x)(x)$$

for a symmetric, positive definite trace class operator $\mathcal{M}^*\mathcal{M}$. From Corollary 3.1 and the subsequent discussion of Subsection 3.2.3 of Chapter 3, there exist a constant $c \in \mathbb{R}$ and symmetric function $\kappa \in L^2(\mathbb{R}_+^2)$ such that

$$\varphi(x) = c^2 + \int_0^x \int_0^x \left(\int_0^\infty \kappa(r,u)\kappa(u,v)\,du\right) w^{-1/2}(r) w^{-1/2}(v)\,dr\,dv.$$

Using that κ is symmetric, we readily find

$$\varphi'(x) = 2\int_0^x \int_0^\infty \kappa(x,u)\kappa(u,v)\,du w^{-1/2}(v)\,dv w^{-1/2}(x).$$

Hence, we estimate using Cauchy-Schwarz' inequality,

$$
\begin{aligned}
&\int_0^\infty w(x)|\varphi'(x)|^2\,dx \\
&= 4\int_0^\infty (\int_0^x \int_0^\infty \kappa(x,u)\kappa(u,v)\,du w^{-1/2}(v)\,dv)^2\,dx \\
&\le 4\int_0^\infty \int_0^x (\int_0^\infty \kappa(x,u)\kappa(u,v)\,du)^2\,dv \int_0^x w^{-1}(v)\,dv\,dx \\
&\le 4\int_0^\infty (\int_0^\infty \kappa^2(x,u)\,du)(\int_0^x \int_0^\infty \kappa^2(u,v)\,du\,dv) \int_0^x w^{-1}(v)\,dv)\,dx \\
&\le 4\int_0^\infty \int_0^\infty \kappa^2(x,u)\,du\,dx \int_0^\infty \int_0^\infty \kappa^2(u,v)\,du\,dv \int_0^\infty w^{-1}(v)\,dv < \infty.
\end{aligned}
$$

In the above we appealed to the condition $w^{-1} \in L^1(\mathbb{R}_+)$ assumed to hold for the Filipović space. Additionally, we have that

$$\varphi(0) = |\mathcal{M}h_0|_w^2 \le \|\mathcal{M}\|_{\text{op}}^2 |h_0|_w^2 < \infty$$

and we conclude that $\varphi \in H_w$. Hence, we have shown that $\mu(t,\cdot) \in H_w$ for the drift condition (6.14).

Next, let us focus on the pure-jump case, i.e., the case when $\mathbf{L}$ does not have any continuous martingale part and therefore $Q_W = 0$ (still restricting the attention to $d = 1$). According to Proposition 6.3, the drift condition in this situation becomes

$$\mu(t) = -\int_{H_w} \{\exp(\Sigma(t)z) - 1 - \Sigma(t)z\}\,\nu(dz) \tag{6.16}$$

where ν is the Lévy measure on H_w of L. As the Filipović space is an algebra according to Proposition 3.1, we have that

$$\exp(\Sigma(t)z) - 1 - \Sigma(t)z \in H_w$$

for any $z \in H_w$. We notice that the constant function 1 is trivially an element of H_w. Now, if

$$\int_{H_w} |\exp(\Sigma(t)z) - 1 - \Sigma(t)z|_w\,\nu(dz) < \infty,$$

then the integral in (6.16) is well-defined as a Bochner integral in H_w, and we conclude that $\mu(t) \in H_w$.

Let us end this Section by exploring a sufficient condition for Bochner integrability of the integral in (6.16). Recall that for two functions $u, v \in H_w$, we have by Proposition 3.1 that $|u \cdot v|_w \leq c|u|_w|v|_w$ for a positive constant c. Thus,

$$|\exp(\Sigma(t)z) - 1 - \Sigma(t)z|_w \leq \sum_{n=2}^{\infty} \frac{1}{n!}|(\Sigma(t)z)^n|_w \leq \sum_{n=2}^{\infty} \frac{c^{n-1}}{n!}\|\Sigma(t)\|_{\text{op}}^n |z|_w^n.$$

It follows that

$$\begin{aligned}
&\int_{H_w} |\exp(\Sigma(t)z) - 1 - \Sigma(t)z|_w \, \nu(dz) \\
&\quad \leq \int_{|z|_w<1} c\|\Sigma(t)\|_{\text{op}}^2 |z|_w^2 \sum_{n=0}^{\infty} \frac{c^n}{(n+2)!}\|\Sigma(t)\|_{\text{op}}^n |z|_w^n \, \nu(dz) \\
&\qquad + \int_{|z|_w\geq 1} c^{-1} \sum_{n=2}^{\infty} \frac{c^n}{n!}\|\Sigma(t)\|_{\text{op}}^n |z|_w^n \, \nu(dz) \\
&\quad \leq c\|\Sigma(t)\|_{\text{op}}^2 \exp(c\|\Sigma(t)\|_{\text{op}}) \int_{|z|_w<1} |z|_w^2 \, \nu(dz) \\
&\qquad + c^{-1} \int_{|z|_w\geq 1} \exp(c\|\Sigma(t)\|_{\text{op}}|z|_w) \, \nu(dz).
\end{aligned}$$

The first integral is finite by the definition of the Lévy measure, and therefore a sufficient condition for Bochner integrability, and thereby the existence of a drift condition, is the exponential integrability of $|z|_w$ with respect to ℓ on $|z|_w \geq 1$ with coefficient $c\|\Sigma(t)\|_{\text{op}}$. This must hold for all t on the horizon of interest, ω-wise as Σ may be stochastic.

6.3 A Stochastic Volatility Model

In this Section we present a class of stochastic volatility models which fits naturally into the specification of arithmetic or geometric HJM models. So far, $\mathbf{\Sigma}$ has been assumed to be a deterministic operator-valued function of time. Our class of stochastic volatility models will be a term structure of a stochastically evolving volatility, which describes the random time-dependency of the variance at each maturity as well as the random correlations across maturities. It is specified as a random scaling of the price increments, and builds on the popular Barndorff-Nielsen and Shephard (BNS) stochastic volatility model first introduced in [18]. Our exposition follows Benth, Rüdiger and Süss [45].

We focus on a Wiener-driven residual noise-component of the dynamics of $\mathbf{f}$, either in the arithmetic model (6.3) or the geometric (6.11), i.e., we consider a model for $\mathbf{\Sigma}$ for

$$R(t) := \int_0^t \mathbf{\Sigma}(s) d\mathbf{W}(s). \tag{6.17}$$

To this end, let $(\mathcal{L}(t))_{t\geq 0}$ be a square-integrable Lévy process in the space of Hilbert-Schmidt operators on H, denoted $L_{\mathrm{HS}}(H)$. We remark that $L_{\mathrm{HS}}(H)$ is a separable Hilbert space. Assume that $\mathfrak{C} \in L(L_{\mathrm{HS}}(H))$, and define the Ornstein–Uhlenbeck process

$$d\mathcal{Y}(t) = \mathfrak{C}\mathcal{Y}(t)dt + d\mathcal{L}(t) \tag{6.18}$$

with $\mathcal{Y}(0) = \mathcal{Y}_0 \in L_{\mathrm{HS}}(H)$. As $\mathfrak{C}$ is a bounded linear operator on $L_{\mathrm{HS}}(H)$, we find that its C_0-semigroup is given by

$$\mathfrak{S}_t := \exp(t\mathfrak{C}) \tag{6.19}$$

and we have a mild solution of (6.18) given by

$$\mathcal{Y}(t) = \mathfrak{S}_t\mathcal{Y}_0 + \int_0^t \mathfrak{S}_{t-s} d\mathcal{L}(s). \tag{6.20}$$

Following the BNS stochastic volatility model, $\mathcal{Y}$ is a stochastic process for the variance, where $\mathfrak{C}$ plays the role of the speed of mean reversion, that is, how fast the volatility is reverting back towards zero after a shock described by $\mathcal{L}$ has occurred. The stochastic volatility is defined as the square-root of the variance, $\mathbf{\Sigma}(t) = \mathcal{Y}(t)^{1/2}$. We next discuss conditions ensuring that $\mathcal{Y}(t)$ is a symmetric, positive definite stochastic process, which, moreover, makes (6.17) well-defined as a stochastic integral.

We define the following class of Lévy processes with an increasing path property:

Definition 6.1 The $L_{\mathrm{HS}}(H)$-valued Lévy process $\mathcal{L}$ has *non-decreasing paths* if $\mathcal{L}(t)$ is symmetric and $t \mapsto \langle \mathcal{L}(t)\mathbf{u}, \mathbf{u}\rangle_H$ is non-decreasing in $t \geq 0$ for every $\mathbf{u} \in H$, *a.s.*

As is evident from the definition, $\mathcal{L}(t)$ is non-negative definite for every $t \geq 0$ since $\mathcal{L}(0) = 0$ by definition of Lévy processes. But even more, we see that for $t \geq s \geq 0$, we have for $\mathbf{u} \in H$

$$\langle (\mathcal{L}(t) - \mathcal{L}(s))\mathbf{u}, \mathbf{u}\rangle_H = \langle \mathcal{L}(t)\mathbf{u}, \mathbf{u}\rangle_H - \langle \mathcal{L}(s)\mathbf{u}, \mathbf{u}\rangle_H$$

which is non-negative *a.s.* by the non-decreasing path property of $t \mapsto \langle \mathcal{L}(t)\mathbf{u}, \mathbf{u}\rangle_H$. Thus, the increments of $\mathcal{L}$ are symmetric and non-negative definite. We have the following:

Lemma 6.1 *Assume* $(\mathcal{L}(t))_{t\geq 0}$ *is an* $L_{HS}(H)$*-valued Lévy process with non-decreasing paths. Then* $t \mapsto \langle \mathcal{L}(t)\mathbf{u}, \mathbf{u}\rangle_H$ *is a real-valued Lévy process with non-decreasing paths, i.e., a subordinator.*

Proof By Peszat and Zabczyk [165, Thm. 4.49(ii)], $t \mapsto \langle \mathcal{L}(t), \mathcal{T}\rangle_{\text{HS}}$ is a real-valued Lévy process for any $\mathcal{T} \in L_{\text{HS}}(H)$. If $\mathbf{u} \in H$, then $\mathbf{u}^{\otimes 2} \in L_{\text{HS}}(H)$. Moreover, for an ONB $(\mathbf{e}_i)_{i\in\mathbb{N}}$ in H, we have by Parseval's identity

$$
\begin{aligned}
\langle \mathcal{L}(t), \mathbf{u}^{\otimes 2}\rangle_{\text{HS}} &= \sum_{i=1}^{\infty} \langle \mathcal{L}(t)\mathbf{e}_i, \mathbf{u}^{\otimes 2}\mathbf{e}_i\rangle_H \\
&= \sum_{i=1}^{\infty} \langle \mathcal{L}(t)\mathbf{e}_i, \mathbf{u}\rangle_H \langle \mathbf{u}, \mathbf{e}_i\rangle_H \\
&= \sum_{i=1}^{\infty} \langle \mathbf{e}_i, \mathcal{L}(t)^*\mathbf{u}\rangle_H \langle \mathbf{u}, \mathbf{e}_i\rangle_H \\
&= \langle \mathcal{L}(t)\mathbf{u}, \mathbf{u}\rangle_H .
\end{aligned}
$$

It follows that $t \mapsto \langle \mathcal{L}(t)\mathbf{u}, \mathbf{u}\rangle_H$ is a Lévy process. Furthermore, by the non-decreasing paths of $\mathcal{L}$, this real-valued Lévy process has non-decreasing paths and is therefore a subordinator. The assertions follow. □

In the BNS stochastic volatility model, real-valued subordinators are the drivers of the variance process. Under some natural positivity preserving conditions on the semigroup $\mathfrak{S}$, we can ensure that $\mathcal{Y}$ is a symmetric and positive definite process in $L_{\text{HS}}(H)$:

Proposition 6.4 *Let* $\mathcal{L}$ *be a Lévy process in* $L_{HS}(H)$ *which has non-decreasing paths. Assume that* $\mathfrak{S}$ *in* (6.19) *is such that for any* $t \geq 0$ *and* $\mathcal{T} \in L_{HS}(H)$, $(\mathfrak{S}_t\mathcal{T})^* = \mathfrak{S}_t\mathcal{T}^*$ *and* $\mathfrak{S}_t\mathcal{T}$ *is non-negative definite whenever* $\mathcal{T}$ *is. Then, if* $\mathcal{Y}_0 \in L_{HS}(H)$ *is symmetric and non-negative definite, the process* $(\mathcal{Y}(t))_{t\geq 0}$ *defined in* (6.20) *is symmetric and non-negative definite a.s.*

Proof From (6.18) it follows for any $t \geq 0$,

$$
\begin{aligned}
\mathcal{Y}(t)^* &= \mathcal{Y}_0^* + (\int_0^t \mathfrak{C}\mathcal{Y}(s)ds)^* + \mathcal{L}(t)^* \\
&= \mathcal{Y}_0 + \int_0^t (\mathfrak{C}\mathcal{Y}(s))^* ds + \mathcal{L}(t).
\end{aligned}
$$

Since $\mathfrak{C}$ is a bounded operator on $L_{\text{HS}}(H)$, for any $\mathcal{T} \in L_{\text{HS}}(H)$ we have

$$
(\mathfrak{C}\mathcal{T})^* = (\lim_{t\downarrow 0} \frac{1}{t}(\mathfrak{S}_t\mathcal{T} - \mathcal{T}))^* = \lim_{t\downarrow 0} \frac{1}{t}(\mathfrak{S}_t\mathcal{T}^* - \mathcal{T}^*) = \mathfrak{C}\mathcal{T}^*
$$

by continuity of the adjoint operation and assumption on the semigroup. Hence,

$$d\mathcal{Y}(t)^* = \mathfrak{C}\mathcal{Y}(t)^* dt + d\mathcal{L}(t)$$

and thus $\mathcal{Y}(t)^*$ solves the same stochastic differential equation as $\mathcal{Y}(t)$. As mild solutions are unique, $\mathcal{Y}(t)^* = \mathcal{Y}(t)$. This shows symmetry.

Recall the mild solution of $\mathcal{Y}(t)$ in (6.20). By assumption on $\mathfrak{S}_t$ and $\mathcal{Y}_0$, $\mathfrak{S}_t\mathcal{Y}_0$ is non-negative definite. Next, we know that $\int_0^t \mathfrak{S}_{t-s} d\mathcal{L}(s)$ is defined as the strong limit of $\sum_{m=1}^M \mathfrak{S}_{t-s_m} \Delta\mathcal{L}(s_m)$ in $L^2(\Omega \times [0,t]; H)$. Here, $(s_m)_{m=1}^M$ is a nested partition of $[0,t]$, and $\Delta\mathcal{L}(s_m) := \mathcal{L}(s_{m+1}) - \mathcal{L}(s_m)$ is an increment of $\mathcal{L}$. But $\Delta\mathcal{L}(s_m)$ is non-negative definite $a.s.$, and therefore each term in the sum above is non-negative definite, $a.s.$, by the assumption on $\mathfrak{S}$. Hence it follows that $\int_0^t \mathfrak{S}_{t-s} d\mathcal{L}(s)$ is non-negative definite $a.s.$, and the proof is complete. □

Let us consider two specifications of $\mathfrak{C}$ which is of particular interest: For some fixed $C \in L(H)$, define

$$\mathfrak{C}^{(1)}\mathcal{T} = C\mathcal{T}C^*, \tag{6.21}$$

and

$$\mathfrak{C}^{(2)}\mathcal{T} = C\mathcal{T} + \mathcal{T}C^*, \tag{6.22}$$

for any $\mathcal{T} \in L_{\mathrm{HS}}(H)$. We show that these two cases of $\mathfrak{C}$ yield semigroups satisfying the required properties:

Lemma 6.2 *Assume* $\mathfrak{C}^{(i)}$ *is defined by either* (6.21) *or* (6.22) *for* $i = 1, 2$, *respectively. Then* $\mathfrak{C}^{(i)} \in L_{HS}(H)$, *and the corresponding semigroups are given by*

$$\mathfrak{S}_t^{(1)}\mathcal{T} = \sum_{n=0}^{\infty} \frac{t^n}{n!} C^n \mathcal{T} C^{*n}$$

and

$$\mathfrak{S}_t^{(2)}\mathcal{T} = \exp(tC)\mathcal{T}\exp(tC^*)$$

respectively, for any $\mathcal{T} \in L_{HS}(H)$. *Moreover, for* $i = 1, 2$ *we find that* $(\mathfrak{S}_t^{(i)}\mathcal{T})^* = \mathfrak{S}_t^{(i)}\mathcal{T}^*$ *and* $\mathfrak{S}_t^{(i)}\mathcal{T}$ *is non-negative definite whenever* $\mathcal{T}$ *is.*

Proof If $\mathcal{S}, \mathcal{T} \in L_{\mathrm{HS}}(H)$, it easily follows that $\mathfrak{C}^{(i)}(\mathcal{S} + \mathcal{T}) = \mathfrak{C}^{(i)}\mathcal{S} + \mathfrak{C}^{(i)}\mathcal{T}$, $i = 1, 2$. Moreover, we find

$$\|\mathfrak{C}^{(1)}\mathcal{T}\|_{\mathrm{HS}} = \|C\mathcal{T}C^*\|_{\mathrm{HS}} \leq \|C\|_{\mathrm{op}}\|\mathcal{T}C^*\|_{\mathrm{HS}} = \|C\|_{\mathrm{op}}\|C\mathcal{T}^*\|_{\mathrm{HS}} \leq \|C\|_{\mathrm{op}}^2\|\mathcal{T}\|_{\mathrm{HS}}$$

where we used that $\|\mathcal{G}^*\|_{\text{HS}} = \|\mathcal{G}\|_{\text{HS}}$ for any $\mathcal{G} \in L_{\text{HS}}(H)$. Hence, $\mathfrak{C}^{(1)} \in L(L_{\text{HS}}(H))$. Appealing to the triangle inequality,

$$\|\mathfrak{C}^{(2)}\mathcal{T}\|_{\text{HS}} \leq \|C\mathcal{T}\|_{\text{HS}} + \|\mathcal{T}C^*\|_{\text{HS}} \leq 2\|C\|_{\text{op}}\|\mathcal{T}\|_{\text{HS}}$$

from which it follows that $\mathfrak{C}^{(2)} \in L(L_{\text{HS}}(H))$ as well.

By induction over $n \geq 1$ it is easily seen that

$$(\mathfrak{C}^{(1)})^n\mathcal{T} = C^n\mathcal{T}C^{*n}$$

and the representation of $\mathfrak{S}_t^{(1)}$ follows. Furthermore, it holds that

$$(C^n\mathcal{T}C^{*n})^* = C^n\mathcal{T}^*C^{*n}$$

and, if $\mathcal{T}$ is non-negative definite,

$$\langle(C^n\mathcal{T}C^{*n}\mathbf{u}, \mathbf{u}\rangle_H = \langle\mathcal{T}(C^{*n}\mathbf{u}), C^{*n}\mathbf{u}\rangle_H \geq 0$$

for any $\mathbf{u} \in H$. Therefore,

$$(\mathfrak{S}_t^{(1)}\mathcal{T})^* = \mathfrak{S}_t^{(1)}\mathcal{T}^*.$$

and $\mathfrak{S}_t^{(1)}\mathcal{T}$ is non-negative definite whenever $\mathcal{T}$ is.

For the case $i = 2$, we notice that

$$\exp(t\mathfrak{C}^{(2)})\mathcal{T} = \sum_{k=0}^{\infty}\frac{t^k}{k!}(\mathfrak{C}^{(2)})^k\mathcal{T}$$

and

$$\exp(tC)\mathcal{T}\exp(tC^*) = \sum_{n,m=0}^{\infty}\frac{t^{n+m}}{n!m!}C^n\mathcal{T}C^{*m}.$$

If we now spell out the definition of $(\mathfrak{C}^{(2)})^n$ and compare the two sums, we show the expression of $\mathfrak{S}_t^{(2)}$. From this it readily follows the required properties of $\mathfrak{S}_t^{(2)}$. □

We mention in passing that specifying the "mean reversion speed" as $\mathfrak{C}^{(2)}$ can be seen as the analogue of the matrix-valued volatility model by Barndorff-Nielsen and Stelzer [19]. By choosing C as a diagonal matrix with bounded operators on the diagonal, we obtain a BNS volatility model for each commodity forward price separately. Blending in operators in the off-diagonal elements of C leads to mixing the term structures of stochastic volatility across commodities.

Under the assumptions that $(\mathfrak{S}_t\mathcal{T})^* = \mathfrak{S}_t\mathcal{T}^*$, $\mathfrak{S}_t\mathcal{T}$ is non-negative definite for $\mathcal{T}$ being non-negative definite, $\mathcal{L}$ is a square-integrable Lévy process with non-decreasing paths and $\mathcal{Y}_0$ is symmetric and non-negative definite, we can define the stochastic volatility process as

$$\boldsymbol{\Sigma}(t) := \mathcal{Y}(t)^{1/2}, \tag{6.23}$$

where $\mathcal{Y}(t)^{1/2}$ is the unique symmetric non-negative definite square root of $\mathcal{Y}(t)$. We assume this given from now on.

In order for R in (6.17) to be well-defined, we must have that the expected value of the Hilbert-Schmidt norm of $\mathcal{Y}(s)^{1/2}Q_W^{1/2}$ is square-integrable on any interval,. where Q_W is the covariance operator of W. We prove that this holds:

Proposition 6.5 *It holds that*

$$\mathbb{E}\left[\int_0^t \|\mathcal{Y}(s)^{1/2}Q_w^{1/2}\|_{HS}^2 ds\right] < \infty$$

for any $0 < t < \infty$.

Proof First, notice that

$$\begin{aligned}
\|\mathcal{Y}(s)^{1/2}Q_w^{1/2}\|_{\mathrm{HS}}^2 &= \sum_{i=1}^{\infty} |\mathcal{Y}(s)^{1/2}Q_W^{1/2}\mathbf{e}_i|_H^2 \\
&= \sum_{i=1}^{\infty} \langle Q_W^{1/2}\mathcal{Y}(s)Q_W^{1/2}\mathbf{e}_i, \mathbf{e}_i\rangle_H \\
&= \mathrm{Tr}(Q_W^{1/2}\mathcal{Y}(s)Q_W^{1/2}),
\end{aligned}$$

for some ONB $(\mathbf{e}_i)_{i\in\mathbb{N}}$ of H. From the mild solution of $\mathcal{Y}(t)$ in (6.20) and linearity of the trace,

$$\mathbb{E}\left[\mathrm{Tr}(Q_W^{1/2}\mathcal{Y}(s)Q_W^{1/2})\right] = \mathrm{Tr}(Q_W^{1/2}\mathfrak{S}_s\mathcal{Y}_0 Q_W^{1/2}) + \mathbb{E}\left[\mathrm{Tr}(Q_W^{1/2}\int_0^s \mathfrak{S}_{s-u}d\mathcal{L}(u)Q_W^{1/2})\right].$$

If $\mathcal{X}$ is an $L_{\mathrm{HS}}(H)$-valued integrable random variable, we derive,

$$\begin{aligned}
\mathbb{E}[\mathrm{Tr}(Q_W^{1/2}\mathcal{X}Q_W^{1/2}] &= \sum_{i=1}^{\infty} \mathbb{E}[\langle Q_W^{1/2}\mathcal{X}Q_W^{1/2}\mathbf{e}_i, \mathbf{e}_i\rangle_H] \\
&= \sum_{i=1}^{\infty} \mathbb{E}[\langle \mathcal{X}(Q_W^{1/2}\mathbf{e}_i), Q_W^{1/2}\mathbf{e}_i\rangle_H]
\end{aligned}$$

$$= \sum_{i=1}^{\infty} \langle \mathbb{E}[\mathcal{X}], (Q_W^{1/2}\mathbf{e}_i)^{\otimes 2}\rangle_{\mathrm{HS}}$$

$$= \sum_{i=1}^{\infty} \langle \mathbb{E}[\mathcal{X}] Q_W^{1/2}\mathbf{e}_i, Q_W^{1/2}\mathbf{e}_i\rangle_H$$

$$= \mathrm{Tr}(Q_W^{1/2}\mathbb{E}[\mathcal{X}]Q_W^{1/2}).$$

Here, $\mathbb{E}[\mathcal{X}]$ is interpreted as the Bochner integral in $L_{\mathrm{HS}}(H)$. If we let $\mathcal{X} = \int_0^s \mathfrak{S}_{s-u} d\mathcal{L}(u)$, it holds by appealing to the characteristic functional of $\mathcal{L}$ (recall theory in Section 2.1 of Chapter 2)

$$\mathbb{E}[\int_0^s \mathfrak{S}_{s-u} d\mathcal{L}(u)] = \int_0^s \mathfrak{S}_u du \mathbb{E}[\mathcal{L}(1)].$$

From the time-continuity and norm-bounds of $\mathfrak{S}_s$, it follows that $\mathbb{E}[\mathrm{Tr}(Q_W^{1/2}\mathcal{Y}(s)$ $Q_W^{1/2}]$ is integrable on $[0, t]$ for any $t < \infty$ with respect to the Lebesgue measure. The claim holds. □

In both the arithmetic and geometric forward price models, we encounter the stochastic integral

$$\int_0^t \mathcal{S}_{t-s}\mathbf{\Sigma}(s)d\mathbf{W}(s)$$

which is well-defined if

$$\mathbb{E}\left[\int_0^t \|\mathcal{S}_{t-s}\mathbf{\Sigma}(s)Q_W^{1/2}\|_{\mathrm{HS}}^2 ds\right] < \infty. \tag{6.24}$$

However, by the Hille-Yosida bound on the operator norm of a C_0-semigroup, it follows

$$\|\mathcal{S}_{t-s}\mathbf{\Sigma}(s)Q_W^{1/2}\|_{\mathrm{HS}} \leq \|\mathcal{S}_{t-s}\|_{\mathrm{op}}\|\mathcal{Y}(s)^{1/2}Q_W^{1/2}\|_{\mathrm{HS}} \leq Ke^{C(t-s)}\|\mathcal{Y}(s)^{1/2}Q_W^{1/2}\|_{\mathrm{HS}}$$

for some positive constants K and C. But then (6.24) is satisfied due to Prop. 6.5.

We can define $\mathcal{L}$ as a compound Poisson process with values in $L_{\mathrm{HS}}(H)$. If N is a Poisson process with values in $\mathbb{N} \cup \{0\}$, and $(\mathcal{X}_i)_{i\in\mathbb{N}}$ is an IID sequence of $L_{\mathrm{HS}}(H)$-valued square-integrable random variables being symmetric and positive definite, we can define the Lévy process $\mathcal{L}$ as

$$\mathcal{L}(t) = \sum_{i=1}^{N(t)} \mathcal{X}_i. \tag{6.25}$$

Indeed, for any $\mathbf{u} \in H$,

$$\langle \mathcal{L}(t)\mathbf{u}, \mathbf{u} \rangle_H = \sum_{i=1}^{N(t)} \langle \mathcal{X}_i \mathbf{u}, \mathbf{u} \rangle_H$$

where, by the IID-property of $(\mathcal{X}_i)_{i \in \mathbb{N}}$ along with the symmetry and positive definiteness, it follows that $(\langle \mathcal{X}_i \mathbf{u}, \mathbf{u} \rangle_H)_{i \in \mathbb{N}}$ is an IID-sequence of $\mathbb{R}_+$-valued random variables. Hence, $t \mapsto \langle \mathcal{L}(t)\mathbf{u}, \mathbf{u} \rangle_H$ is a compound Poisson process with positive jumps, and henceforth $\mathcal{L}$ is a Lévy process which has non-decreasing paths.

For example, we can define $\mathcal{X}_i = Z_i^{\otimes 2}$ for an IID-sequence $(Z_i)_{i \in \mathbb{N}}$ of centered and square-integrable H-valued random variables. Obviously,

$$\langle Z^{\otimes 2}\mathbf{u}, \mathbf{u} \rangle_H = \langle Z, \mathbf{u} \rangle^2 \geq 0$$

for any $\mathbf{u} \in H$, and the non-negativity of $Z^{\otimes 2}$ follows. Furthermore,

$$\mathbb{E}[Z^{\otimes 2}] = Q_Z$$

where Q_Z is a symmetric, positive definite trace class covariance operator of Z. Also, $\|Z^{\otimes 2}\|_{\mathrm{HS}}^2 = |Z|^4 < \infty \; a.s.$ The variable Z could for example be chosen as a centered Gaussian random variable in H.

Let us end this Section with some brief remarks concerning the Samuelson effect. The stochastic volatility model introduced here does not account for the Samuelson effect. A quick remedy to fix this is to appeal to multiplication operators as we recall from Section 3.2.4 in Chapter 3. Indeed, if we assume that H is the product space of Filipović spaces, we can define a $d \times d$-dimensional diagonal matrix operator $\mathcal{M} \in L(H)$, with multiplication operators on the diagonal. These operators can be $\mathcal{M}_{g_i}, i = 1, \ldots, d$, where in a simplistic situation $\mathcal{M}_{g_i}(u)(x) = \beta_i \exp(-\alpha_i x) u(x)$ for $u \in H_w, \alpha_i, \beta_i \in \mathbb{R}_+$. Here we have assumed that w, the scaling function in the Filipović space, is such that $x \mapsto \beta_i \exp(-\alpha_i x) \in H_w$. Then we define a stochastic volatility model with Samuelson effect as

$$\boldsymbol{\Sigma}(t) = \mathcal{M}\mathcal{Y}(t)^{1/2}. \tag{6.26}$$

The integrability condition in (6.24) with $\mathcal{M}\mathcal{Y}(s)^{1/2}$ instead of $\mathcal{Y}(s)^{1/2}$ remains to hold since $\mathcal{M}$ is a bounded linear operator. In this specification of the Samuelson effect, each commodity in the d-variate market model has its own decay rate α_i.

By introducing an exponentially decaying function for the Samuelson effect, we are essentially killing the randomness in the long end of the forward curve. However, recalling the discussion in Remark 5.3, there is clear empirical evidence that also prices of contracts which mature far into the future are randomly varying, i.e., are volatile. For example, yearly power swaps in the Nordic electricity market maturing in two years have an estimated volatility above 20% (annually) (see Benth *et al.* [25, Sect. 8.6.2]). This estimate is based on a data analysis of all daily swap prices on the Nordic power market ranging over the years from 2001 to 2006. Rather than choosing a generic family of functions $g(x) = \beta \exp(-\alpha x)$ as the model for the Samuelson effect, one could choose

$$g(x) = a + \beta \exp(-\alpha x)$$

for $a, \beta, \alpha \in \mathbb{R}_+$. We focus here on the univariate case where $d = 1$. Then, asymptotically, we reach the fixed level a as a volatility scale in the long end of the market. As the constant a disappears after differentiation, we also find that $g \in H_w$ for appropriately chosen weight functions w. Since the noise $\mathcal{Y}^{1/2}(t)dW(t) \in H_w$ is constant in the long end, we will have that for large values of x, the forward prices will randomly fluctuate according to the variance of $a\delta_x\mathcal{Y}^{1/2}(t)dW(t)$ for x large. Without stochastic volatility, say, this becomes $a^2\langle Q_W h_\infty, h_\infty\rangle_w dt > 0$ (see definition of h_∞ in (3.5) and Lemma 3.4).

6.4 Markovian HJM Models

Rather than specifying explicit HJM-models, we can formulate a class of stochastic partial differential equations (SPDEs) for the forward price $\mathbf{f}(t)$. To this end, consider the SPDE

$$d\mathbf{f}(t) = \nabla_x \mathbf{f}(t)dt + \mu(t, \mathbf{f}(t))dt + \mathbf{\Sigma}(t, \mathbf{f}(t))d\mathbf{W}(t) + \int_H \mathbf{\Gamma}(t, \mathbf{f}(t-), \mathbf{z})\widetilde{\mathbf{N}}(d\mathbf{z}, dt) \tag{6.27}$$

where $\mathbf{f}(0) = \mathbf{f}_0 \in H$. Here, $\mathbf{W}$ is a Q_W-Wiener process on H, while $\widetilde{\mathbf{N}}$ is a compensated Poisson random measure on H with Lévy measure denoted by $\nu(d\mathbf{z})$. Furthermore, we have measurable coefficient functions $\mu : \mathbb{R}_+ \times H \to H$, $\mathbf{\Sigma} : \mathbb{R}_+ \times H \to L_{\text{HS}}(H)$ and $\mathbf{\Gamma} : \mathbb{R}_+ \times H \times H \to H$. The coefficient functions are assumed to satisfy the following conditions:

$$t \mapsto |\mu(t, 0)|_H \in L^2_{\text{loc}}(\mathbb{R}_+)$$

$$t \mapsto \|\mathbf{\Sigma}(t, 0)\|_{\text{HS}} \in L^2_{\text{loc}}(\mathbb{R}_+)$$

$$t \mapsto \int_H |\mathbf{\Gamma}(t, 0, \mathbf{z})|^2_H \nu(d\mathbf{z}) \in L^1_{\text{loc}}(\mathbb{R}_+)$$

and, for some function $K \in L^2_{\mathrm{loc}}(\mathbb{R}_+)$,

$$|\mu(t,\mathbf{g})-\mu(t,\mathbf{h})|_H \leq K(t)|\mathbf{g}-\mathbf{h}|_H$$

$$\|\mathbf{\Sigma}(t,\mathbf{g})-\mathbf{\Sigma}(t,\mathbf{h})\|_{\mathrm{HS}} \leq K(t)|\mathbf{g}-\mathbf{h}|_H$$

$$\int_H |\mathbf{\Gamma}(t,\mathbf{g},\mathbf{z})-\mathbf{\Gamma}(t,\mathbf{h},\mathbf{z})|_H^2 \nu(d\mathbf{z}) \leq K(t)^2|\mathbf{g}-\mathbf{h}|_H^2$$

for any $\mathbf{g},\mathbf{h} \in H$.

We refer to (6.27) as a *Markovian* model for the forward price dynamics, as the SPDE is explicitly depending on the state $\mathbf{f}(t)$ as well as time.

Notice that the Poisson random measure can come from a square-integrable Lévy process $\mathbf{L}$ on H, with $\mathbf{W}$ being the continuous martingale part. Indeed, we may pose the equation

$$d\mathbf{f}(t) = \nabla_x \mathbf{f}(t)dt + \mu(t,\mathbf{f}(t))dt + \mathbf{\Sigma}(t,\mathbf{f}(t-))d\mathbf{L}(t)$$

which then becomes, after re-writing using the compensated Poisson random measure (see discussion after Definition 4.1)

$$d\mathbf{f}(t) = \nabla_x \mathbf{f}(t)dt + \mu(t,\mathbf{f}(t))dt + \mathbf{\Sigma}(t,\mathbf{f}(t-))d\mathbf{W}(t) + \int_H \mathbf{\Sigma}(t,\mathbf{f}(t-))\mathbf{z}\widetilde{\mathbf{N}}(d\mathbf{z},dt).$$

In this case, we have that $\mathbf{\Gamma}(t,\mathbf{g},\mathbf{z}) := \mathbf{\Sigma}(t,\mathbf{g})\mathbf{z}$. It is easy to see that the conditions on $\mathbf{\Gamma}$ is verified from the assumptions on $\mathbf{\Sigma}$ and the square-integrability of $\mathbf{L}$ in this case. In passing, we also note that we could choose another state space for the Lévy process (or, the Wiener process and the compensated Poisson random measure), only at the expense of slightly more notation.

We remark that we have based our formulation on the set-up in Section 4.3 of Chapter 4, where the noise is split into a Wiener process and a pure-jump compensated Poisson measure. The C_0-semigroup $\mathcal{S}_t$ associated with ∇_x is given as a diagonal matrix operator in (6.5), where each element on the diagonal is a quasi-contractive semigroup by assumption (recall Section 6.1 of this Chapter). We have the following simple result:

Lemma 6.3 *The C_0-semigroup $(\mathcal{S}_t)_{t\geq 0}$ is quasi-contractive on H.*

Proof As each $\mathcal{S}_t^i$ is quasi-contractive, it holds that $\|\mathcal{S}_t^i\|_{\mathrm{op}} \leq \exp(c_i t)$ for constants $c_i \in \mathbb{R}, i = 1,\dots,d$. Define $c := \max_{1\leq i\leq d} c_i$, and observe that for a given $\mathbf{g} = (g_1,\dots,g_d) \in H$ with $|\mathbf{g}|_H = 1$ it holds,

$$|\mathcal{S}_t\mathbf{g}|_H^2 = \sum_{i=1}^d |\mathcal{S}_t^i g_i|_i^2 \leq \sum_{i=1}^d \|\mathcal{S}_t^i\|_{\mathrm{op}}^2 |g_i|_i^2 \leq e^{2ct}|\mathbf{g}|_H^2 \leq e^{2ct}.$$

The assertion follows. □

Hence, from Theorem 4.3, there exists a unique mild mean-square continuous adapted RCLL (cádlág) solution $(\mathbf{f}(t))_{t\geq 0}$ to (6.27), where

$$\mathbf{f}(t) = \mathcal{S}_t\mathbf{f}_0 + \int_0^t \mathcal{S}_{t-s}\mu(s,\mathbf{f}(s)))ds + \int_0^t \mathcal{S}_{t-s}\mathbf{\Sigma}(s,\mathbf{f}(s))d\mathbf{W}(s) + \int_0^t \int_H \mathcal{S}_{t-s}\mathbf{\Gamma}(s,\mathbf{f}(s-),\mathbf{z})\widetilde{\mathbf{N}}(d\mathbf{z},ds) \tag{6.28}$$

and such that

$$\mathbb{E}[\sup_{t\in[0,T]} |\mathbf{f}(t)|_H^2] < \infty \tag{6.29}$$

for all $T > 0$.

We have the following necessary martingale drift condition on $\mathbf{f}$:

Proposition 6.6 *Assume that* $(\mathbf{f}(t))_{t\geq 0}$ *is the unique mild mean-square continuous adapted RCLL (cádlág) solution to* (6.27) *with* (6.29). *If* $\mu(t,\mathbf{x}) = 0$ *for every* $(t,\mathbf{x}) \in \mathbb{R}_+ \times H$, *then* $t \mapsto \mathbf{F}(t,T), t \leq T$ *is a* $\mathbb{Q}$*-martingale.*

Proof First of all, by (6.29) and the fact that $\delta_{T-t} \in H^*$, $\mathbf{F}(t,T) := \delta_{T-t}\mathbf{f}(t)$ has finite variance for all $t \leq T$, and hence is a square-integrable process. Since, by assumption on μ and the fact that $\delta_x\mathcal{S}_y = \delta_{x+y}$,

$$\begin{aligned}\mathbf{F}(t,T) &= \delta_{T-t}\mathcal{S}_t\mathbf{f}_0 + \int_0^t \delta_{T-t}\mathcal{S}_{t-s}\mu(s,\mathbf{f}(s))ds + \int_0^t \delta_{T-t}\mathcal{S}_{t-s}\mathbf{\Sigma}(s,\mathbf{f}(s))d\mathbf{W}(s)\\ &\quad + \int_0^t \int_H \delta_{T-t}\mathcal{S}_{t-s}\mathbf{\Gamma}(s,\mathbf{f}(s-),\mathbf{z})\widetilde{\mathbf{N}}(d\mathbf{z},ds)\\ &= \delta_T\mathbf{f}_0 + \int_0^t \delta_{T-s}\mathbf{\Sigma}(s,\mathbf{f}(s))d\mathbf{W}(s) + \int_0^t \int_H \delta_{T-s}\mathbf{\Gamma}(s,\mathbf{f}(s-))\widetilde{\mathbf{N}}(d\mathbf{z},ds),\end{aligned}$$

it follows that $t \mapsto \mathbf{F}(t,T), t \leq T$ is a true $\mathbb{Q}$-martingale. Thus, the assertion is proved. □

We note from the proof above that $\mathbf{F}(0,T) = \delta_T\mathbf{f}_0$. But, in general we do not have that $\delta_{T-s}\mathbf{\Sigma}(s,\mathbf{f}(s))$ (and likewise for the jump term) is a function of $\mathbf{F}(s,T)$ and therefore the dynamics of $\mathbf{F}(t,T)$ is not a Markovian real-valued process. Indeed, the dynamics of $\mathbf{F}(t,T)$ depends on the whole curve of $\mathbf{f}(t)$ in general.

6.5 Swaps: Forwards with Delivery Period

In many forward markets the spot is not delivered at a fixed point in the future, but over an agreed period. In electricity markets, say, power is delivered to the buyer of a forward contract over a period which can be the coming month, next year or a specific week. Delivery of forward contracts on gas are also settled over specified periods. There are obvious physical reasons for trading in contracts with a delivery period rather than a fixed delivery time in these markets.

Forwards traded on temperature in a given city is another interesting market where delivery takes place over agreed time periods. The accumulation of heating or cooling degree days (HDD or CDD) over a specific month, say, is "delivered" when holding a long position in the temperature futures. Of course, temperature as such cannot be delivered, so the market converts measured temperature into currency.

Sometimes, forward contracts with a delivery period are called *swaps*, as they may be viewed as swapping a fixed price for a floating one (or vice versa). We refer to Benth *et al.* [25] and Benth and Šaltytė Benth [49] for more on the specifics of such contracts in various markets.

We consider swap prices, and show that these can be expressed in terms of bounded linear operators on $\mathbf{f}$, the fixed-delivery forward price specified in the Musiela parametrization. To this end, let $\overline{\mathbf{F}}(t, T_1, T_2)$ be the swap price at time $t \geq 0$ of $d \in \mathbb{N}$ swap contracts with delivery in the period $[T_1, T_2]$. It is supposed that $t \leq T_1 < T_2$.

The swap price can be expressed as the aggregation of fixed delivery forward prices $\mathbf{F}(t, T)$ over the delivery period $T \in [T_1, T_2]$. Hence,

$$\overline{\mathbf{F}}(t, T_1, T_2) = \int_{T_1}^{T_2} \widetilde{\mathbf{a}}(T; T_1, T_2)^\top \mathbf{F}(t, T) dT \tag{6.30}$$

for some measurable vector-valued *aggregation function* $\widetilde{\mathbf{a}}(\cdot; T_1, T_2) : [T_1, T_2] \to \mathbb{R}^d$. Leaving precise assumptions for the well-definedness of (6.30) aside for a moment, we introduce the Musiela notation $x := T_1 - t \geq 0$ for *time to start of delivery* and define

$$\overline{\mathbf{f}}_\lambda(t, x) := \overline{\mathbf{F}}(t, t + x, t + x + \lambda) \tag{6.31}$$

with $\lambda := T_2 - T_1 > 0$ being the *length of delivery*. A simple change-of-variables in (6.30) reveals that

$$\overline{\mathbf{f}}_\lambda(t, x) = \int_x^{x+\lambda} \widetilde{\mathbf{a}}(t + y; t + x, t + x + \lambda)^\top \mathbf{f}(t, y)\, dy.$$

As we shall see in examples below, the relevant aggregation functions $\widetilde{\mathbf{a}}$ are of a form such that

$$\mathbf{a}_\lambda(y - x) := \widetilde{\mathbf{a}}(t + y; t + x, t + x + \lambda)\,, \tag{6.32}$$

for $y \geq x$, i.e., they are homogeneous in $y - x$ as well as not being dependent on current time t. Hence,

$$\bar{\mathbf{f}}_\lambda(t, x) = \int_0^\infty 1_{[0,\lambda]}(y - x)\mathbf{a}_\lambda(y - x)^\top \mathbf{f}(t, y)\, dy, \tag{6.33}$$

where we have used that $1_{[x,x+\lambda]}(y) = 1_{[0,\lambda]}(y - x)$. With (6.33) we have realized the swap price dynamics $\bar{\mathbf{f}}_\lambda(t)$ in a Musiela parametrization as an integral operator acting on $\mathbf{f}(t)$. If the integral operator in (6.33) defines a bounded linear operator on H, we have that $\bar{\mathbf{f}}_\lambda(t) \in H$ for any $\mathbf{f}(t) \in H$.

Let us consider some examples of aggregation functions, focusing on the one-dimensional case $d = 1$. For temperature forwards, one is simply aggregating up the fixed-delivery forward prices leading to

$$\widetilde{a}(T; T_1, T_2) = 1.$$

This implies that $a_\lambda(z) = 1$ as well. In energy markets, like gas and power, one defines the swap price by averaging the fixed-delivery prices. In the forward case, one has

$$\widetilde{a}(T; T_1, T_2) = \frac{1}{T_2 - T_1}$$

while for futures-style swaps there is a margin account leading to

$$\widetilde{a}(T; T_1, T_2) = \frac{e^{-rT}}{\int_{T_1}^{T_2} e^{-rs}\, ds}$$

where $r > 0$ is the risk-free constant interest rate. In the former case, we get $a_\lambda(z) = 1/\lambda$, whereas in the latter futures-style we find

$$a_\lambda(z) = \frac{r}{1 - e^{-r\lambda}} e^{-rz}.$$

We observe that all the examples of a_λ are measurable, bounded and positive functions.

Let us analyse the integral operator in (6.33) when the state space H is formed by the Filipović spaces. For simplicity of the exposition, we let $d = 1$. We have the following proposition:

Proposition 6.7 *Assume $a_\lambda : \mathbb{R}_+ \to \mathbb{R}$ defined in* (6.32) *for $d = 1$ is a bounded and measurable function. Consider the operator $\mathcal{D}_\lambda$ on H_w given by*

$$\mathcal{D}_\lambda(g) = \int_0^\infty 1_{[0,\lambda]}(y - \cdot) a_\lambda(y - \cdot) g(y)\, dy.$$

It holds that $\mathcal{D}_\lambda \in L(H_w)$.

Proof The proof first rewrites the operator $\mathcal{D}_\lambda$ into a multiplication operator and an integral operator on g'. Next, we show that these two linear operators are closed.

Introduce the function A_λ on $\mathbb{R}_+$

$$A_\lambda(z) := \int_0^z a_\lambda(v)\,dv$$

which is well-defined since a_λ is assumed bounded and measurable. Integration-by-parts and the fundamental theorem of calculus for elements in H_w (recall Lemma 3.1) yield, for any $g \in H_w$:

$$\begin{aligned}\mathcal{D}_\lambda(g)(x) &= \int_0^\lambda a_\lambda(z)g(z+x)\,dz \\ &= A_\lambda(\lambda)g(x+\lambda) - \int_0^\lambda A_\lambda(z)g'(z+x)\,dz \\ &= A_\lambda(\lambda)g(x) + \int_0^\infty q_\lambda(y-x)g'(y)\,dy,\end{aligned}$$

where

$$q_\lambda(z) := (A_\lambda(\lambda) - A_\lambda(z))1_{[0,\lambda]}(z)$$

for $z \geq 0$.

Obviously, $g \mapsto A_\lambda(\lambda)g$ is a linear bounded operator on H_w. We show that the linear integral operator $g \mapsto \int_0^\infty q_\lambda(y-\cdot)g'(y)\,dy$ is bounded on H_w by appealing to Proposition 3.2.

We start with proving that part 1 of Proposition 3.2 holds: we need to show that $q_\lambda(\cdot - x)/\sqrt{w} \in L^2(\mathbb{R}_+)$, where we recall w to be the weight function in H_w. But using the assumed boundedness of a_λ and $w^{-1} \in L^1(\mathbb{R}_+)$ (recall Assumption 3.1 in Chapter 3 for the latter)

$$\begin{aligned}\int_0^\infty w^{-1}(y)(q_\lambda(y-x))^2\,dy &= \int_x^{x+\lambda} w^{-1}(y)(\int_{y-x}^\lambda a_\lambda(u)\,du)^2\,dy \\ &\leq c^2 \int_x^{x+\lambda} w^{-1}(y)(\lambda - (y-x))^2\,dy \\ &\leq c^2\lambda^2 \int_0^\infty w^{-1}(y)\,dy < \infty,\end{aligned}$$

for some positive constant c. This shows the first part.

To demonstrate that part 2 of Proposition 3.2 holds, we define the function

$$\xi(x) = \int_x^{x+\lambda} (A_\lambda(\lambda) - A_\lambda(y-x))g'(y)\,dy$$

for $x \geq 0$. Its weak derivative is given by

$$\begin{aligned}\xi'(x) &= \frac{d}{dx}\int_x^{x+\lambda}(A_\lambda(\lambda)-A_\lambda(y-x))g'(y)\,dy \\ &= (A_\lambda(\lambda)-A_\lambda(\lambda))g'(x+\lambda)-(A_\lambda(\lambda)-A_\lambda(0))g'(x) \\ &\quad +\int_x^{x+\lambda}(-A'_\lambda(y-x))(-1)g'(y)\,dy \\ &= -A_\lambda(\lambda)g'(x)+\int_x^{x+\lambda}a_\lambda(y-x)g'(y)\,dy.\end{aligned}$$

We show that the H_w-norm of ξ is finite: by again appealing to the boundedness of a_λ, we find by Cauchy-Schwarz' inequality

$$\begin{aligned}\xi^2(0) &= (\int_0^\lambda\int_y^\lambda a_\lambda(u)\,du g'(y)\,dy)^2 \\ &\leq (\int_0^\lambda\int_y^\lambda |a_\lambda(u)|\,du|g'(y)|\,dy)^2 \\ &\leq c^2\lambda^2(\int_0^\lambda |g'(y)|\,dy)^2 \\ &= c^2\lambda^2(\int_0^\lambda w^{1/2}(y)|g'(y)|w^{-1/2}(y)\,dy)^2 \\ &\leq c^2\lambda^2\int_0^\lambda w^{-1}(y)\,dy|g|_w^2 < \infty.\end{aligned}$$

Next, considering the weak derivative of ξ we find from the triangle inequality that

$$\xi'(x)^2 \leq 2A_\lambda(\lambda)^2 g'(x)^2 + 2(\int_x^{x+\lambda} a_\lambda(y-x)g'(y)\,dy)^2$$

Henceforth, restricting the attention to the second term, we have from boundedness of a_λ, the Cauchy-Schwarz inequality and w being an increasing function,

$$\begin{aligned}\int_0^\infty w(x)(\int_x^{x+\lambda} a_\lambda(y-x)g'(y)\,dy)^2\,dx &\leq \int_0^\infty w(x) \\ &\quad \times (\int_x^{x+\lambda}|a_\lambda(y-x)||g'(y)|\,dy)^2\,dx \\ &\leq c^2\int_0^\infty w(x)(\int_x^{x+\lambda}|g'(y)|\,dy)^2\,dx\end{aligned}$$

$$
\begin{aligned}
&\leq c^2\lambda \int_0^\infty w(x) \int_x^{x+\lambda} g'(y)^2\, dy\, dx \\
&\leq c^2\lambda \int_0^\infty \int_x^{x+\lambda} w(y)g'(y)^2\, dy\, dx \\
&= c^2\lambda \int_0^\infty \int_0^\infty \\
&\quad \times 1_{[x,x+\lambda]}(y)\, dx\, w(y)g'(y)^2\, dy \\
&\leq c^2\lambda^2 |g|_w^2 < \infty.
\end{aligned}
$$

Here, we used Tonelli's Theorem together with the fact that $1_{[x,x+\lambda]}(y) = 1_{[y-\lambda,y]}(x)$ in the two last estimations above. This shows that $\xi \in H_w$, and the proof is complete. □

We recall from the discussion above that typical aggregation functions yield bounded measurable functions a_λ. Additionally, they are positive, a property that is not required by the analysis behind Proposition 6.7. We note that if $\mathbf{a}_\lambda$ is a diagonal matrix with elements being functions a_λ on the diagonal which are bounded and measurable, we can canonically define a bounded linear operator on $H_w \times \cdots \times H_w$ such that $\bar{\mathbf{f}}_\lambda(t)$ is an $H_w \times \cdots \times H_w$-valued dynamics given by $\mathbf{f}(t)$. Indeed, we can use different aggregation functions in the different coordinates, to build up a joint multivariate forward dynamics including power, gas and temperature, say. With a bit of additional notational effort, we can also include different delivery periods across the different coordinates.

We show that $\mathcal{D}_\lambda$ commutes with the shift semigroup $\mathcal{S}_t$ in the case of H as above:

Lemma 6.4 *For any $t \geq 0$ it holds that $\mathcal{D}_\lambda \mathcal{S}_t = \mathcal{S}_t \mathcal{D}_\lambda$.*

Proof Fix $t \geq 0$. For any $x \in \mathbb{R}_+$, we have for $\mathbf{g} \in H$

$$
\begin{aligned}
\mathcal{D}_\lambda(\mathcal{S}_t\mathbf{g})(x) &= \int_0^\infty 1_{[0,\lambda]}(y-x)\mathbf{a}_\lambda(y-x)\mathbf{g}(t+y)dy \\
&= \int_0^\infty 1_{[0,\lambda]}(y-(t+x))\mathbf{a}_\lambda(y-(t+x))\mathbf{g}(y)dy \\
&= \mathcal{S}_t\mathcal{D}_\lambda\mathbf{g}(x).
\end{aligned}
$$

Hence, the Lemma follows. □

Now, recalling the Markovian forward dynamics $\mathbf{f}(t)$ with the martingale drift condition in Prop. 6.6, we find

$$\begin{aligned}
\bar{\mathbf{f}}_\lambda(t) &= \mathcal{D}_\lambda(\mathbf{f}(t)) \\
&= \mathcal{D}_\lambda \mathcal{S}_t \mathbf{f}_0 + \int_0^t \mathcal{D}_\lambda \mathcal{S}_{t-s} \boldsymbol{\Sigma}(s, \mathbf{f}(s)) d\mathbf{W}(s) \\
&\quad + \int_0^t \int_H \mathcal{D}_\lambda \mathcal{S}_{t-s} \boldsymbol{\Gamma}(s, \mathbf{f}(s-), \mathbf{z}) \widetilde{\mathbf{N}}(d\mathbf{z}, ds) \\
&= \mathcal{S}_t \mathcal{D}_\lambda \mathbf{f}_0 + \int_0^t \mathcal{S}_{t-s} \mathcal{D}_\lambda \boldsymbol{\Sigma}(s, \mathbf{f}(s)) d\mathbf{W}(s) \\
&\quad + \int_0^t \int_H \mathcal{S}_{t-s} \mathcal{D}_\lambda \boldsymbol{\Gamma}(s, \mathbf{f}(s-), \mathbf{z}) \widetilde{\mathbf{N}}(d\mathbf{z}, ds).
\end{aligned}$$

Hence, $\bar{\mathbf{f}}_\lambda(t)$ is a mild solution of

$$d\bar{\mathbf{f}}_\lambda(t) = \nabla_x \bar{\mathbf{f}}_\lambda(t) dt + \mathcal{D}_\lambda \boldsymbol{\Sigma}(t, \mathbf{f}(t)) d\mathbf{W}(t) + \int_H \mathcal{D}_\lambda \boldsymbol{\Gamma}(t, \mathbf{f}(t-), \mathbf{z}) \widetilde{\mathbf{N}}(d\mathbf{z}, dt). \tag{6.34}$$

This also shows, not unexpectedly, that the swap price process $t \mapsto \overline{\mathbf{F}}(t, T_1, T_2), t \leq T_1$ is a $\mathbb{Q}$-martingale. Indeed, it is a true one as $\mathbf{f}(t)$ is square-integrable and all the operators involved (δ_x and $\mathcal{D}_\lambda$) are bounded.

6.6 Finite Factor HJM Models

The most straightforward and easiest to implement way to specify an HJM model for a single forward price is via a finite factor model. We introduce and analyse such models in this section.

After specifying a finite dimensional stochastic process $\mathbf{X} = (X_1, \ldots, X_n)^\top$ and finitely many continuous curves $g_1, \ldots, g_n : \mathbb{R}_+ \to \mathbb{R}$ as well as a continuous base curve $g_0 : \mathbb{R}_+ \to \mathbb{R}$, the model is given as (note that we model only one forward here, so that $d = 1$):

$$f(t, x) = g_0(t + x) + \sum_{i=1}^{n} X_i(t) g_i(x), \quad t, x \geq 0. \tag{6.35}$$

This has an obvious extension to several forward prices where the dynamics of the factor processes would be allowed to be cross-dependent. However, the generalisation results into the same conditions for each forward curve separately with no extra condition between forward curves of different assets. Thus, to simplify the exposition we continue with the single forward case.

Basically, the main result of this section is that a factor model (6.35) is risk neutral if and only if it satisfies

$$\mathbf{g}(x) = \exp(xA)\mathbf{c}, \quad x \geq 0$$

where $A \in \mathbb{R}^{n\times n}$, $\mathbf{c} \in \mathbb{R}^n$, $\mathbf{g}(x) := (g_1(x), \dots, g_n(x))^\top$ and the process $\mathbf{X}$ satisfies

$$\mathbf{X}(t) = \mathbf{X}(0) + \int_0^t (A^\top \mathbf{X}(s) + \boldsymbol{\gamma}(s))ds + \mathbf{M}(t), \quad t \geq 0$$

where $\mathbf{M}$ is a martingale and $\boldsymbol{\gamma}$ any $\mathbb{R}^n$-valued progressively measurable process with $\sum_{j=1}^n \gamma_j(t)g_j(x) = 0$ for any $t, x \geq 0$ and g_0 is simply any continuous curve. Usually, the latter condition on $\boldsymbol{\gamma}$ enforces that it is constant zero and if it does not, then it enforces that $\boldsymbol{\gamma}$ has no effect on $f(t)$ (cf. Remark 6.2 below).

The corresponding spot model is obtained via

$$S(t) := f(t,0) = g_0(t) + \sum_{j=1}^n X_j(t)c_j, \quad t \geq 0 \tag{6.36}$$

which is slightly more general than the arithmetic spot model discussed in Section 5.1 of Chapter 5, see Equation (5.4). In Benth and Krühner [36] it has essentially been shown that these models are dense in the set of all (possibly infinite dimensional) models.

Throughout this section we assume that $X_1, \dots, X_n$ are driven by a common mean zero and square integrable Lévy process (or Brownian motion) $\mathbf{L}$ of some dimension m and that

$$dX_i(t) = \beta_i(t)dt + \boldsymbol{\sigma}_i(t)d\mathbf{L}(t), \quad i = 1, \dots, n \tag{6.37}$$

where β_i is an $\mathbb{R}$-valued progressively measurable with locally integrable paths and $\boldsymbol{\sigma}_i$ is an $\mathbb{R}^m$-valued progressively measurable processes with $\int_0^t \mathbb{E}[|\boldsymbol{\sigma}_i(t)|^2]dt < \infty$. Here, $|\cdot|$ is the canonical Euclidean norm in $\mathbb{R}^m$.

We start by recovering the forward price dynamics for contracts with delivery periods and the spot dynamics. The spot price dynamics is immediately recovered as the arithmetic spot model. The forwards with delivery period turn out to be of relatively simple structure as well.

Proposition 6.8 *Given the forward price dynamics in* (6.35). *We have*

$$S(t) := f(t,0) = g_0(t) + \sum_{i=1}^n X_i(t)g_i(0), \quad t \geq 0$$

and

$$
\begin{aligned}
F(t, T_1, T_2) &:= \frac{1}{T_2 - T_2} \int_{T_1}^{T_2} f(t, u-t) du \\
&= F_0(T_1, T_2) + \sum_{j=1}^{n} X_j(t) F_j(T_1 - t, T_2 - t), \quad t \in [0, T_1] \qquad (6.38)
\end{aligned}
$$

where

$$
F_j(T_1, T_2) := \frac{1}{T_2 - T_1} \int_{T_1}^{T_2} g_j(x) dx
$$

for any $j = 0, \dots, n$, $0 < T_1 < T_2$.

Proof $x \mapsto f(t, x)$ is continuous (hence locally integrable) for any fixed $t \geq 0$ by assumption. Linearity of the integral yields the claims. □

Remark 6.1 In the previous proposition it is possible that $g_i(0) = 0$ for some i. In that case, the corresponding factors X_i do not play any role for the spot price dynamics but they do play a role for futures price dynamics. Their increments are corresponding to market changes that do not have an immediate effect on operations right now, and hence no immediate effect on the spot, but are latent in the system.

Björk [58] observed that arbitrage-free factor term structure models for interest rate markets have curves of a very specific structure. The same holds by similar arguments in the present setup under some technical assumptions. In particular, we will see that the curves $g_1, \dots, g_n$ are infinitely differentiable.

Proposition 6.9 *Assume that the functions* $g_0, \dots, g_n$ *are absolutely continuous and denote by* $g_0', \dots g_n'$ *versions of the respective corresponding derivatives. Assume that for any* $x \geq 0$ *we have that* f *is a solution to the SDE*

$$
df(t, x) = \partial_x f(t, x) dt + dM^x(t)
$$

for some local martingale M^x. *Finally, we assume the following non-degeneracy condition, i.e. there is no affine space* $V \subsetneq \mathbb{R}^n$ *with* $P(\forall t \geq 0 : X(t) \in V) = 1$.

Then the following holds

(i) $\mathbf{g}(x) = \exp(Ax)\mathbf{g}(0)$ *for some* $A \in \mathbb{R}^{n \times n}$ *and any* $x \geq 0$, *where* $\mathbf{g} = (g_1, \dots, g_n)$. *(A does not depend on x).*

(ii) g_i *is infinitely differentiable for any* $i = 1, \dots, d$.

(iii) There is an $\mathbb{R}^n$*-valued progressively measurable process* $\boldsymbol{\gamma}$ *such that*

$$
\boldsymbol{\beta}(t) = A^\top \mathbf{X}(t) + \boldsymbol{\gamma}(t)
$$

and $\sum_{i=1}^{n} \gamma_i(t) g_i(x) = 0$ *for any* $t, x \geq 0$. *(A is the same as in (i).)*

(iv) $(F(t, T_1, T_2))_{t \in [0, T_1]}$ *is a martingale for any* $0 < T_1 < T_2$.

Proof By definition of f in (6.35) and (6.37) we have

$$df(t,x) = g_0'(t+x)dt + \sum_{i=1}^{n} g_i(x)\beta_i(t)dt + \sum_{i=1}^{n} g_i(x)\boldsymbol{\sigma}_i(t)d\mathbf{L}(t).$$

On the other hand,

$$\partial_x f(t,x) = g_0'(x+t) + \sum_{i=1}^{n} g_i'(x)X_i(t).$$

Comparing with the assumed dynamics for $f(t,x)$ we find that

$$\sum_{i=1}^{n} g_i(x)\beta_i(t) = \sum_{i=1}^{n} g_i'(x)X_i(t),$$

$$\sum_{i=1}^{n} g_i(x)\boldsymbol{\sigma}_i(t)d\mathbf{L}(t) = dM^x(t).$$

From the first equality we see that g_j is the solution to an implicit linear first order differential equation. The non-degeneracy condition allows to rewrite the linear ODE as an explicit equation and, hence, $\mathbf{g}$ is of the form claimed in Claim (i) for some matrix $A \in \mathbb{R}^{n\times n}$. Claim (ii) follows from Claim (i).

Denote by $\langle\cdot,\cdot\rangle$ the scalar product in $\mathbb{R}^n$. We have with $\mathbf{g}'(x) := (g_1'(x),\ldots,g_n'(x))$ that

$$\langle A^\top \mathbf{X}(t), \mathbf{g}(x)\rangle = \langle \mathbf{X}(t), A\mathbf{g}(x)\rangle = \langle \mathbf{X}(t), \mathbf{g}'(x)\rangle = \langle \boldsymbol{\beta}(t), \mathbf{g}(x)\rangle$$

for any $t, x \geq 0$ and, hence, for $\boldsymbol{\gamma}(t) := \boldsymbol{\beta}(t) - A^\top\mathbf{X}(t)$ that $\langle\boldsymbol{\gamma}(t), \mathbf{g}(x)\rangle = 0$. Claim (iii) follows. Finally, we conclude the proof by using equation (6.38) and our findings above to get Claim (iv). □

Remark 6.2 If we define $\mathbf{c} := \mathbf{g}(0) \in \mathbb{R}^n$ and the sub-vector space

$$V := \mathrm{Span}(\mathbf{c}, A\mathbf{c}, \ldots, A^{n-1}\mathbf{c}),$$

then we see that $\mathbf{g}(x) \in V$ for any $x \geq 0$. It is known that in fact $V = \mathrm{Span}(\mathbf{g}(x) : x \geq 0)$. The condition for $\boldsymbol{\gamma}$ in the previous proposition is therefore equivalent to

$$\boldsymbol{\gamma}(t) \in V^\perp, \quad t \geq 0.$$

If $\pi_V : \mathbb{R}^n \to V$ is the orthogonal projection, $\mathbf{Y}(t) := \pi_V(\mathbf{X}(t))$ the orthogonal projection of $\mathbf{X}(t)$ to V and $\mathbf{Z}(t) := \mathbf{X}(t) - \mathbf{Y}(t)$ the compliment, then, with $\boldsymbol{\sigma}$ being the $n \times n$-matrix built by the vectors $\boldsymbol{\sigma}_i$,

$$d\mathbf{Y}(t) = A\mathbf{Y}(t)dt + \pi_V \boldsymbol{\sigma}(t)d\mathbf{L}(t),$$
$$d\mathbf{Z}(t) = \boldsymbol{\gamma}(t)dt + \pi_V^c \boldsymbol{\sigma}(t)d\mathbf{L}(t),$$

and hence,

$$f(t,x) = g_0(x+t) + \sum_{i=1}^{n} X_i(t)g_i(x) = g_0(x+t) + \sum_{i=1}^{n} Y_i(t)g_i(x).$$

This shows that the additional drift $\boldsymbol{\gamma}$ has no direct effect for f, however, components of $\mathbf{Z}$ could possibly feed into the diffusion coefficient.

By possibly choosing a new basis of $\mathbb{R}^n$ one may assume that A is in normal form which simplifies the calculation of the matrix exponential $\exp(Ax)$. The initial function g_0 and values of $g_1(0), \ldots, g_n(0)$ are up to choice as well as the driving noise $\mathbf{L}$ and the coefficient $\boldsymbol{\sigma}$. We are going to see this in Lemma 6.5 below.

Consider next an example of a specification of a factor model:

Example We make the explicit choice that $\mathbf{L}$ is an n-dimensional Brownian motion, $\sigma = I$ and $A = -\lambda I + N$ where I is the identity matrix on $\mathbb{R}^{n\times n}$, $N \in \mathbb{R}^{n\times n}$ is the matrix with entries 1 in the upper side diagonal and 0 entries otherwise and $\lambda \in \mathbb{R}$. In this case, one has

$$\exp(xA) = e^{-\lambda x} \sum_{i=0}^{n-1} \frac{x^i}{i!} N^i.$$

Consequently,

$$g_i(x) = e^{-\lambda x} p_i(x)$$

where p_i is a polynomial of order equal to $i-1$ for $i = 1, \ldots, n$. With the choice $\gamma = 0$ we find that

$$\beta_i(t) = -\lambda X_i(t) + 1X_{i+1}(t), \quad X_{n+1}(t) := 0.$$

The process $\mathbf{X}$ turns out to be a multi-dimensional OU-process where the factor X_i has mean level X_{i+1} and reversion rate λ.

Lemma 6.5 *Let $A \in \mathbb{R}^{n\times n}$, $\mathbf{c} \in \mathbb{R}^n$, $g_0 : \mathbb{R}_+ \to \mathbb{R}$ be continuous, $\mathbf{X}(0) \in \mathbb{R}^n$, $\mathbf{L}$ a mean-zero, square-integrable Lévy process in $\mathbb{R}^m$ and $\boldsymbol{\sigma}_j$ progressively measurable process in $\mathbb{R}^m$ with $\int_0^t |\boldsymbol{\sigma}_j(s)|^2 ds < \infty$ for any $j = 1, \dots, n$, $t \geq 0$. We define for the $n \times m$-matrix $\boldsymbol{\sigma} := (\boldsymbol{\sigma}_1, \dots, \boldsymbol{\sigma}_n)$*

$$\begin{aligned}
\mathbf{g}(x) &:= \exp(xA)\mathbf{c}, \quad x \geq 0, \\
\mathbf{X}(t) &:= \exp(tA^\top)\mathbf{X}(0) + \int_0^t \exp((t-s)A^\top)\boldsymbol{\sigma}(s)d\mathbf{L}(s), \quad t \geq 0, \\
f(t,x) &:= g_0(x+t) + \sum_{i=1}^n X_i(t)g_i(x), \\
\partial_x f(t,x) &:= g_0'(x+t) + \sum_{i=1}^n X_i(t)g_i'(x),
\end{aligned}$$

for $t, x \geq 0$. Then for any $x \geq 0$ we have

$$df(t,x) = \partial_x f(t,x)dt + dM_t^x$$

for some martingale M^x.

Proof Note that $d\mathbf{X}(t) = A^\top \mathbf{X}(t)dt + \boldsymbol{\sigma}(t)d\mathbf{L}(t)$, $t \geq 0$. Consequently, we have

$$\begin{aligned}
df(t,x) &= g_0'(x+t)dt + \langle d\mathbf{X}(t), \mathbf{g}(t)\rangle \\
&= g_0'(x+t)dt + \langle \mathbf{X}(t), \mathbf{g}'(t)\rangle dt + \mathbf{g}(x)\boldsymbol{\sigma}(t)d\mathbf{L}(t) \\
&= \partial_x f(t,x)dt + \mathbf{g}(x)\boldsymbol{\sigma}(t)d\mathbf{L}(t),
\end{aligned}$$

as claimed. □

We like to close this section with a very explicit 5-dimensional factor model for the forward dynamics.

Example We build the model that comes from the previous Lemma with

$$A := \begin{pmatrix} 0 & 0 & 0 & 0 & 0 \\ 0 & -\alpha & 1 & 0 & 0 \\ 0 & 0 & -\alpha & 0 & 0 \\ 0 & 0 & 0 & -\beta & 1 \\ 0 & 0 & 0 & 0 & -\beta \end{pmatrix},$$

(continued)

for some $\alpha, \beta > 0$, $\mathbf{c} := (1, 0, 1, 0, 1)$, $g_0 : \mathbb{R}_+ \to \mathbb{R}$ continuous, $\mathbf{L}$ a 5-dimensional Brownian motion, $\boldsymbol{\sigma}$ the diagonal matrix $\boldsymbol{\sigma} = \operatorname{diag}(\sigma_1, \ldots, \sigma_5)$. Note that

$$\exp(xA) = \begin{pmatrix} 1 & 0 & 0 & 0 & 0 \\ 0 & e^{-x\alpha} & xe^{-x\alpha} & 0 & 0 \\ 0 & 0 & e^{-x\alpha} & 0 & 0 \\ 0 & 0 & 0 & e^{-x\beta} & xe^{-x\beta} \\ 0 & 0 & 0 & 0 & e^{-x\beta} \end{pmatrix},$$

and consequently,

$$\mathbf{g}(x) := \exp(xA)\mathbf{c} = (1, xe^{-\alpha x}, e^{-\alpha x}, xe^{-\beta x}, e^{-\beta x}). \tag{6.39}$$

Also, the 5-dimensional process $\mathbf{X}$ is given as

$$\begin{aligned} dX_1(t) &= \sigma_1 dB_1(t), \\ dX_2(t) &= -\alpha X_2(t) + \sigma_2 dB_2(t), \\ dX_3(t) &= (X_2(t) - \alpha X_3(t))dt + \sigma_3 dB_3(t), \\ dX_4(t) &= -\beta X_4(t) + \sigma_4 dB_4(t), \\ dX_5(t) &= (X_4(t) - \beta X_5(t))dt + \sigma_5 dB_5(t), \end{aligned}$$

for $t \geq 0$. Hence, we have the forward price dynamics

$$\begin{aligned} f(t,x) = g_0(t+x) &+ X_1(t) + X_2(t)xe^{-\alpha x} + X_3(t)e^{-\alpha x} \\ &+ X_4(t)xe^{-\beta x} + X_5(t)e^{-\beta x} \end{aligned} \tag{6.40}$$

for $t, x \geq 0$.

In Figure 6.2 we have plotted the simulated time evolution of this 5-factor model for the HJM-dynamics of power forward prices. In the plot, we have assumed that g_0 is a constant equal to 80, signifying a flat level of 80 Euro/MWh. The two constants α and β are exponential decay rates which we can relate to the speed of mean reversion of the spot price. The half-life, that is, the time the spot reverts on average to half its long-term value, is known to be $\ln(2)/\alpha$ for a speed of mean reversion (see Clewlow and Strickland [72], say, for the definition of the half-life.). In the modeling, we chose $\alpha = 0.5$ and $\beta = 0.005$, corresponding to a half-life of roughly 1.5 days and 138 days, respectively. For the factor processes X_2 and X_3, we assumed a high volatility being 50% annually, and for the slower mean reverting

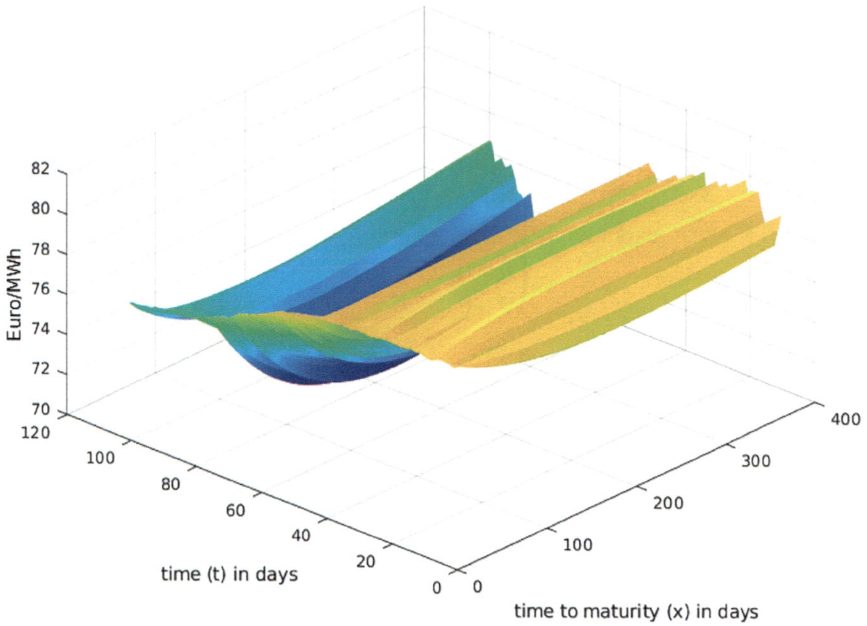

Fig. 6.2 Simulation of the time evolution of forward prices in a 5-factor model

factors X_4 and X_5 20% annually. The level factor X_1 varied with a volatility of 10% annually. We simulated the one year term structure of forward prices for 100 days.

It is worth noticing that the factor model in (6.40) is closely related to the so-called Svensson model for smooth forward rate curves in fixed-income theory (see Svensson [181]). Indeed, the exact match is obtained if X_5 would be zero. The factor model that we consider here a stochastic evolution of the given shape-functions in **g**. We remark that Diebold and Rudebusch [86] have proposed to model the dynamics of forward rates in fixed income by stochastic processes for the parameters in front of the shape functions. This is in line with our model above, and we emphasise that in our framework the dynamics is arbitrage-free. The Nelson-Siegel shape factors $(1, e^{-\alpha x}, xe^{-\alpha x})$ is seen as the three first in our dynamics (see Nelson and Siegel [158] for their smooth forward rate curve model). Additionally, the function g_0 may model seasonal effects in the dynamics.

6.7 Modelling Under the Market Probability and Seasonality

Forward and swap contracts are traded assets in commodity markets. In some, like for example power, these contracts may be both physical and financial, where, in the latter case, delivery is in terms of the money-equivalent of the commodity. Although we have so far only focused on the risk-neutral price dynamics of the forward and swaps, the market dynamics (i.e., the dynamics under the market/objective probability denoted $\mathbb{P}$) is also of interest. In this Section we make some brief remarks on HJM-modeling under the market probability.

To simplify matters, we concentrate our presentation on simple HJM-models of arithmetic type, as defined in (6.3), however, formulated under the market probability $\mathbb{P}$,

$$d\mathbf{f}(t) = \nabla_x \mathbf{f}(t)dt + \boldsymbol{\mu}(t)dt + \boldsymbol{\Sigma}(t)d\mathbf{L}(t). \tag{6.41}$$

I.e., we assume that $\mathbf{L}$ is an H-valued Lévy process under $\mathbb{P}$ which has mean zero and is square-integrable. Furthermore, the integrability condition on the volatility operator $\boldsymbol{\Sigma}$ in (6.2) holds under $\mathbb{P}$, that is, $(\boldsymbol{\Sigma}(t))_{t\geq 0}$ is a predictable stochastic process with values in $L(H)$, satisfying

$$\mathbb{E}\left[\int_0^T \|\boldsymbol{\Sigma}(t)Q^{1/2}\|_{\mathrm{HS}}^2 dt\right] < \infty \tag{6.42}$$

for all $T < \infty$. The drift $(\boldsymbol{\mu}(t))_{t\geq 0}$ is a predictable stochastic process, satisfying the local integrability condition

$$\mathbb{E}[\int_0^T |\boldsymbol{\mu}(t)|_H^2 dt] < \infty, \tag{6.43}$$

for all $T < \infty$.

To this end, a common approach to model seasonal variations consistent with the analysis in Chapter 5 is to define the forward price as

$$\mathbf{f}(t,x) := \boldsymbol{\Lambda}(t+x) + \widetilde{\mathbf{f}}(t,x) \tag{6.44}$$

where $\boldsymbol{\Lambda}$ is a deterministic seasonality function and $\widetilde{\mathbf{f}}(t)$ is the *deseasonalized* forward price. Since, for $x = 0$, $\mathbf{f}(t,0) = \mathbf{S}(t)$ is the spot price of the commodity, $\boldsymbol{\Lambda}$ is the seasonality function of the spot. Indeed, in a market like power, say, one expects a seasonally varying mean spot price. Thinking of Norway, the winter spot prices are on average much higher than the summer, and a simple spot price seasonality function may take the form as a Fourier series of sine and cosine functions with yearly frequency (and possibly other frequencies to explain mean fluctuations on other scales) (see, e.g., Weron [188]). On the other hand, the function $x \mapsto \boldsymbol{\Lambda}(t+x)$ will in this case *not* be an element of the Filipović space, say, as its derivative will also become a Fourier series of sine and cosine functions. For example, in the context of one market, if $\Lambda(t) = \sin(2\pi t/365)$, then $\Lambda'(t+x) = (2\pi/365)\cos(2\pi(t+x)/365)$ and its square is obviously not integrating $w(x)dx$ over $\mathbb{R}_+$ for any function $w(x) \geq 1$.

On the other hand, the deseasonalized forward price dynamics $(\widetilde{\mathbf{f}}(t))_{t\geq 0}$ may be formulated as a stochastic dynamics of the form (6.41),

$$d\widetilde{\mathbf{f}}(t) = \nabla_x \widetilde{\mathbf{f}}(t)dt + \boldsymbol{\mu}(t)dt + \boldsymbol{\Sigma}(t)d\mathbf{L}(t).$$

where the drift $(\boldsymbol{\mu}(t))_{t\geq 0}$ and volatility $(\boldsymbol{\Sigma}(t))_{t\geq 0}$ satisfy the conditions posed above. Note that

$$\mathbf{f}(0, x) = \boldsymbol{\Lambda}(x) + \widetilde{\mathbf{f}}(0, x).$$

So, although we may assume that $\widetilde{\mathbf{f}}(0) = \widetilde{\mathbf{f}}_0 \in H$, $\mathbf{f}(0)$ may fail to be an element in H if $\boldsymbol{\Lambda} \notin H$. From a practical perspective, this still does not pose any real problem as we model the seasonality function $\boldsymbol{\Lambda}$ first, and then focus on the deseasonalized forward price afterwards. As long as the seasonality function is differentiable, we find that the dynamics of $\mathbf{f}$ is given by (6.41), but with an initial condition $\mathbf{f}(0)$ that may not be an element in H. We remark in passing that as long as $\boldsymbol{\Lambda}$ is a continuous function, it makes sense to apply both the shift semigroup and the evaluation operator on $\boldsymbol{\Lambda}$.

With this in view, we have that

$$\mathbf{F}(t, T) = \boldsymbol{\Lambda}(T) + \widetilde{\mathbf{f}}(t, T - t) \tag{6.45}$$

for $t \leq T$. To have an arbitrage-free dynamics, we recall from the arbitrage theory of mathematical finance (see, e.g., Duffie [88]) that there must exist a probability $\mathbb{Q} \sim \mathbb{P}$ such that $t \mapsto \mathbf{F}(t, T)$ for $t \leq T$ is a (local) $\mathbb{Q}$-martingale. But this means that $t \mapsto \widetilde{\mathbf{f}}(t, T - t)$ must be a (local) $\mathbb{Q}$-martingale.

We discuss here the so-called Esscher transform, which is a way to construct measures $\mathbb{Q} \sim \mathbb{P}$ preserving the Lévy property of the process $\mathbf{L}$ in H. It was first introduced by Esscher [95] in risk theory in insurance mathematics, and has become a popular class of measure changes applied in finance (see, e.g., Gerber and Shiu [115]) as well as energy and commodity markets (see Benth, Šaltytė Benth and Koekebakker [25]). For $\theta \in H$, assume that

$$\mathbb{E}\left[\exp\left(\langle \theta, \mathbf{L}(1)\rangle_H\right)\right] < \infty \tag{6.46}$$

i.e., that the real-valued random variable $\langle \theta, \mathbf{L}(1)\rangle_H$ has finite exponential moment. This is equivalent to assuming

$$\int_{\langle \mathbf{z}, \theta\rangle_H > 1} e^{\langle \mathbf{z}, \theta\rangle_H}\, \nu(d\mathbf{z}) < \infty \tag{6.47}$$

for the Lévy measure ν of $\mathbf{L}$. Introduce the real-valued process

$$Z(t) := \exp\left(\langle \theta, \mathbf{L}(t)\rangle_H - \psi(-\mathrm{i}\theta)t\right) \tag{6.48}$$

for $t \leq T < \infty$. Here, ψ is the characteristic exponent of $\mathbf{L}$, which we recall from Peszat and Zabczyk [165, Thm. 4.27] to be

$$\psi(\mathbf{u}) = -\frac{1}{2}|Q_W^{1/2}\mathbf{u}|_H^2 + \int_H e^{\mathrm{i}\langle \mathbf{z},\mathbf{u}\rangle_H} - 1 - \mathrm{i}\langle \mathbf{z},\mathbf{u}\rangle_H \, \nu(d\mathbf{z}) \tag{6.49}$$

for $\mathbf{u} \in H$ in our case of a mean-zero square-integrable Lévy process. By the exponential integrability condition of $\mathbf{L}$, it follows that $Z(t)$ in (6.48) is a strictly positive martingale process with mean 1. Hence, we define a probability $\mathbb{Q} \sim \mathbb{P}$ on $\mathcal{F}_T$ with Radon-Nikodym density process

$$\left.\frac{d\mathbb{Q}}{d\mathbb{P}}\right|_{\mathcal{F}_t} = Z(t) \tag{6.50}$$

for $0 \leq t \leq T < \infty$. This change of probability is called the *Esscher transform*, and in a financial context the parameter θ is referred to as *the market price of risk*. Recall the discussion of the market price of risk in Chapter 5 (Section 5.4), where it was introduced in connection with forward prices derived from spot models.

We next show that $t \mapsto \mathbf{L}(t), t \leq T$ is a Lévy process under the probability $\mathbb{Q}$:

Proposition 6.10 *For $\theta \in H$, assume that* (6.46) *holds for the Lévy process* $\mathbf{L}$. *Then* $(\mathbf{L}(t))_{t\leq T}$ *is a Lévy process with respect to the probability $\mathbb{Q}$ defined in* (6.50). *Moreover, with respect to $\mathbb{Q}$,* $\mathbf{L}$ *has characteristic triplet $(\gamma, Q_W, \nu_\theta)$, where the drift $\gamma \in H$ is*

$$\gamma = Q_W\theta + \int_H \mathbf{z}(e^{\langle \mathbf{z},\theta\rangle_H} - 1)\nu(d\mathbf{z})$$

and

$$\nu_\theta(d\mathbf{z}) = e^{\langle \mathbf{z},\theta\rangle_H}\nu(d\mathbf{z})$$

its Lévy measure.

Proof We consider the conditional characteristic exponent of $\mathbf{L}$ with respect to $\mathbb{Q}$: for $T \geq t > s \geq 0$ and $\mathbf{u} \in H$, it follows from the Bayes formula for conditional expectation and the independent increment property of $\mathbf{L}$ that

$$\begin{aligned}
&\mathbb{E}_{\mathbb{Q}}\left[\exp\left(\mathrm{i}\langle \mathbf{u}, \mathbf{L}(t)-\mathbf{L}(s)\rangle_H\right) | \mathcal{F}_s\right] \\
&= \mathbb{E}\left[\exp\left(\mathrm{i}\langle \mathbf{u}, \mathbf{L}(t)-\mathbf{L}(s)\rangle_H\right)\frac{Z(t)}{Z(s)} | \mathcal{F}_s\right] \\
&= \mathbb{E}\left[\exp\left(\mathrm{i}\langle \mathbf{u}-\mathrm{i}\theta, \mathbf{L}(t)-\mathbf{L}(s)\rangle_H\right) | \mathcal{F}_s\right]\exp\left(\psi(-\mathrm{i}\theta)(t-s)\right) \\
&= \exp\left((\psi(\mathbf{u}-\mathrm{i}\theta) - \psi(-\mathrm{i}\theta)(t-s)\right) \\
&= \mathbb{E}_{\mathbb{Q}}\left[\exp\left(\mathrm{i}\langle \mathbf{u}, \mathbf{L}(t)-\mathbf{L}(s)\rangle_H\right)\right]
\end{aligned}$$

where Z is defined in (6.48) and ψ in (6.49). This shows that the increments $\mathbf{L}(t) - \mathbf{L}(s)$ are independent of $\mathcal{F}_s$ and that their distributions are stationary, hence a Lévy process. Moreover, since we have

$$\psi(\mathbf{u} - \mathrm{i}\theta) - \psi(-\mathrm{i}\theta) = \mathrm{i}\langle Q_W\theta, \mathbf{u}\rangle_H + \mathrm{i}\int_H \langle \mathbf{z}, \mathbf{u}\rangle_H (e^{\langle \mathbf{z},\theta\rangle_H} - 1)\nu(d\mathbf{z})$$
$$- \frac{1}{2}|Q_W^{1/2}\mathbf{u}|_H^2 + \int_H (e^{\mathrm{i}\langle \mathbf{z},\mathbf{u}\rangle_H} - 1 - \mathrm{i}\langle \mathbf{z}, \mathbf{u}\rangle_H)e^{\langle \mathbf{z},\theta\rangle_H}\nu(d\mathbf{z})$$

the result follows. □

The Esscher transform is parametric in θ, the market price of risk. As θ is an element of H, we may define a market price of risk which varies with time to maturity, that is, a term structure of the market price of risk. On the other hand, in our framework θ is assumed not to depend on time. A time-dependent θ would violate the stationarity of increments property under $\mathbb{Q}$, as seen in the proof of Prop. 6.10 above. This would mean that $\mathbf{L}$ is not a Lévy process under $\mathbb{Q}$, but more generally a so-called independent increment process (see Jacod and Shiryaev [128] and Sato [175] for a general treatment of such processes in finite dimensional case, and Benth, Šaltytė Benth and Koekebakker [25] for an application to energy markets). We do not go into more detail on independent increment processes here, and focus on time-constant market prices of risk for general Esscher transforms.

From Prop. 6.10 we see that when $\mathbf{L}$ is a Wiener process under $\mathbb{P}$, i.e., there are no jumps and thus $\nu = 0$, $\mathbf{L}$ is a drifted Wiener process with respect to $\mathbb{Q}$. In fact, there is a Q_W-Wiener process $\widetilde{W}$ with respect to $\mathbb{Q}$ such that

$$\mathbf{W}(t) = \widetilde{\mathbf{W}}(t) + tQ_W\theta.$$

The Esscher transform therefore reduces to the Girsanov transform for Wiener processes. In Da Prato and Zabczyk [81, Thm. 10.14], a general Girsanov transform for H-valued Wiener processes are proven, with the market price of risk θ being a predictable stochastic process with values in H, satisfying the integrability condition

$$\mathbb{E}\left[\exp\left(\int_0^T \langle\theta(s), d\mathbf{W}(s)\rangle_H - \frac{1}{2}\int_0^T |Q_W^{1/2}\theta(s)|_H^2 ds\right)\right] = 1$$

This condition ensures the martingale property of the Radon-Nikodym density process. Thus, in the no-jump case we may easily expand the market price of risk beyond a non-random, time-independent case. The Esscher transform is structure-preserving for Lévy processes in general for constant market prices of risk only, as we have considered above.

Let us return to the question of an arbitrage-free dynamics of (6.45), which we recall is related to $t \mapsto \widetilde{\mathbf{f}}(t, T-t)$, $t \leq T$ being a $\mathbb{Q}$-martingale for some probability $\mathbb{Q} \sim \mathbb{P}$. We find the following sufficient condition:

Lemma 6.6 *For $\theta \in H$, assume that the exponential integrability condition* (6.46) *holds for* $\mathbf{L}$ *and let* $\mathbb{Q} \sim \mathbb{P}$ *be defined as the Esscher transform with market price of risk* $\theta \in H$ *and density process in* (6.50). *If*

$$\boldsymbol{\mu}(t) + \boldsymbol{\Sigma}(t)\mathbb{E}_{\mathbb{Q}}[\mathbf{L}(1)] = 0$$

for all $t \leq T$, *then* $t \mapsto \widetilde{\mathbf{f}}(t, T-t)$ *is a* $\mathbb{Q}$*-martingale.*

Proof It holds for $t \leq T$ that $\delta_{T-t}\widetilde{\mathbf{f}} = \widetilde{\mathbf{f}}(t, T-t)$ and using $\delta_t \mathcal{S}_x = \delta_0 \mathcal{S}_{t+x}$ it follows

$$\delta_{T-t}\widetilde{\mathbf{f}}(t) = \delta_T \widetilde{\mathbf{f}}_0 + \delta_0 \int_0^t \mathcal{S}_{T-v}\boldsymbol{\mu}(v)dv + \delta_0 \int_0^t \mathcal{S}_{T-v}\boldsymbol{\Sigma}(v)d\mathbf{L}(v).$$

By the condition of exponential integrability, the Esscher transform is well-defined and so is the conditional expectation under $\mathbb{Q}$ of $\delta_{T-t}\widetilde{\mathbf{f}}(t)$ with respect to $\mathcal{F}_s$, $T \geq t \geq s \geq 0$. But, by the Lévy property of $\mathbf{L}$ under the probability $\mathbb{Q}$,

$$\mathbb{E}_{\mathbb{Q}}[\delta_{T-t}\widetilde{\mathbf{f}}(t) \,|\, \mathcal{F}_s] = \delta_{T-s}\widetilde{\mathbf{f}}(s) + \delta_0 \int_s^t \mathcal{S}_{T-v}\boldsymbol{\mu}(v)dv + \delta_0 \int_s^t \mathcal{S}_{T-v}\boldsymbol{\Sigma}(v)\mathbb{E}_{\mathbb{Q}}[\mathbf{L}(1)]dv$$

and the result follows. □

We find from the $\mathbb{Q}$-characteristics of $\mathbf{L}$ in Proposition 6.10 that

$$\mathbb{E}_{\mathbb{Q}}[\mathbf{L}(1)] = Q_W \theta + \int_H \mathbf{z}(e^{\langle \mathbf{z}, \theta \rangle_H} - 1)\nu(d\mathbf{z})$$

Thus, the relationship between the drift μ and the volatility $\boldsymbol{\Sigma}$ must be

$$\boldsymbol{\mu}(t) = -\boldsymbol{\Sigma}(t)\left(Q_E \theta + \int_H \mathbf{z}(e^{\langle \mathbf{z}, \theta \rangle_H} - 1)\nu(d\mathbf{z})\right)$$

for some $\theta \in H$.

One purpose of having the dynamics of $\mathbf{f}(t)$ available for the risk-neutral probability $\mathbb{Q}$ is option pricing. If we allow for a seasonality in the model, we find for a call option with strike K at exercise time $\tau \leq T$ the payoff

$$\max(F(\tau, T) - K, 0) = \max(\widetilde{f}(\tau, T-\tau) - (K - \Lambda(T)), 0)$$

I.e., the seasonality $\Lambda(T)$ can be moved over to the strike, and we can study the option price of a contract on $\widetilde{f}$. We analyse option prices in detail in Chapter 7.

Let us discuss the *risk premium* implied by the forward models. The risk premium, as briefly discussed in Section 5.2 of Chapter 5, is defined as the difference between the forward price and the predicted spot price at delivery, i.e., the risk premium $\mathbf{R}_p(t,T)$ at time $t \geq 0$ for a contract delivering at time $T \geq t$ is given by

$$\mathbf{R}_p(t,T) := \mathbf{F}(t,T) - \mathbb{E}[\mathbf{F}(T,T)\,|\,\mathcal{F}_t] \tag{6.51}$$

where we recall that $\mathbf{F}(T,T) = \mathbf{S}(T)$, the spot price at time T. This definition gives us a d-dimensional vector of risk premia, one for each market under consideration. But, using the representation in (6.45), we need to compute the conditional expectation of $\delta_0\widetilde{\mathbf{f}}(T)$ given $\mathcal{F}_t$. From the $\mathbb{P}$-dynamics of $\widetilde{\mathbf{f}}(t)$, we find

$$\begin{aligned}\delta_0\widetilde{\mathbf{f}}(T) &= \delta_0\mathcal{S}_{T-t}\widetilde{\mathbf{f}}(t) + \int_t^T \delta_0\mathcal{S}_{T-s}\boldsymbol{\mu}(s)ds + \int_t^T \delta_0\mathcal{S}_{T-s}\boldsymbol{\Sigma}(s)d\mathbf{L}(s)\\ &= \widetilde{\mathbf{f}}(t, T-t) + \int_t^T \delta_{T-s}\boldsymbol{\mu}(s)ds + \int_t^T \delta_{T-s}\boldsymbol{\Sigma}(s)d\mathbf{L}(s).\end{aligned}$$

The stochastic integral has zero conditional mean by its martingale property, and hence by the Fubini Theorem

$$\mathbf{R}_p(t,T) = -\int_t^T \delta_{T-s}\mathbb{E}[\boldsymbol{\mu}(s)\,|\,\mathcal{F}_t]ds. \tag{6.52}$$

The drift μ models the risk premium, and from the negative sign in front it follows that for $\mathbf{R}_p(t,T) < 0$ whenever $\delta_x\boldsymbol{\mu}(s) \geq 0$, i.e., the drift is positive (this is a sufficient, but not necessary condition for a negative risk premium, of course). A negative premium is the typical situation in commodity markets, as this refers to producers inducing a hedging pressure to secure the prices of their production, and is therefore willing to pay a premium for this. The premium is in terms of accepting lower prices for their production compared to what they expect to get in the spot market. As earlier indicated, in the power market also the consumers might want to hedge future price risk, in particular upward spikes, and thereby induce a *positive premium*. This change in sign across the term structure of the risk premium in power market was first observed and explained by Geman and Vašíček [114] for the PJM power market in the US, and has later been investigated empirically and theoretically from various perspectives (see, e.g., Bessembinder and Lemon [57] and Benth, Cartea and Kiesel [26] and Veraart and Veraart [186]).

Moreover, in power market as well as gas and also markets for temperature derivatives, swaps are the traded objects. The risk premium will naturally be defined as the difference between the swap price and the expected payment from spot over the delivery period. The definition of the risk premium is in this case changed to

$$\widehat{\mathbf{R}}(t,T_1,T_2) = \frac{1}{T_2 - T_1}\int_{T_1}^{T_2} \mathbf{R}_p(t,T)dT.$$

This is the definition when the swap has aggregation function $\widetilde{a}(T; T_1, T_2) = 1/(T_2 - T_1)$ as we discussed in Section 6.5 of this Chapter. Other cases are obtained by an appropriate change of the aggregation function.

We end this section with a brief remark on geometric models and seasonality under the market probability $\mathbb{P}$. In this case it would be natural to model the seasonality as

$$\mathbf{f}(t, x) = \mathbf{\Lambda}(t + x)\widetilde{\mathbf{f}}(t, x)$$

where $\widetilde{\mathbf{f}}(t) = \exp(\widetilde{\mathbf{g}})$ and the multiplication between $\mathbf{\Lambda}(t + x)$ and $\widetilde{\mathbf{f}}(t, x)$ is coordinate-wise. We model $\widetilde{\mathbf{g}}$ as in (6.11), with the same conditions except formulated under $\mathbb{P}$. Also in this geometric model case we see that the condition to have a no-arbitrage model becomes a (local) martingale condition on $\exp(\widetilde{\mathbf{g}}(t, T - t))$ for some probability $\mathbb{Q} \sim \mathbb{P}$. The risk premium may be more conveniently stated on a logarithmic scale, as the difference $\ln \mathbf{F}(t, T) - \ln \mathbb{E}[\mathbf{S}(T) \,|\, \mathcal{F}_t]$. This brings the study of the risk premium essentially back to the arithmetic case.

6.8 The Initial Forward Curve: Smoothing of the Term Structure

The initial forward curve f_0 is an ingredient in the HJM-models we have proposed and analysed in this Chapter. What we observe in the market is the term structure of forward prices, i.e., $F_0(T_1), \ldots, F_0(T_k)$ for $k \in \mathbb{N}$ maturities $T_1 < T_2 < \ldots < T_k$. In this Section we discuss how one can create a smooth curve $x \mapsto f_0(x)$ from these observed forward prices.[1]

The classical approach to obtain a smooth initial forward curve is to fit a parametric function, with the Nelson-Siegel function as a typical choice (see Nelson and Siegel [158]). We recall the Nelson-Siegel function from (3.2) as

$$f_{\mathrm{NS}}(x) = \beta_0 + \beta_1 e^{-\gamma x} + \beta_2 x e^{-\gamma x}$$

for $x \geq 0$, and parameters $\beta_0, \beta_1, \beta_2 \in \mathbb{R}$ and $\gamma \in \mathbb{R}_+$. For the initial time $t = 0$, we observe that time to maturity coincides with time of maturity, i.e., $x = T$. The Nelson-Siegel function can be fitted to actual forward price observations by a nonlinear regression of the data, meaning that we minimize the distance between the function and the data. This will result in a representation of the initial forward curve by the fitted Nelson-Siegel function by $f_0(x) = f_{\mathrm{NS}}(x)$ for $x \geq 0$. However, typically $f_{\mathrm{NS}}(T_k) \neq F_0(T_k)$.

Indeed, consider 25 futures prices of crude oil observed at NYMEX on December 30, 2022.[2] The $k = 25$ maturities are monthly, starting with delivery in February

[1] To simplify the exposition, we let $d = 1$ here.

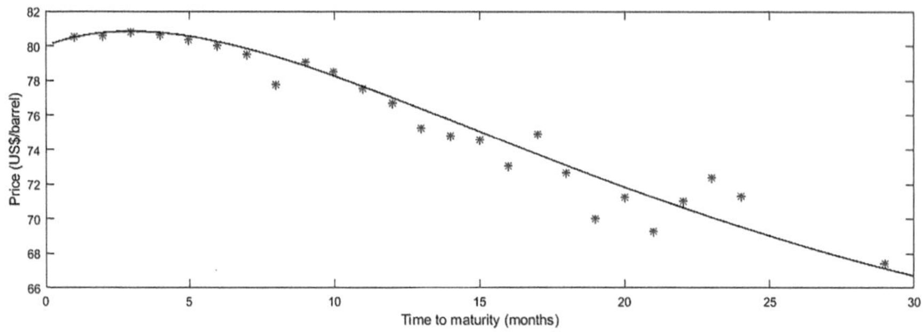

Fig. 6.3 Futures prices of NYMEX crude oil on December 30 2022 with the fitted Nelson-Siegel curve (solid line)

2023 and ranging to January 2025. The last maturity T_{25} is July 2025. The prices denoted in US$/barrel are plotted (as red stars) in Figure 6.3. In the same figure we have also included a Nelson-Siegel curve with parameters $\beta_0 = 60$, $\beta_1 = 20$, $\beta_2 = 2.333$ and $\gamma = 0.0864$. These parameters where chosen by matching the hump, spot and long-term level. As is evident from the figure, the Nelson-Siegel curve provides a reasonable representation of the term structure of prices, however, not matching the actual prices observed in the market.

From the definition of the Nelson-Siegel function $f_{\mathrm{NS}}(x)$, we have that

$$f_{\mathrm{NS}}(0) = \beta_0 + \beta_1$$

whereas

$$\lim_{x \to \infty} f_{\mathrm{NS}}(x) = \beta_0 .$$

For the crude oil prices, this results in a spot price equal to $\beta_0 + \beta_1 = 80$US$/barrel and long-term price level of $\beta_0 = 60$US$/barrel. A simple calculation reveals that the function has a hump located at maturity

$$\widehat{T} = \frac{1}{\gamma}\left(1 - \frac{\beta_1 \gamma}{\beta_2}\right)$$

when $0 < \beta_1 \gamma < \beta_2$. In our empirical example this results in a hump at maturity equal to three months, where the price is given as 80.8US$/barrel.

If one has available swap prices, which is the case in gas and power markets, say, one can modify the Nelson-Siegel function by simply averaging over the delivery periods in the calibration. For instance, suppose we have available $k \in \mathbb{N}$ observed

[2] The data are downloaded from finance.yahoo.com. We do not make any distinction between forward and futures here, but treat them as the same asset class.

power swap prices

$$F_0(T_1^1, T_2^2), F_0(T_2^1, T_2^2), \ldots, F_0(T_k^1, T_k^2)$$

for delivery over the time intervals $[T_j^1, T_j^2]$, $T_j^1 < T_j^2$, $j = 1, \ldots, k \in \mathbb{N}$. We recover $f_0(x)$ in this case by, say, minimizing the squared distances

$$\sum_{j=1}^{k} |F_0(T_j^1, T_j^2) - \overline{f_{\mathrm{NS}}}(T_j^1, T_j^2)|^2$$

over the parameters $\beta_0, \beta_1, \beta_2$ and γ, where

$$\overline{f_{\mathrm{NS}}}(T_j^1, T_j^2) = \frac{1}{T_j^2 - T_j^1} \int_{T_j^1}^{T_j^2} f_{\mathrm{NS}}(u)\, du.$$

Notice that the integral can be analytically calculated.

We remark in passing that the Svensson model [181], being a very popular choice in fixed-income market modeling, is an extension of the Nelson-Siegel function. The Svensson model is adding a second exponential term and thereby provides more flexibility in capturing various shapes of the term structure (cf. Section 6.6 in this chapter).

As mentioned, the Nelson-Siegel curve only approximates the observed prices, but does not (in general) *interpolate* the data. This may be an undesirable property. Benth, Koekebakker and Ollmar [33] propose a method based on polynomial spline interpolation of electricity forward prices (swaps) observed in the Nordic electricity market. Their approach is assuming a maximally smooth forward price curve, which is interpolating the data, and it is shown that polynomial splines of fourth order solve the problem. The idea is to split the representation of the forward curve into a seasonal term structure and an adjustment term which is given by the spline. In view of the case study of oil prices above, one can substitute the seasonality function by the Nelson-Siegel function. We refer to Kiesel, Paraschiv and Sætherø [139] for a recent application to hourly price forward curves.

Splitting the smooth initial forward curve into a seasonality or mean function and an adjustment term which ensures data interpolation, is also the starting point for the so-called *kriging* technique. Kriging is a popular method in geostatistical prediction (see Cressie [77] and Cressie and Wikle [78]), but has also been applied to forward curve smoothing (see Benth [23] for applications to oil and gas markets). The method assumes that the observed forward price data are observations from a Gaussian random field, where the covariance operator is known. By conditioning on the data, one obtains a new conditional Gaussian random field where the data are interpolated. We now explain the kriging technique in more detail.

The following small exposition is based on so-called *simple kriging* (see Cressie [77, Ch. 3]). To this end, assume that f is a Gaussian random field on $x \in \mathbb{R}$, given by

$$f(x) = m(x) + Y(x) \tag{6.53}$$

where m is some deterministic mean function and Y is a zero-mean stationary Gaussian random field. The mean function can be taken as the Nelson-Siegel function fitted to the data, $m(x) := f_{\text{NS}}(x)$. The kriged forward curve is now given as

$$f_{\text{K}}(x) := \mathbb{E}[f(x) \mid f(T_1) = F_0(T_1), \ldots, f(T_k) = F_0(T_k)] \tag{6.54}$$

which, after some calculations appealing to standard results on conditioning Gaussian random variables, has the explicit expression (see Cressie [77] for details)

$$f_{\text{K}}(x) = m(x) + c_x^\top C^{-1}\widehat{y}. \tag{6.55}$$

Here, C is the $k \times k$ correlation matrix with elements $(C_{ij})_{i,j=1}^k$

$$C_{ij} = \text{Corr}(Y(T_i), Y(T_j))$$

and the vector $c_x \in \mathbb{R}^k$ has coordinates given by $\text{Corr}(Y(x), Y(T_i))$, $i = 1, \ldots, k$. Finally, $\widehat{y}$ is the k-dimensional (column) vector with coordinates $F_0(T_i) - m(T_i)$, $i = 1, \ldots, k$. We notice that $f_{\text{K}}(T_i) = F_0(T_i)$ for all $i = 1, \ldots, k$, and therefore the kriged forward curve indeed interpolates the data.

The correlation function for the Gaussian random field Y on $\mathbb{R}$ determines the smoothness of the predicted forward curve. Common choices in spatial statistics are stationary correlation functions which only depend on the distance between locations, here being maturities. For example, exponential or Gaussian correlation functions can be applied, being, respectively, of the form

$$\rho(h) = \exp(-\zeta h)$$

and

$$\rho(h) = \exp(-\zeta h^2)$$

where $\zeta \in \mathbb{R}_+$ is a constant called the *smoothness* parameter. Thus, we have that $\text{Corr}(Y(x_i), Y(x_j)) = \rho(|x_i - x_j|)$. The higher value of ζ, the smaller is the correlation between two points, and hence, the weaker they are related in the kriged term structure. Therefore, lower values of ζ gives smoother curves.

Returning to our crude oil futures price data example, we now apply the kriging method for obtaining a smooth initial curve f_0 that interpolates the data. As mean function $m(x)$, we use the Nelson-Siegel curve specified above. Furthermore, we assume a Gaussian correlation function, with smoothness parameter $\zeta = 1$. The resulting kriged forward curve is plotted as a blue line in Fig. 6.4. In the figure we

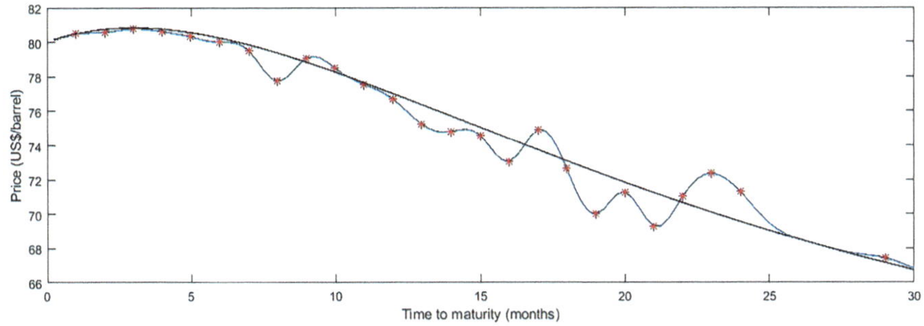

Fig. 6.4 Futures prices of NYMEX crude oil on December 30 2022 with the fitted Nelson-Siegel curve (black line) and the kriged curve (blue line)

have included the Nelson-Siegel curve and the observed data for reference. If we increase the value of ζ, the kriged surface will become less smooth and tend back to the mean inbetween observation points resembling a saw-tooth curve.

In our discussion, we have adopted the spatial statistics approach without any reference to an underlying Hilbert space structure. The correlation function of the random field Y can be seen as the kernel function in an integral operator representing the covariance operator in an appropriately defined Hilbert space. This leads us to reproducing kernel Hilbert spaces. Recently, Filipović, Pelger and Ye [101] have developed smoothing methods for conditional factor models applied to treasury bond prices.

We can extend the kriging technique to swap data in the following way. Suppose we know the mean function $m(x)$ (for example, by fitting a Nelson-Siegel function to swap data as described above). Then we find that

$$f_{\mathrm{K}}(x) := \mathbb{E}\Big[f(x) \mid F_0(T_1^1, T_1^2), \dots, F_0(T_k^1, T_k^2)\Big] = m(x) + c_x^\top C^{-1} \widehat{y}\,.$$

Here, C is the correlation matrix with elements

$$\frac{1}{(T_i^2 - T_i^1)(T_j^2 - T_j^1)} \int_{T_i^1}^{T_i^2} \int_{T_j^1}^{T_j^2} \mathrm{Corr}(Y(u), Y(v))\,du\,dv\,,$$

the vector c_x has coordinates

$$\frac{1}{T_i^2 - T_i^1} \int_{T_i^1}^{T_i^2} \mathrm{Corr}(Y(x), Y(u))\,du$$

and finally the vector $\widehat{y}$ has coordinates

$$F_0(T_i^1, T_i^2) - \frac{1}{T_i^2 - T_i^1} \int_{T_i^1}^{T_i^2} m(u)\,du\,.$$

The indices i and j range from 1 to k. With an exponential correlation function, the above expressions can be analytically integrated. If we choose a Gaussian correlation function, numerical integration must be performed. In Benth [23], an empirical example of Henry Hub gas futures prices from NYMEX is studied.

6.9 Bibliographical Notes

Already in Jamshidian [130] we find a term structure model of commodity futures prices. An account on early papers using an HJM-approach to commodities, and in particular energy markets, can be found in Clewlow and Strickland [72]. Audet *et al.* [9] proposed and analysed a term structure model for the Nordic electricity forward market driven by an infinite dimensional Gaussian noise. Such an infinite dimensional forward curve model was later extended by Barth and Benth [20] to non-Gaussian Lévy noise, where a numerical method for simulation is defined. Benth and Paraschiv [42] fitted an HJM term structure model with infinite dimensional noise to German power futures prices.

A general study of affine models and convenience yield in HJM factor model with Gaussian and jump noise is performed by Björk and Landen [59]. Hinderks and Wagner [124], Latini, Piccirilli, and Vargiolu [147] and Piccirilli, Schmeck, and Vargiolu [166] study factor models in view of power market futures, while Goatte, Oudjane and Russo [117] use factor models driven by independent increment processes in hedging. We remark in passing that independent increment processes are also known as additive processes by Sato [175]. Callegaro, Mazzoran and Sgarra [66] model jump clustering by introducing a branching or Hawkes process into the noise driver of a HJM-term structure model. They apply their model to the French power market.

Ambit fields have been suggested by Bardorff-Nielsen, Benth and Veraart [14] as a class of HJM-models for forward prices in energy markets. We refer to Barndorff-Nielsen, Benth and Veraart [15] for an extensive introduction to ambit stochastics and further discussions on their applications to term structure modelling. Di Persio and Perin [85] apply ambit fields to the German electricity futures market.

The analysis of swaps and integral operators in Section 6.5 is based on the papers Benth and Krühner [34, 37]. We refer to Benth and Koekebakker [32] for a discussion on the challenges in modeling the swap dynamic directly in a HJM-framework. A multivariate discrete forward curve model for swaps is proposed by Andersen, Koekebakker and Westgaard [5], estimated to swap price data from the Nordic market.

Benth and Simonsen [53] suggested a stochastic volatility model generalising the Heston model to infinite dimensions. This is an alternative to the infinite dimensional Barndorff-Nielsen and Shephard stochastic volatility model considered in Section 6.3. Cox, Karbach and Khedher [76] generalise this class of infinite dimensional stochastic volatility models to affine processes taking values in the positive cone of Hilbert-Schmidt operators on the state space of the HJM-forward model. Schmidt, Tappe and Yu [176] present an extensive analysis of affine processes in Hilbert space, while and Cox, Karbach and Khedher [75] study positive affine pure-jump processes on the space of Hilbert-Schmidt operators. A rough volatility model is suggested by Benth and Harang [29], where the dynamics is a fractional Ornstein–Uhlenbeck process in Hilbert space. We mention in passing that a framework for volatility estimation based on realized quadratic variation and law of large numbers is found in Benth, Schroers and Veraart [50].

In finance, leverage is usually thought of as asset prices being negatively correlated with volatility. In turbulent markets, prices tend to fall with increased volatility. In energy markets, the opposite effect of prices increasing with increasing volatility is often observed (see Eydeland and Wolyniec [97] and Geman [111]). This is coined the *inverse leverage effect*. Nomikos and Andriosopoulos [159] reveal both direct and inverse leverage for eight different commodity futures on NYMEX. Kristoufek [143] analyses empirically front month futures prices, showing that natural gas has an inverse leverage effect, while various oil prices follow the behaviour in asset markets. Benth and Sgarra [52] extends the HJM-dynamics with the stochastic volatility model of Section 6.3 to allow for a leverage effect.

The Esscher transform for infinite dimensional processes introduced in Section 6.7 follows the exposition in Benth and Sgarra [52], who adopted the result by Quiao and Wu [170].

Chapter 7
Pricing of Commodity and Energy Options

Abstract Risk-neutral prices of options on infinite-dimensional forward price models are derived for general payoff functions. In the case of Gaussian forward price models, expressions for the prices are computed in terms of integrals over the normal density function. Several particular examples are analysed, where we recover the Black-76 formula for plain-vanilla call options and the Margrabe formula for spread options. Using the density method together with Gateaux differentiation, we derive expressions for the delta of option prices. A general pricing approach is developed based on the Fourier transform of the payoff function and the characteristic functional related to the price dynamics, where we include stochastic volatility in our analysis. The delta of the option price is also discussed in this case. Finally, we focus on Markovian forward models, and discuss stability of the prices when the payoff function is Lipschitz continuous.

In this Chapter we derive arbitrage-free prices for options on forwards frequently encountered in commodity and energy markets. In organised market-places, typical options are plain vanilla call and put options on forwards with a given delivery time or period. OTC markets trade in a number of different other contracts, for example spread options, which may be cross-commodity based or calendar spreads, say.

We present two approaches, the first rather general one is appealing to the Fourier transform following the idea proposed by Carr and Madan [71]. This method establishes the price as a Fourier integral in terms of the Fourier transform of the payoff function and the characteristic exponent of the price process (or some functional thereof). The second method is tailor-made to the Gaussian situation, where we can derive analytic expressions closely related to the classical Black-76 formula (see, e.g., Geman [111, Sect. 4.7]).

F. E. Benth, P. Krühner, *Stochastic Models for Prices Dynamics in Energy and Commodity Markets*, Springer Finance,
https://doi.org/10.1007/978-3-031-40367-5_7

7.1 Arbitrage-free Pricing of Options

Let the payoff at time $\tau \geq 0$ from entering an option be

$$p(G(\mathbf{f}(\tau))) \tag{7.1}$$

where $G \in H^*$ and $p : \mathbb{R} \to \mathbb{R}$ is a measurable function. Here, we have placed ourselves in the set-up of Chapter 6, where $\mathbf{f} \in H := H_1 \times \cdots \times H_d$ is a multivariate (a d-variate) forward price dynamics. We refer to the functional G as the *gathering* functional.

Let us consider some examples. The most commonly traded derivatives in commodity markets are call and put options on forwards with given delivery times. I.e., a call option with strike K and exercise time τ on a forward delivering at time $T \geq \tau$ has payoff $\max(F(\tau, T) - K, 0)$. Since

$$F(\tau, T) = f(\tau, T - \tau) = \delta_{T-\tau} f(\tau)$$

we find $G := \delta_{T-\tau} \in H^*$. Moreover, $p(x) = \max(x - K, 0)$. If we have a call option on a forward with delivery period $[T_1, T_2]$, $T_1 < T_2$, paying $\max(\overline{F}(\tau, T_1, T_2) - K, 0)$ at time $\tau \leq T_2$, the gathering functional G will involve the operator $\mathcal{D}_\lambda$ in Proposition 6.7. Indeed, for a given aggregation function $\widetilde{a}$ introduced in (6.30), we find

$$\overline{F}(\tau, T_1, T_2) = \delta_{T_1-\tau} \mathcal{D}_{T_2-T_1}(f(\tau)).$$

That is, $G := \delta_{T_1-\tau} \mathcal{D}_{T_2-T_1}$, which is a bounded linear functional on H_w by Proposition 6.7 and $\delta_x \in H_w^*$.

We may also represent basket and spread options in terms of payoffs of the form (7.1). For example, a put option with strike K and exercise time τ on a basket of d forwards with delivery period $[T_1, T_2]$ has payoff $\max(K - \mathbf{v}^\top \overline{\mathbf{F}}(\tau, T_1, T_2), 0)$, with $\mathbf{v} \in \mathbb{R}^d$ is the vector yielding the weights in each forward in the basket. Hence, the gathering functional in this case becomes $G = \delta_{T_1-\tau} \mathbf{v}^\top \mathcal{D}_{T_2-T_1}(\mathbf{f}(\tau))$. Here, $\mathcal{D}_{T_2-T_1}$ operates on each coordinate of $\mathbf{f} \in H$, where H now is a product of Filipović spaces, and we have from Proposition 6.7 that $G \in H^*$. Moreover, $p(x) = \max(K - x, 0)$, obviously. A cross-commodity spread option is included here, by choosing $d = 2$ and $\mathbf{v} = (1, -h_r)^\top$, with $h_r > 0$ being the constant heat rate converting the second commodity into the nomination of the first. Calendar spread options can be formulated by selecting $G = \delta_{T_1^a-\tau} \mathcal{D}_{T_2^a-T_1^a} \mathcal{P}_1 - \delta_{T_1^b-\tau} \mathcal{D}_{T_2^b-T_1^b} \mathcal{P}_2$. Here, $[T_1^a, T_2^a]$ and $[T_1^b, T_2^b]$ are the delivery periods of the two respective forwards in the calendar spread and $\mathcal{P}_i$, $i = 1, 2$ are the projection operators from H into H_i. Again it follows that $G \in H^*$.

From the arbitrage pricing theory (see, e.g., Björk [58]) we have that the price of an option at time $t \geq 0$, paying $p(G(\mathbf{f}(\tau)))$ at time $\tau \geq t$, is given by

$$C(t;\tau) = \exp(-r(\tau - t))\mathbb{E}_{\mathbb{Q}}[p(G(\mathbf{f}(\tau)))\,|\mathcal{F}_t]. \tag{7.2}$$

Here, r is the risk-free rate of return which we assume to be a constant. The expectation is taken with respect to the risk-neutral probability measure $\mathbb{Q}$, and it is therefore convenient to assume that $\mathbf{f}$ is modelled already under this probability (see martingale drift conditions in Proposition 6.2 for simple HJM models, and Proposition 6.3 for the geometric case).

In all the examples above we notice that the gathering functional G is linearly transforming the infinite dimensional forward price $\mathbf{f}(\tau)$ into a real-valued price at exercise time τ. The function p is next giving the actual payoff of the option in question.

For (7.2) to be well-defined, we need to have that $p(G(\mathbf{f}(\tau))) \in L^1(\mathbb{Q})$. The following Lemma presents a sufficient condition for this moment condition.

Lemma 7.1 *Assume p is Lipschitz continuous. If $|\mathbf{f}(\tau)|_H \in L^1(\mathbb{Q})$, then (7.2) is well-defined.*

Proof By the assumption on p, we have

$$|p(x)| \leq |p(x) - p(0)| + |p(0)| \leq C|x| + |p(0)|,$$

for some constant $C > 0$. Hence,

$$\mathbb{E}_{\mathbb{Q}}[|p(G(\mathbf{f}(\tau)))|] \leq C\mathbb{E}_{\mathbb{Q}}[|G(\mathbf{f}(\tau))|] + |p(0)| \leq C\|G\|_{\text{op}}\mathbb{E}_{\mathbb{Q}}[|\mathbf{f}(\tau)|] + |p(0)|.$$

The result follows. □

The simple HJM models considered in Subsection 6.1.1 of Chapter 6 satisfies this first-moment condition (see Proposition 6.2). The simple geometric HJM models introduced in Section 6.2 of Chapter 6 become integrable whenever $\mathbf{g}(t)$ has finite exponential moments. This holds under additional conditions on the parameters of the model and the Lévy process. For example, if the volatility $\boldsymbol{\Sigma}$ is deterministic and bounded, $|\mathbf{f}(t)|_H \in L^1(\mathbb{Q})$ whenever $\mathbf{L}$ has finite exponential moments. A discussion analogous to the analysis at the end of Section 6.2 of Chapter 6 gives conditions on the Lévy measure for this to hold. We notice that put and call option payoff functions p satisfy the Lipschitz condition.

7.2 Gaussian HJM-Models and the Black-76 Formula

We first recall the simple HJM-model $\mathbf{f}(t)$ in (6.7), modelled under the risk-neutral probability $\mathbb{Q}$, as we introduced it in Chapter 6, with $\mathbf{L}$ being a Wiener process in H with covariance operator Q_W,

$$\mathbf{f}(t) = \mathcal{S}_t\mathbf{f}_0 + \int_0^t \mathcal{S}_{t-s}\boldsymbol{\Sigma}(s)d\mathbf{W}(s). \tag{7.3}$$

We assume that $\boldsymbol{\Sigma}$ is non-random, i.e., a measurable function from $\mathbb{R}_+$ into $L(H)$ such that $\int_0^T \|\boldsymbol{\Sigma}(t)Q_W^{1/2}\|_{\text{HS}}^2 dt < \infty$ for all $T < \infty$. Then, $\mathbf{f}(t)$ is a Gaussian process with values in H:

Lemma 7.2 *The H-valued stochastic process $(\mathbf{f}(t))_{t\geq 0}$ defined in* (7.3) *is a Gaussian process.*

Proof For arbitrary $n \in \mathbb{N}$, we need to show that for any selection of times $0 \leq t_1 < \cdots < t_n < \infty$ and arbitrary elements $g_1, \ldots, g_n \in H$, $\langle \mathbf{f}(t_1), g_1\rangle_H, \ldots, \langle \mathbf{f}(t_n), g_n\rangle_H$ is an n-variate Gaussian random variable. Notice that

$$\begin{aligned}\int_0^{t_{k+1}} \mathcal{S}_{t_{k+1}-s}\boldsymbol{\Sigma}(s)d\mathbf{W}(s) &= \mathcal{S}_{t_{k+1}-t_k}\int_0^{t_k} \mathcal{S}_{t_k-s}\boldsymbol{\Sigma}(s)d\mathbf{W}(s)\\ &\quad + \int_{t_k}^{t_{k+1}} \mathcal{S}_{t_{k+1}-s}\boldsymbol{\Sigma}(s)d\mathbf{W}(s)\end{aligned}$$

and therefore by the independent increment property of the Wiener process along with $\boldsymbol{\Sigma}$ being deterministic, any $\mathbf{f}(t_{k+1})$ can be expressed by $\mathbf{f}(t_k)$ and an independent random variable. Hence, it suffices to show that $\int_{t_k}^{t_{k+1}} \mathcal{S}_{t_{k+1}-s}\boldsymbol{\Sigma}(s)d\mathbf{W}(s)$ is Gaussian in H for any k. For $g \in H$, the characteristic function becomes

$$\begin{aligned}&\mathbb{E}_{\mathbb{Q}}[\exp(\mathrm{i}\langle \int_{t_k}^{t_{k+1}} \mathcal{S}_{t_{k+1}-t_k}\boldsymbol{\Sigma}(s)d\mathbf{W}(s), g\rangle)_H]\\ &\quad = \exp\left(-\frac{1}{2}\int_{t_k}^{t_{k+1}} |Q_W^{1/2}\boldsymbol{\Sigma}^*(s)\mathcal{S}_{t_{k+1}-s}^* g|_H^2 ds\right).\end{aligned}$$

This is the characteristic function of a Gaussian random variable on $\mathbb{R}$, and we are done. □

We can calculate an expression of the option price in terms of the standard normal distribution:

Proposition 7.1 *Assume that p is Lipschitz. Then the option price in* (7.2) *with the simple HJM model in* (7.3) *is given by*

$$C(t;\tau) = e^{-r(\tau-t)}\int_{\mathbb{R}} p(G\mathcal{S}_{\tau-t}\mathbf{f}(t) + \sigma_{t,\tau}z)\varphi(z)dz$$

where

$$\sigma_{t,\tau}^2 = \int_t^\tau |Q_W^{1/2}\mathbf{\Sigma}(s)^*\mathcal{S}_{\tau-s}^*G^*1|_H^2 ds$$

and φ is the density of the standard normal distribution function.

Proof Since $\mathbf{f}(\tau)$ is Gaussian by Lemma 7.2, it has finite variance (see Peszat and Zabczyk [165, Thm. 3.31]). Together with the assumed Lipschitz continuity of p, it follows from Lemma 7.1 that the payoff is $\mathbb{Q}$-integrable.

By (7.3), we derive

$$\mathbf{f}(\tau) = \mathcal{S}_{\tau-t}\mathbf{f}(t) + \int_t^\tau \mathcal{S}_{\tau-s}\mathbf{\Sigma}(s)d\mathbf{W}(s)$$

using the semigroup property of $\mathcal{S}_t$ and linearity of the stochastic integral. where, by the independent increment property of $\mathbf{W}$ the Gaussian random variable $\int_t^\tau \mathcal{S}_{\tau-s}\mathbf{\Sigma}(s)d\mathbf{W}(s)$ is independent of $\mathcal{F}_t$. Moreover, we find

$$G\int_t^\tau \mathcal{S}_{\tau-s}\mathbf{\Sigma}(s)d\mathbf{W}(s) = \langle\int_t^\tau \mathcal{S}_{\tau-s}\mathbf{\Sigma}(s)d\mathbf{W}(s), G^*1\rangle_H,$$

which, by Lemma 7.2 is a Gaussian random variable on $\mathbb{R}$ with mean zero and variance given by $\sigma_{t,\tau}^2$. Appealing to the $\mathcal{F}_t$-measurability of $\mathbf{f}(t)$, the result follows. □

For example, if $p(x) = \max(x - K)$, the payoff of a call option, a straightforward calculation using the properties of the standard normal distribution function reveals that

$$C(t;\tau) = e^{-r(\tau-t)}\sigma_{t,\tau}\Psi\left(\frac{G\mathcal{S}_{\tau-t}\mathbf{f}(t) - K}{\sigma_{t,\tau}}\right)$$

where $\Psi(x) = x\Phi(x) + \varphi(x)$, or $\Psi'(x) = \Phi(x)$, with Φ being the cumulative standard normal distribution function. If we have a call option on a forward maturing at time T, we have seen that $G := \delta_{T-\tau}$. In that case, we have $G\mathcal{S}_{\tau-t}f(t) = \delta_{T-t}f(t) = F(t,T)$ using that $\delta_x\mathcal{S}_t = \delta_{x+t}$. For the volatility $\sigma_{t,\tau}$, we find,

$$\sigma_{t,\tau}^2 = \int_t^\tau |Q_W^{1/2}\mathbf{\Sigma}(s)^*\delta_{T-s}^*1|_H^2 ds.$$

The presence of the covariance operator Q_W and $\mathbf{\Sigma}(s)$ lead to a mixing of all the volatilities across maturities when building up $\sigma_{t,\tau}$.

On the other hand, if the we have a call option with exercise time τ on a swap contract delivering over the period $[T_1, T_2]$, $\tau \le T_1 < T_2$, we recall from Section 7.1 of this chapter that $G = \delta_{T_1-\tau}\mathcal{D}_{T_2-T_1}$. Here, $\mathcal{D}_\lambda$ is the bounded linear operator from

Proposition 6.7 mapping a fixed-delivery forward into a swap. Hence, for a single commodity swap (i.e., one-dimensional case),

$$
\begin{aligned}
G\mathcal{S}_{\tau-t} f(t) &= \delta_{T_1-\tau}\mathcal{D}_{T_2-T_1}\mathcal{S}_{\tau-t} f(t) \\
&= \delta_{T_1-\tau}\int_{\cdot}^{\cdot+T_2-T_1} a_{T_2-T_1}(y-\cdot) f(t, y+\tau-t)dy \\
&= \int_{T_1-\tau}^{T_2-\tau} a_{T_2-T_1}(y-(T_1-\tau)) f(t, y+\tau-t)dy \\
&= \int_{T_1-t}^{T_2-t} a_{T_2-T_1}(z-(T_1-t)) f(t, z)dz \\
&= \overline{F}(t, T_1, T_2),
\end{aligned}
$$

where the function a_λ appears in the definition of $\mathcal{D}_\lambda$ from the aggregation function $\widetilde{a}$. So, as for the case of a fixed-delivery forward, we also get for swaps an option price which is a function of the initial price of the underlying, here being $\overline{F}(t, T_1, T_2)$.

We can treat geometric simple HJM-models Gaussian type in a similar manner. From Section 6.2 we recall that when H is an algebra we can define the forward price to be $\mathbf{f}(t) = \exp(\mathbf{g}(t))$, where, in a Gaussian set-up, we have from (6.12),

$$
\mathbf{g}(t) = \mathcal{S}_t \mathbf{g}_0 + \int_0^t \mathcal{S}_{t-s}\boldsymbol{\mu}(s)ds + \int_0^t \mathcal{S}_{t-s}\boldsymbol{\Sigma}(s)d\mathbf{W}(s) \tag{7.4}
$$

By Proposition 6.3, the martingale drift conditions yields that

$$
\boldsymbol{\mu}(t) = -\frac{1}{2}\|Q_W^{1/2}\boldsymbol{\Sigma}(t)^*\mathbf{h}_\cdot\|^2,
$$

and we remind that $\boldsymbol{\Sigma}$ is assumed to be deterministic in our Gaussian framework. By Lemma 7.2, we have that $\mathbf{g}$ is a Gaussian process. Moreover, it is easily seen that

$$
\mathbf{g}(\tau) = \mathcal{S}_{\tau-t}\mathbf{g}(t) + \int_t^\tau \mathcal{S}_{\tau-s}\boldsymbol{\mu}(s)ds + \int_t^\tau \mathcal{S}_{\tau-s}\boldsymbol{\Sigma}(s)d\mathbf{W}(s) \tag{7.5}
$$

for any $\tau \geq t \geq 0$.

Now, it is hard to exploit the Gaussianity of $\mathbf{g}$ to compute option prices except in some special cases. For example, for options on swaps, we have that $G\mathbf{f}(\tau)$ is an integral operator on $\exp(\mathbf{g}(\tau))$, which is not expressible as $\exp(\widetilde{G}\mathbf{g}(\tau))$ for some operator $\widetilde{G} \in L(H)$. Heuristically speaking, we are facing the computation of a nonlinear function of a sum of lognormally distributed variables which do not have a closed-form expression in general. On the other hand, if $G = \delta_{T-\tau}$, the evaluation operator, the situation is much simpler. In this case, we can derive a formula for the option price:

Proposition 7.2 *Assume that p is Lipschitz and $G := \sum_{i=1}^d b_i \delta_{T_i-\tau}\mathcal{P}_i$ for $T_i \geq \tau$, $a_i \in \mathbb{R}$ and the projections $\mathcal{P}_i : H \to H_i$, $i = 1, \ldots, d$. Then, the option price in (7.2) with the simple exponential HJM model in (7.4), is*

$$C(t;\tau) = e^{-r(\tau-t)} \int_{\mathbb{R}^d} p\left(\sum_{i=1}^d b_i F_i(t, T_i) e^{\int_t^\tau \delta_{T_i-s}\mathcal{P}_i \boldsymbol{\mu}(s)ds + x_i}\right) \varphi_d(\mathbf{x}) d\mathbf{x},$$

where φ_d is the density function of a d-variate normal distribution function with zero mean and covariance matrix $K(t, \tau)$ having elements

$$K_{ij}(t,\tau) = \int_t^\tau \delta_{T_j-s}\mathcal{P}_j \boldsymbol{\Sigma}(s) Q_W \boldsymbol{\Sigma}(s)^* \mathcal{P}_i^* \delta_{T_i-s}^* 1 ds$$

for $i, j = 1, \ldots, d$.

Proof Since $\mathbf{g}(\tau)$ is Gaussian by Lemma 7.2, it has finite exponential moment: Indeed, let X denote the Wiener integral of $\mathbf{g}(t)$,

$$X := \int_0^t \mathcal{S}_{t-s} \boldsymbol{\Sigma}(s) d\mathbf{W}(s),$$

which is a zero-mean square integrable random variable in H. by the elementary inequality $2xy \leq x^2 + y^2$ it follows that

$$|X|_H = \frac{1}{\sqrt{2s}} |\sqrt{2s} X|_H \leq \frac{1}{4s} + s|X|_H^2$$

for any $s > 0$. Imposing Fernique's Theorem (see Peszat and Zabczyk [165, Thm. 3.31]), it holds that $\exp(s|X|_H^2) \in L^1(\mathbb{Q})$ for any $s < 1/2\mathbb{E}_{\mathbb{Q}}[|X|_H^2]$. This shows that $\exp(|X|_H) \in L^1(\mathbb{Q})$ and therefore $\mathbf{g}(t)$ has finite exponential moment. Together with the assumed Lipschitz continuity of p, it follows from Lemma 7.1 that the payoff is $\mathbb{Q}$-integrable.

Since $\delta_x(fg) = \delta_x f \cdot \delta_x g$ in H, we have

$$\begin{aligned}
\delta_{T_i-\tau}\mathcal{P}_i \mathbf{f}(\tau) &= \exp(\delta_{T_i-\tau}\mathcal{P}_i \mathbf{g}(\tau)) \\
&= \exp\left(\delta_{T_i-\tau}\mathcal{S}_{\tau-t}\mathcal{P}_i \mathbf{g}(t) + \int_t^\tau \delta_{T_i-\tau}\mathcal{S}_{\tau-s}\mathcal{P}_i \boldsymbol{\mu}(s)ds \right. \\
&\quad \left. + \int_t^\tau \delta_{T_i-\tau}\mathcal{S}_{\tau-s}\mathcal{P}_i \boldsymbol{\Sigma}(s) d\mathbf{W}(s)\right) \\
&= \exp\left(\delta_{T_i-t}\mathcal{P}_i \mathbf{g}(t) + \int_t^\tau \delta_{T_i-s}\mathcal{P}_i \boldsymbol{\mu}(s)ds \right. \\
&\quad \left. + \int_t^\tau \delta_{T_i-s}\mathcal{P}_i \boldsymbol{\Sigma}(s) d\mathbf{W}(s)\right).
\end{aligned}$$

Now, let

$$X_i := \int_t^\tau \delta_{T_i - s} \mathcal{P}_i \mathbf{\Sigma}(s) d\mathbf{W}(s)$$

for $i = 1, \ldots, d$, and deduce from Lemma 7.2 that $(X_1, \ldots, X_d)$ is a d-variate Gaussian random variable with zero mean. We find the covariance between X_i and X_j as follows (using $h_0^i \in H_i$ as the representative of $\delta_0 \in H_i^*$):

$$\begin{aligned}
\mathbb{E}_{\mathbb{Q}}[X_i X_j] &= \mathbb{E}_{\mathbb{Q}}\left[\langle \int_t^\tau \mathcal{S}_{T_i - s} \mathcal{P}_i \mathbf{\Sigma}(s) d\mathbf{W}(s), h_0^i \rangle_i \langle \int_t^\tau \mathcal{S}_{T_j - s} \mathcal{P}_j \mathbf{\Sigma}(s) d\mathbf{W}(s), h_0^j \rangle_j \right] \\
&= \int_t^\tau \langle \mathcal{Q}_W \mathbf{\Sigma}(s)^* \mathcal{P}_i^* \mathcal{S}_{T_i - s}^* h_0^i, \mathbf{\Sigma}(s)^* \mathcal{P}_j^* \mathcal{S}_{T_j - s}^* h_0^j \rangle_H ds \\
&= \int_t^\tau \delta_{T_j - s} \mathcal{P}_j \mathbf{\Sigma}(s) \mathcal{Q}_W \mathbf{\Sigma}(s)^* \mathcal{P}_i^* \delta_{T_i - s}^* 1 ds.
\end{aligned}$$

The result follows. □

Letting $d = 1$, we easily derive the famous Black-76-formula for a call option with strike K and exercise time τ on a forward delivering at time T (with $T \geq \tau \geq t \geq 0$). The payoff function is $p(x) = \max(x - K)$, and gathering functional is $G = \delta_{T-\tau}$. From Proposition 7.2 we see that the variance becomes

$$\sigma_{t,\tau,T}^2 := \int_t^\tau |\mathcal{Q}_W^{1/2} \Sigma(s) h_{T-s}|_H^2 ds = -2 \int_t^\tau \delta_{T-s} \mu(s) ds, \tag{7.6}$$

and a straightforward calculation using the properties of the normal distribution reveals the price as

$$C(t; \tau) = e^{-r(\tau - t)} \left(F(t, T) \Phi(d_1) - K \Phi(d_2) \right) \tag{7.7}$$

where Φ is the standard cumulative normal distribution function, $d_1 = d_2 + \frac{1}{2}\sigma_{t,\tau,T}^2$ and

$$d_2 = \frac{\ln F(t, T) - \ln K}{\sigma_{t,\tau,T}} \tag{7.8}$$

We have recovered the Black-76 formula.

Next, we consider a spread option and recover the : To this end, let $d = 2$ consider a payoff function $\max(x_1 - x_2, 0)$. Hence, we have a call option on the spread $F_1(\tau, T_1) - F_2(\tau, T_2)$, i.e., an option on the spread between two forwards with two different maturities. With $b_1 = 1$ and $b_2 = -1$ in Proposition 6.7, the price can be expressed as

$$C(t;\tau) = e^{-r(\tau-t)}\mathbb{E}_{\mathbb{Q}}[\max(F_1(t,T_1)e^{\int_t^\tau \delta_{T_1-s}\mathcal{P}_1\boldsymbol{\mu}(s)ds+X_1} \\ - F_2(t,T_2)e^{\int_t^\tau \delta_{T_2-s}\mathcal{P}_2\boldsymbol{\mu}(s)ds+X_2}, 0)]$$

for two normally distributed random variables X_1, X_2 having mean zero, variances $\sigma_i^2, i = 1, 2$, with

$$\sigma_i^2 = \int_t^\tau |Q_W^{1/2}\boldsymbol{\Sigma}(s)^*\mathcal{P}_i^* h^i_{T_i-s}|_H^2 ds$$

and covariance $c_{1,2}$ with

$$c_{1,2} = \int_t^\tau \langle Q_W^{1/2}\boldsymbol{\Sigma}(s)^*\mathcal{P}_1^* h^1_{T_1-s}, Q_W^{1/2}\boldsymbol{\Sigma}(s)^*\mathcal{P}_2^* h^2_{T_2-s}\rangle_H ds.$$

Recall in the above notation that $h^i_x \in H_i, i = 1, 2$ is the representative of $\delta_x \in H_i^*$. From the drift condition on the forward prices (see Proposition 6.3), it holds that

$$\int_t^\tau \delta_{T_i-s}\mathcal{P}_i\boldsymbol{\mu}(s)ds = -\frac{1}{2}\sigma_i^2$$

for $i = 1, 2$. Now, defining the random variables $Y_i := X_i/\sigma_i, i = 1, 2$, we find that (Y_1, Y_2) are bivariate Gaussian with mean zero and variance 1, with the same correlation as (X_1, X_2). If we denote this correlation as ρ, we can express $Y_1 = \rho Y_2 + \sqrt{1-\rho^2}Z$, where Z is standard normally distributed random variable being independent of Y_2. Hence, after factorizing and changing probability to $\widetilde{\mathbb{Q}} \sim \mathbb{Q}$ with Radon-Nikodym derivative $d\widetilde{\mathbb{Q}} = \exp(-0.5\sigma_2^2 + \sigma_2 Y_2)d\mathbb{Q}$, we find

$$\begin{aligned}
&e^{r(\tau-t)}C(t;\tau) \\
&\quad = F_2(t,T_2)\mathbb{E}_{\mathbb{Q}}\Bigg[e^{-\frac{1}{2}\sigma_2^2+\sigma_2 Y_2} \\
&\qquad \times \max\left(\frac{F_1(t,T_1)}{F_2(t,T_2)}e^{-\frac{1}{2}\sigma_1^2+\frac{1}{2}\sigma_2^2}e^{(\sigma_1\rho-\sigma_2)Y_2+\sigma_1\sqrt{1-\rho^2}Z} - 1, 0\right)\Bigg] \\
&\quad = F_2(t,T_2)\mathbb{E}_{\widetilde{\mathbb{Q}}}\left[\max\left(\frac{F_1(t,T_1)}{F_2(t,T_2)}e^{-\frac{1}{2}\sigma_1^2+\frac{1}{2}\sigma_2^2}e^{(\sigma_1\rho-\sigma_2)(\widetilde{Y}_2+\sigma_2)+\sigma_1\sqrt{1-\rho^2}Z} - 1, 0\right)\right]
\end{aligned}$$

where $\widetilde{Y}_2$ is a standard normal random variable independent of Z under $\widetilde{\mathbb{Q}}$. Therefore, a straightforward calculation using the properties of the normal distribution leads to the option price

$$C(t,\tau) = e^{-r(\tau-t)}\left(F_1(t,T_1)\Phi(d_1) - F_2(t,T_1)\Phi(d_2)\right) \tag{7.9}$$

where $d_1 = d_2 + \widetilde{\sigma}^2_{t,\tau,T_1,T_2}$,

$$d_2 = \frac{\ln(F_1(t,T_1)/F_2(t,T_2)) - \frac{1}{2}\widetilde{\sigma}^2_{t,\tau,T_1,T_2}}{\widetilde{\sigma}_{t,\tau,T_1,T_2}} \tag{7.10}$$

and the *total variance* is given by

$$\begin{aligned}\widetilde{\sigma}^2_{t,\tau,T_1,T_2} &= \int_t^\tau |Q_W^{1/2}\mathbf{\Sigma}(s)^*\mathcal{P}_i^* h^1_{T_1-s}|^2_H ds \\ &\quad - 2\int_t^\tau \langle Q_W^{1/2}\mathbf{\Sigma}(s)^*\mathcal{P}_1^* h^1_{T_1-s}, Q_W^{1/2}\mathbf{\Sigma}(s)^*\mathcal{P}_2^* h^2_{T_2-s}\rangle_H ds \\ &\quad + \int_t^\tau |Q_W^{1/2}\mathbf{\Sigma}(s)^*\mathcal{P}_2^* h^2_{T_2-s}|^2_H ds \end{aligned} \tag{7.11}$$

The price of a spread option on two forwards in (7.9) is known as the Margrabe formula (see Margrabe [155]). Notice that in our context we are pricing a calendar spread as the two maturity times may be different, as well as a cross-commodity spread as the two forwards may be in different markets.

In fact, the infinite dimensional term structure dynamics impose a dependency structure on plain vanilla call and put option prices across different exercise times and forward maturities. This can be valuable information in selecting the right options for given risk management purposes. For instance, one can answer questions related to diversification of risk when entering an option deal. We refer the reader to Benth and Krühner [37] for a numerical study of correlation structures. Recalling back to the Margrabe option above, we can indeed use the correlation structure imposed from the term structure dynamics to obtain the optimal calendar times T_1 and T_2 in the spread (here, of course, "optimal" refers to some given risk goal).

Next, we discuss the sensitivity of the option price with respect to the current forward curve. First, we consider the set-up given in Proposition 7.2, where it is simple to derive the *delta* of the option price, that is, the gradient of the option price with respect to the current prices $F_1(t,T_1), \ldots, F_d(t,T_d)$: we have

Proposition 7.3 *Assume the conditions of Proposition 7.2. Then*

$$\begin{aligned}\nabla_{\mathbf{F}} C(t;\tau) = -e^{-r(\tau-t)} \int_{\mathbb{R}^d} p\left(\sum_{i=1}^d b_i F_i(t,T_i) e^{\int_t^\tau \delta_{T_i-s}\mathcal{P}_i \boldsymbol{\mu}(s)ds + x_i}\right) \varphi_d(\mathbf{x}) \\ \times\, diag\left(F_i^{-1}(t,T_i)\right) K^{-1}(t,\tau)\mathbf{x}d\mathbf{x},\end{aligned}$$

where $\mathbf{F} := (F_1(t,T_1), \ldots, F_d(t,T_d)) \in \mathbb{R}^d$ *and* $diag(a_i)$ *is a diagonal matrix with elements* $a_1, \ldots, a_d$ *on the diagonal.*

Proof In the argument, we let $r = 0$ for simplicity, and collapse the notation to

$$C(t;\tau) = \int_{\mathbb{R}^d} p\left(\sum_{i=1}^{d} b_i e^{\mu_i + \ln F_i + x_i}\right) \varphi_d(\mathbf{x}) d\mathbf{x}$$
$$= \int_{\mathbb{R}^d} p\left(\sum_{i=1}^{d} b_i e^{\mu_i + y_i}\right) \varphi_d(\mathbf{y} - \ln \mathbf{F}) d\mathbf{y}$$

with the obvious meaning of μ_i, F_i and with $\ln \mathbf{F} := (\ln F_1, \ldots, \ln F_d)^\top$. It holds that

$$\nabla_{\mathbf{F}} \varphi_d(\mathbf{y} - \ln \mathbf{F}) = -\frac{1}{2} \varphi_d(\mathbf{y} - \ln \mathbf{F}) \nabla_{\mathbf{F}} \{(\mathbf{y} - \ln \mathbf{F})^\top K^{-1}(t,\tau)(\mathbf{y} - \ln \mathbf{F})\}$$
$$= -\varphi(\mathbf{y} - \ln \mathbf{F}) \mathrm{diag}\left(\frac{1}{F_i}\right) K^{-1}(t,\tau)(\mathbf{y} - \ln \mathbf{F})$$

Observe that since p is assumed Lipschitz continuous,

$$|p(\exp(x_i))| \leq |p(1)| + c|\exp(x_i) - 1|$$

for some constant $c > 0$, and thus $p(\exp(x_i))x_i$ is at most of growth $x_i \exp(x_i)$, which is integrable with respect to the normal density function. Hence, by Theorem 2.27 in Folland [104], the claimed result follows since we can move the gradient operator inside the integral. □

The delta of the option is a vector, where each element i is the sensitivity with respect to the forward $F_i(t, T_i)$. In the expression of this vector, we observe that the payoff function is not differentiated, and that we scale by the inverse of the covariance matrix. The method applied in the proof is referred to as the *density method*, as the dependency on the current forward prices in the (possibly non-differentiable) payoff function is moved into the Gaussian density function (which is smooth).

Let us re-visit the case of an option on a swap (i.e., assuming $d = 1$, a univariate market), where the swap dynamics is defined by a simple geometric Gaussian HJM-model. If the swap delivers over the period $[T_1, T_2]$, and the option has exercise time $\tau \leq T_1$, then the price is given by

$$C(t;\tau) = e^{-r(\tau - t)} \mathbb{E}_{\mathbb{Q}}[p(\delta_{T_1 - \tau} \mathcal{D}_{T_2 - T_1} f(\tau)) \,|\, \mathcal{F}_t]$$

with $\mathcal{D}_\lambda$ being the operator from Proposition 6.7. We recall that $f(\tau) = \exp(g(\tau))$ and for $\tau \geq t$,

$$g(\tau) = \mathcal{S}_{\tau - t} g(t) + X_{t,\tau}$$

with

$$X_{t,\tau} := \int_t^\tau \mathcal{S}_{\tau-s}\mu(s)ds + \int_t^\tau \mathcal{S}_{\tau-s}\Sigma(s)dW(s)$$

and μ given by the drift condition in Proposition 6.3. Evidently, $X_{t,\tau}$ is an H_w-valued Gaussian random variable independent of $\mathcal{F}_t$. Thus, the option price becomes

$$C(t;\tau) = e^{-r(\tau-t)}\mathbb{E}_{\mathbb{Q}}\left[p\left(\delta_{T_1-\tau}\mathcal{D}_{T_2-T_1}\exp(\mathcal{S}_{\tau-t}g(t) + X_{t,\tau})\right)\right]$$

We find,

$$\begin{aligned}
&\delta_{T_1-\tau}\mathcal{D}_{T_2-T_1}\exp(\mathcal{S}_{\tau-t}g(t) + X_{t,\tau}) \\
&\quad = \int_{T_1-\tau}^{T_2-\tau} a_{T_2-T_1}(y-(T_1-\tau))e^{g(t,y+\tau-t)}e^{X_{t,\tau}(y)}dy \\
&\quad = \int_{T_1-t}^{T_2-t} a_{T_2-T_1}(z-(T_1-t))f(t,z)e^{X_{t,\tau}(z-(\tau-t))}dz,
\end{aligned}$$

where the function a_λ appears in the definition of $\mathcal{D}_\lambda$ from the aggregation function $\widetilde{a}$. But this reveals that the option price $C(t;\tau)$ is in fact dependent on the forward curve $z \mapsto f(t,z)$, and *not* a function of the current swap price $\overline{F}(t,T_1,T_2)$. Hence, we see that in general prices of options on forwards with delivery period will depend on the initial curve and not on particular contracts, and are therefore real-valued nonlinear functionals on the Hilbert space. I.e., we can express the price $C(t;\tau)$ as $\overline{C}(t;\tau,g(t))$, with $\overline{C}(t;\tau,\cdot)$ being a real-valued nonlinear functional on H.

Indeed, we can extend these considerations to the general case, and derive that $C(t;\tau) = \overline{C}(t;\tau,\mathbf{g}(t))$ where from (7.5) and the definition of the option price in (7.2)

$$\begin{aligned}
\overline{C}(t;\tau,\mathbf{g}) := e^{-r(\tau-t)}\mathbb{E}_{\mathbb{Q}}\Bigg[p\Bigg(&G(\exp(\mathcal{S}_{\tau-t}\mathbf{g} + \int_t^\tau \mathcal{S}_{\tau-t}\boldsymbol{\mu}(s)ds \\
&+ \int_t^\tau \mathcal{S}_{\tau-s}\boldsymbol{\Sigma}(s)d\mathbf{W}(s)))\Bigg)\Bigg] \qquad (7.12)
\end{aligned}$$

Here, we have used that the stochastic integral is independent of $\mathcal{F}_t$. It holds that

$$\exp(\mathcal{S}_{\tau-t}\mathbf{g}) = \mathcal{S}_{\tau-t}\exp(\mathbf{g})$$

and so we may also express the option price by $C(t;\tau) = \overline{C}(t;\tau,\mathbf{f}(t))$ where

$$\overline{C}(t;\tau,\mathbf{f}) = e^{-r(\tau-t)}\mathbb{E}_{\mathbb{Q}}\left[p\left(G((\mathcal{S}_{\tau-t}\mathbf{f})\exp(\int_t^\tau \mathcal{S}_{\tau-t}\boldsymbol{\mu}(s)ds + \int_t^\tau \mathcal{S}_{\tau-s}\boldsymbol{\Sigma}(s)d\mathbf{W}(s)))\right)\right]. \tag{7.13}$$

We have slightly abused the meaning of $\overline{C}$ with the faith that the reader understands from the context which one we have in mind.

We step forward with analysing the delta of general options by extending the density method presented above to the infinite dimensional case. We consider the Gateaux derivative of $\overline{C}(t;\tau,\mathbf{f})$ in a direction $\mathbf{h}\in H$, and have the following:

Proposition 7.4 *Assume that p is Lipschitz continuous. Then, for $\mathbf{f} = \exp(\mathbf{g})$, $\mathbf{g}\in H$ and $\mathbf{h}\in H$, such that $\mathcal{S}_{\tau-t}(\mathbf{h}\mathbf{f}^{-1})\in Dom(Q_{t,\tau}^{-1})$, it holds that*

$$D_{\mathbf{h}}\overline{C}(t;\tau,\mathbf{f}) = e^{-r(\tau-t)}\langle Q_{t,\tau}^{-1}\mathcal{S}_{\tau-t}(\mathbf{h}\mathbf{f}^{-1}), \mathbf{v}_{t,\tau}\rangle_H$$

Here,

$$\mathbf{v}_{t,\tau} = \mathbb{E}_{\mathbb{Q}}\left[p\left(G((\mathcal{S}_{\tau-t}\mathbf{f}))\exp(\int_t^\tau \mathcal{S}_{\tau-s}\boldsymbol{\mu}(s)ds + \mathbf{X}_{t,\tau})\right)\mathbf{X}_{t,\tau}\right]$$

where the expectation is interpreted in the Bochner sense as an object in H, with $\mathbf{X}_{t,\tau} := \int_t^\tau \mathcal{S}_{\tau-s}\boldsymbol{\Sigma}(s)d\mathbf{W}(s)$ being an H-valued mean-zero Gaussian random variable with

$$Q_{t,\tau} = \int_t^\tau \mathcal{S}_{\tau-s}\boldsymbol{\Sigma}(s)Q_W\boldsymbol{\Sigma}(s)^*\mathcal{S}_{\tau-s}^*ds$$

as covariance operator.

Proof First, we have that

$$\overline{C}(\mathbf{f}) = \mathbb{E}_{\mathbb{Q}}\left[p\left(G\left(\exp(\int_t^\tau \mathcal{S}_{\tau-s}\boldsymbol{\mu}(s)ds)\exp(\mathcal{S}_{\tau-t}\ln\mathbf{f} + \mathbf{X}_{t,\tau})\right)\right)\right]$$

where we have collapsed the notation slightly and assumed without any loss of generality that the interest rate $r = 0$. Above we have used explictly that $\mathbf{f} = \exp(\mathbf{g})$ which provides us with the definition of $\ln\mathbf{f} = \mathbf{g}\in H$. We recall that we work in Hilbert spaces of functions on $\mathbb{R}_+$, giving the natural interpretation of exponentials and logarithms as pointwise operations. In the Banach algebra H, it is easily shown that the multiplication operator

$$\mathbf{u}\mapsto\exp(\int_t^\tau \mathcal{S}_{\tau-s}\boldsymbol{\mu}(s)ds)\mathbf{u}$$

defines a bounded linear operator in H (see Proposition 3.5 for the case when H is specified by the Filipović spaces). Hence, for an appropriately defined function $q : H \to \mathbb{R}$ we can write

$$\overline{C}(\mathbf{f}) = \mathbb{E}_{\mathbb{Q}}\left[q\left(\exp(\mathcal{S}_{\tau-t} \ln \mathbf{f} + \mathbf{X}_{t,\tau})\right)\right]$$

where from the Lipschitz continuity of p it follows that

$$|q(\mathbf{u})| \leq c_1 + c_2 |\mathbf{u}|_H$$

for positive constants c_1, c_2. Denoting

$$\widehat{C}(\mathbf{g}) = \mathbb{E}_{\mathbb{Q}}\left[q\left(\exp(\mathcal{S}_{\tau-t}\mathbf{g} + \mathbf{X}_{t,\tau})\right)\right]$$

it follows from the chain rule of the Gateaux derivative that

$$D_{\mathbf{h}}\overline{C}(\mathbf{f}) = D_{D_{\mathbf{h}} \ln \mathbf{f}}\widehat{C}(\ln \mathbf{f}).$$

Hence, we must find $D_{\mathbf{h}} \ln \mathbf{f}$ and $D_{\overline{\mathbf{h}}}\widehat{C}(\mathbf{g})$ for an arbitrary $\overline{\mathbf{h}} \in H$:

First, by definition of the Gateaux derivative, it follows,

$$D_{\mathbf{h}} \ln \mathbf{f}(x) = \frac{d}{d\epsilon} \ln(\mathbf{f}(x) + \epsilon \mathbf{h}(x))|_{\epsilon=0} = \mathbf{h}(x)\mathbf{f}(x)^{-1}$$

where we use that $\mathbf{f}^{-1} = \exp(-\mathbf{g}) \in H$ and thus $\mathbf{h}\mathbf{f}^{-1} \in H$ by the algebraic structure of H.

Next we study $D_{\overline{\mathbf{h}}}\widehat{C}(\mathbf{g})$ for $\overline{\mathbf{h}} \in H$: to this end, we recall that $Q_{t,\tau}$ is a positive definite symmetric trace class operator on H, and hence there exists an ONB $(e_i)_{i\in\mathbb{N}}$ in H of eigenvectors of $Q_{t,\tau}$ with corresponding positive eigenvalues $(\lambda_i)_{i\in\mathbb{N}}$. Moreover, $Y_i := \lambda_i^{-1/2}\langle X_{t,\tau}, e_i\rangle_H$ for $i = 1, 2, \ldots$, becomes an $i.i.d.$-sequence of standard normally distributed random variables and we have the representation

$$X_{t,\tau} = \sum_{i=1}^{\infty} \sqrt{\lambda_i} Y_i e_i.$$

Let $\mathcal{P}_n$ be the projection in H onto the span of $(e_i)_{i=1}^n$. We show that

$$\widehat{C}(\mathbf{g}) = \lim_{n\to\infty} \mathbb{E}_{\mathbb{Q}}[q(\exp(\mathcal{P}_n(\mathcal{S}_{\tau-t}\mathbf{g} + \mathbf{X}_{t,\tau})))].$$

Obviously, $\mathcal{P}_n(\mathcal{S}_{\tau-t}\mathbf{g} + \mathbf{X}_{t,\tau}) \to \mathcal{S}_{\tau-t}\mathbf{g} + \mathbf{X}_{t,\tau}$ on H for every ω, and by continuity of $q \circ \exp()$ in H it follows

$$q(\exp(\mathcal{P}_n(\mathcal{S}_{\tau-t}\mathbf{g} + \mathbf{X}_{t,\tau}))) \to q(\exp(\mathcal{S}_{\tau-t}\mathbf{g} + \mathbf{X}_{t,\tau}))$$

as $n \to \infty$. The growth bound on q yields

$$
\begin{aligned}
|q(\exp(\mathcal{P}_n(\mathcal{S}_{\tau-t}\mathbf{g} + \mathbf{X}_{t,\tau})))| &\leq c_1 + c_2 |\exp(\mathcal{P}_n(\mathcal{S}_{\tau-t}\mathbf{g} + \mathbf{X}_{t,\tau}))|_H \\
&\leq c_1 + c_2 \exp(|\mathcal{S}_{\tau-t}\mathbf{g}|_H + |\mathbf{X}_{t,\tau}|_H)
\end{aligned}
$$

where we used that $|\exp(\mathbf{g})|_H \leq \exp(|\mathbf{g}|_H)$ in the Banach algebra H along with the triangle inequality and $\|\mathcal{P}_n\|_{\mathrm{op}} \leq 1$. Fernique's Theorem (see Peszat and Zabczyk [165, Thm. 3.31] and the argument in the first part of the proof of Proposition 7.2 above) yields that $\exp(|\mathbf{X}_{t,\tau}|_H) \in L^1(\mathbb{Q})$, and by the dominated convergence theorem (see Folland [104, Thm. 2.24]) we find

$$
\begin{aligned}
\widehat{C}(\mathbf{g}) &= \mathbb{E}_{\mathbb{Q}}[\lim_{n\to\infty} q(\exp(\mathcal{P}_n(\mathcal{S}_{\tau-t}\mathbf{g} + \mathbf{X}_{t,\tau})))] \\
&= \lim_{n\to\infty} \mathbb{E}_{\mathbb{Q}}[q(\exp(\mathcal{P}_n(\mathcal{S}_{\tau-t}\mathbf{g} + \mathbf{X}_{t,\tau})))]
\end{aligned}
$$

which is the result we desired to show. In particular, for any $\epsilon \geq 0$, we find analogously

$$
\widehat{C}(\mathbf{g} + \epsilon\overline{\mathbf{h}}) = \lim_{n\to\infty} \mathbb{E}_{\mathbb{Q}}[q(\exp(\mathcal{P}_n(\mathcal{S}_{\tau-t}\mathbf{g} + \epsilon\mathcal{S}_{\tau-t}\overline{\mathbf{h}} + \mathbf{X}_{t,\tau})))]
$$

We use this approximation to analyse the derivative.

Since,

$$
\mathcal{P}_n(\mathcal{S}_{\tau-t}\mathbf{g} + \epsilon\mathcal{S}_{\tau-t}\overline{\mathbf{h}} + \mathbf{X}_{t,\tau}) = \sum_{i=1}^{n} \left(\langle \mathcal{S}_{\tau-t}\mathbf{g}, e_i \rangle_H + \epsilon \langle \mathcal{S}_{\tau-t}\overline{\mathbf{h}}, e_i \rangle_H + \sqrt{\lambda_i} Y_i \right) e_i
$$

we can express the expected value as

$$
\begin{aligned}
&\mathbb{E}_{\mathbb{Q}}[q(\exp(\mathcal{P}_n(\mathcal{S}_{\tau-t}\mathbf{g} + \epsilon\mathcal{S}_{\tau-t}\overline{\mathbf{h}} + \mathbf{X}_{t,\tau})))] \\
&= \int_{\mathbb{R}^n} q\left(\exp\left(\sum_{i=1}^{n} (\langle \mathcal{S}_{\tau-t}\mathbf{g}, e_i \rangle_H + \epsilon \langle \mathcal{S}_{\tau-t}\overline{\mathbf{h}}, e_i \rangle_H + \sqrt{\lambda_i} y_i) e_i\right)\right) \\
&\quad \times \varphi(y_1) \cdots \varphi(y_n) dy_1 \cdots dy_n \\
&= \int_{\mathbb{R}^n} q(\exp(\sum_{i=1}^{n} z_i e_i)) \prod_{i=1}^{n} \varphi\left(\frac{z_i - \langle \mathcal{S}_{\tau-t}\mathbf{g}, e_i \rangle_H - \epsilon \langle \mathcal{S}_{\tau-t}\overline{\mathbf{h}}, e_i \rangle_H}{\sqrt{\lambda_i}}\right) \frac{dz_1 \cdots dz_n}{\sqrt{\lambda_1 \cdots \lambda_n}}
\end{aligned}
$$

where φ is the density of the standard normal distribution function. It is straightforward to show that $\varphi'(y) = -y\varphi(y)$. Moreover, $q(\exp(z))$ is bounded by $\exp(z)$ (the norm of the basis functions is one), which is integrable with respect to $z\varphi(z)dz$. Thus, by Theorem 2.27 in Folland [104] we can move the derivative $d/d\epsilon$ inside the integral, to obtain

$$
\begin{aligned}
&\frac{d}{d\epsilon}\mathbb{E}_{\mathbb{Q}}[q(\exp(\mathcal{P}_n(\mathcal{S}_{\tau-t}\mathbf{g}+\epsilon\mathcal{S}_{\tau-t}\overline{\mathbf{h}}+\mathbf{X}_{t,\tau})))] \\
&= \int_{\mathbb{R}^n} q(\exp(\sum_{i=1}^n z_i e_i))\frac{d}{d\epsilon}\prod_{i=1}^n \varphi\left(\frac{z_i-\langle \mathcal{S}_{\tau-t}\mathbf{g}, e_i\rangle_H - \epsilon\langle \mathcal{S}_{\tau-t}\overline{\mathbf{h}}, e_i\rangle_H}{\sqrt{\lambda_i}}\right) \\
&\quad\times \frac{dz_1\cdots dz_n}{\sqrt{\lambda_1\cdots\lambda_n}} \\
&= \int_{\mathbb{R}^n} q(\exp(\sum_{i=1}^n z_i e_i))\frac{d}{d\epsilon}\prod_{i=1}^n \varphi\left(\frac{z_i-\langle \mathcal{S}_{\tau-t}\mathbf{g}, e_i\rangle_H - \epsilon\langle \mathcal{S}_{\tau-t}\overline{\mathbf{h}}, e_i\rangle_H}{\sqrt{\lambda_i}}\right) \\
&\qquad\times\left(\sum_{i=1}^n \frac{\langle \mathcal{S}_{\tau-t}\overline{\mathbf{h}}, e_i\rangle_H}{\sqrt{\lambda_i}}\frac{z_i-\langle \mathcal{S}_{\tau-t}\mathbf{g}, e_i\rangle_H - \epsilon\langle \mathcal{S}_{\tau-t}\overline{\mathbf{h}}, e_i\rangle_H}{\sqrt{\lambda_i}}\right)\frac{dz_1\cdots dz_n}{\sqrt{\lambda_1\cdot\lambda_n}} \\
&= \int_{\mathbb{R}^n} q\left(\exp\left(\sum_{i=1}^n(\langle \mathcal{S}_{\tau-t}\mathbf{g}, e_i\rangle_H+\epsilon\langle \mathcal{S}_{\tau-t}\overline{\mathbf{h}}, e_i\rangle_H+\sqrt{\lambda_i}y_i)e_i\right)\right) \\
&\quad\times\left(\sum_{i=1}^n y_i\frac{\langle \mathcal{S}_{\tau-t}\overline{\mathbf{h}}, e_i\rangle_H}{\sqrt{\lambda_i}}\right)\varphi(y_1)\cdots\varphi(y_n)dy_1\cdots dy_n \\
&= \mathbb{E}_{\mathbb{Q}}\left[q\left(\exp(\mathcal{P}_n(\mathcal{S}_{\tau-t}\mathbf{g}+\epsilon\mathcal{S}_{\tau-t}\overline{\mathbf{h}}+\mathbf{X}_{t,\tau}))\right)\sum_{i=1}^n\frac{1}{\lambda_i}\langle \mathcal{S}_{\tau-t}\overline{\mathbf{h}}, e_i\rangle\langle \mathbf{X}_{t,\tau}, e_i\rangle\right].
\end{aligned}
$$

By the linear growth of q and an application of Fernique's Theorem again (see Peszat and Zabczyk [165, Thm. 3.31]) we can bound the expression inside the expectation above by $|X_{t,\tau}|_h\exp(|X_{t,\tau}|_H) \in L^1(\mathbb{Q})$ uniformly in n for ϵ in a neighborhood of zero. Thus, we can interchange limit in n and the derivative with respect to ϵ in a neighborhood around zero,

$$
\begin{aligned}
&\frac{d}{d\epsilon}\mathbb{E}_{\mathbb{Q}}[q(\exp(\mathcal{S}_{\tau-t}\mathbf{g}+\epsilon\mathcal{S}_{\tau-t}\overline{\mathbf{h}}+\mathbf{X}_{t,\tau}))] \\
&\quad= \lim_{n\to\infty}\frac{d}{d\epsilon}\mathbb{E}_{\mathbb{Q}}[q(\exp(\mathcal{P}_n(\mathcal{S}_{\tau-t}\mathbf{g}+\epsilon\mathcal{S}_{\tau-t}\overline{\mathbf{h}}+\mathbf{X}_{t,\tau})))] \\
&\quad= \lim_{n\to\infty}\mathbb{E}_{\mathbb{Q}}\left[q\left(\exp(\mathcal{P}_n(\mathcal{S}_{\tau-t}\mathbf{g}+\epsilon\mathcal{S}_{\tau-t}\overline{\mathbf{h}}+\mathbf{X}_{t,\tau}))\right)\right. \\
&\qquad\left.\times\sum_{i=1}^n\frac{1}{\lambda_i}\langle \mathcal{S}_{\tau-t}\overline{\mathbf{h}}, e_i\rangle\langle \mathbf{X}_{t,\tau}, e_i\rangle\right].
\end{aligned}
$$

$Q_{t,\tau}$ is a nuclear operator and thus not invertible. But assuming $\overline{\mathbf{h}} \in \mathrm{Dom}(Q_{t,\tau}^{-1}) \subset H$ we get

$$\sum_{i=1}^{n} \frac{1}{\lambda_i} \langle \overline{\mathbf{h}}, e_i \rangle_H \langle \mathbf{X}_{t,\tau}, e_i \rangle_H = \langle Q_{t,\tau}^{-1} \overline{\mathbf{h}}, \mathcal{P}_n \mathbf{X}_{t,\tau} \rangle_H$$

Now, using the same arguments as above we can show that

$$\begin{aligned} &\frac{d}{d\epsilon} \mathbb{E}_{\mathbb{Q}}[q(\exp(\mathcal{S}_{\tau-t}\mathbf{g} + \epsilon \mathcal{S}_{\tau-t}\overline{\mathbf{h}} + \mathbf{X}_{t,\tau}))] \\ &= \lim_{n\to\infty} \left\langle Q_{t,\tau}^{-1}\overline{\mathbf{h}}, \mathbb{E}_{\mathbb{Q}}\left[q\left(\exp(\mathcal{P}_n(\mathcal{S}_{\tau-t}\mathbf{g} + \epsilon \mathcal{S}_{\tau-t}\overline{\mathbf{h}} + \mathbf{X}_{t,\tau}))\right) \mathcal{P}_n \mathbf{X}_{t,\tau} \right]\right\rangle_H \\ &= \left\langle Q_{t,\tau}^{-1}\overline{\mathbf{h}}, \mathbb{E}_{\mathbb{Q}}\left[q\left(\exp(\mathcal{S}_{\tau-t}\mathbf{g} + \epsilon \mathcal{S}_{\tau-t}\overline{\mathbf{h}} + \mathbf{X}_{t,\tau})\right) \mathbf{X}_{t,\tau} \right]\right\rangle_H \end{aligned}$$

The claim follows after choosing $\epsilon = 0$ above. □

To select $\mathbf{h} \in H$ such that $\mathcal{S}_{\tau-t}(\mathbf{h}\mathbf{f}^{-1}) \in \mathrm{Dom}(Q_{t,\tau}^{-1})$ is obviously not simple. Leaving this question aside, let us make some notes on the implication of the Gateaux derivative. It provides a measure for how sensitive the option price is to small changes in the current term structure of forward prices. For example, constant functions are elements in the Filipović space, and working with that state space we may find the sensitivity of the option price with respect to parallel shifts in the current forward curve. Notice that the condition changes to $\mathcal{S}_{\tau-t}(\mathbf{f}^{-1}) \in \mathrm{Dom}(Q_{t,\tau}^{-1})$ when $\mathbf{h}$ is a constant. Further, looking at functions $\mathbf{h}$ which has a support in specific intervals of maturities, we can find the sensitivity of the option price with respect to change in those parts of the term structure.

We end this Section by remarking that in Section 7.5 below, we study continuity of option prices with respect to the initial curve in more generality, based on a Markovian forward price dynamics.

7.3 Fourier-Based Pricing of Options on Simple HJM-Models

In this Section we focus on deriving expressions of the option price $C(t;\tau)$ in (7.2) using the Fourier transform. Carr and Madan [71] suggested the Fourier transform as an approach to price options on non-Gaussian models, where one obtains prices expressed in terms of the integrals over the (conditional) characteristic exponent of the price dynamics of the underlying and the Fourier transform of the payoff function. Indeed, the price can be computed numerically based on fast Fourier transform, and allows for a splitting of the payoff from the dynamics of the underlying.

To this end, recall from Folland [104] that the Fourier transform of a function $v \in L^1(\mathbb{R})$ is defined by

$$\widehat{v}(y) = \int_{\mathbb{R}} v(x)e^{-\mathrm{i}xy}dx. \tag{7.14}$$

Moreover, if $\widehat{v} \in L^1(\mathbb{R})$, then the inverse Fourier transform holds, and yields (see Folland [104])

$$v(x) = \frac{1}{2\pi}\int_{\mathbb{R}} \widehat{v}(y)e^{\mathrm{i}xy}dy. \tag{7.15}$$

Notice the sign in the exponent of the exponential of the Fourier transform and its inverse, which is chosen to compare with the characteristic exponent of random variables.

Assume now that the forward price dynamics $\mathbf{f}(t)$ follow the dynamics in (6.7), but where we assume that $t \mapsto \mathbf{\Sigma}(t)$ is a measurable function with values in $L(H)$ satisfying $\int_0^T \|\mathbf{\Sigma}(t)Q^{1/2}\|_{\mathrm{HS}}^2 dt < \infty$ for any $T < \infty$. Here, Q is the covariance operator of the mean-zero square-integrable H-valued Lévy process $\mathbf{L}(t)$. We have the general Fourier-based option price:

Proposition 7.5 *Assume that $p \in L^1(\mathbb{R})$ with a Fourier transform $\widehat{p} \in L^1(\mathbb{R})$. Then,*

$$C(t;\tau) = \frac{1}{2\pi}e^{-r(\tau-t)}\int_{\mathbb{R}} \widehat{p}(y)\exp \\ \times\left(\mathrm{i}yG\mathcal{S}_{\tau-t}\mathbf{f}(t) + \int_t^\tau \varphi_L(y\mathbf{\Sigma}(s)^*\mathcal{S}_{\tau-s}^*G^*1)ds\right)dy$$

where φ_L is the Lévy exponent of $\mathbf{L}$ *(defined in Section 2.1 of Chapter 2).*

Proof First, by the inverse Fourier transform we find, after appealing to the Fubini theorem (which is valid since $|\exp(\mathrm{i}xy)| = 1$ and $\widehat{p} \in L^1(\mathbb{R})$),

$$\mathbb{E}_{\mathbb{Q}}[p(G\mathbf{f}(\tau))\,|\,\mathcal{F}_t] = \frac{1}{2\pi}\int_{\mathbb{R}} \widehat{p}(y)\mathbb{E}_{\mathbb{Q}}[\exp(\mathrm{i}yG\mathbf{f}(\tau))\,|\,\mathcal{F}_t]dy.$$

It holds for $\tau \geq t \geq 0$,

$$\mathbf{f}(\tau) = \mathcal{S}_{\tau-t}\mathbf{f}(t) + \int_t^\tau \mathcal{S}_{\tau-s}\mathbf{\Sigma}(s)d\mathbf{L}(s)$$

where we find that the stochastic integral on the right-hand side is independent of $\mathcal{F}_t$ due to the independent increment property of $\mathbf{L}$. In view of Proposition 2.1, we have,

$$\mathbb{E}_{\mathbb{Q}}[\exp(\mathrm{i}yG\int_t^\tau \mathcal{S}_{\tau-s}\mathbf{\Sigma}(s)d\mathbf{L}(s))\,|\,\mathcal{F}_t] = \mathbb{E}_{\mathbb{Q}}[\exp(\mathrm{i}yG\int_t^\tau \mathcal{S}_{\tau-s}\mathbf{\Sigma}(s)d\mathbf{L}(s))]$$
$$= \exp(\int_t^\tau \varphi_L(y\mathbf{\Sigma}(s)^*\mathcal{S}^*_{\tau-s}G^*1)ds)$$

where $G^*1 \in H$ is the representative of the linear functional $G \in H^*$ such that $Gh = \langle h, G^*1\rangle_H$ for any $h \in H$. The result follows. □

We notice that the payoff function of a call option $p(x) = \max(x - K, 0)$ is not integrable on $\mathbb{R}$. But an exponential damping of it will be integrable, i.e., $p_\alpha(x) := \exp(-\alpha x)\max(x - K, 0) \in L^1(\mathbb{R})$ for any $\alpha > 0$. A simple calculation reveals that the Fourier transform of p_α becomes

$$\widehat{p}_\alpha(y) = (\alpha + \mathrm{i}y)^{-2}\exp(-(\alpha + \mathrm{i}y)K).$$

Since,

$$|\widehat{p}_\alpha(y)| = e^{-\alpha K}(\alpha^2 + y^2)^{-1}$$

it follows that $|\widehat{p}_\alpha(y)| \in L^1(\mathbb{R})$. Thus, by the inverse Fourier transform,

$$\max(x - K, 0) = e^{\alpha x}p_\alpha(x) = \frac{1}{2\pi}\int_{\mathbb{R}} \widehat{p}_\alpha(y)e^{ixy+\alpha x}dy.$$

Following the arguments in the proof of Proposition 7.5, we need to compute the conditional expectation of $\exp((iy+\alpha)G\mathbf{f}(\tau))$. This requires exponential integrability of the real-valued random variable $\alpha\int_t^\tau G\mathcal{S}_{\tau-s}\mathbf{\Sigma}(s)d\mathbf{L}(s)$. As Gaussian random variables are exponentially integrable, we focus on pure-jump Lévy processes $\mathbf{L}$. If $\alpha\int_t^\tau G\mathcal{S}_{\tau-s}\mathbf{\Sigma}(s)d\mathbf{L}(s)$ is exponentially integrable, we find that

$$\mathbb{E}_{\mathbb{Q}}[\exp(\alpha\int_t^\tau G\mathcal{S}_{\tau-s}\mathbf{\Sigma}(s)d\mathbf{L}(s))] = \exp(\int_t^\tau \varphi_L(-\mathrm{i}\alpha\mathbf{\Sigma}(s)^*\mathcal{S}^*_{\tau-s}G^*1)ds).$$

But in the pure-jump (square-integrable) case, we recall from Section 2.1 in Chapter 2 that the Lévy exponent is

$$\varphi_L(\mathbf{u}) = \int_H \{e^{\mathrm{i}\langle\mathbf{u},\mathbf{z}\rangle} - 1 - \langle\mathbf{u},\mathbf{z}\rangle_H\}\,\nu(d\mathbf{z})$$

for the Lévy measure ν of $\mathbf{L}$ on H. Thus, denoting by $\mathbf{h}(s) := \mathbf{\Sigma}(s)^*\mathcal{S}^*_{\tau-s}G^*1 \in H$, we find

$$\varphi_L(-\mathrm{i}\alpha\mathbf{h}(s)) = \int_H \{e^{\alpha\langle\mathbf{h}(s),\mathbf{z}\rangle_H} - 1 - \alpha\langle\mathbf{h}(s),\mathbf{z}\rangle_H\}\,\nu(d\mathbf{z}).$$

Hence, a sufficient condition for the well-definedness of this integral, and therefore the exponential integrability of $\alpha \int_t^\tau G\mathcal{S}_{\tau-s}\mathbf{\Sigma}(s)\,d\mathbf{L}(s)$, is

$$\int_0^\tau \int_{|z|_H>1} \exp(\alpha \|\mathbf{\Sigma}(s)\|_{\text{op}} \|\mathcal{S}_{\tau-s}\|_{\text{op}} \|G\|_{\text{op}} |\mathbf{z}|_H)\nu(d\mathbf{z})ds < \infty. \tag{7.16}$$

In deriving this sufficient condition, we made use of the integrability of $|z|_H^2$ with respect to the Lévy measure $\nu(d\mathbf{z})$ when $|z|_H < 1$.

We can also derive an expression in terms of a Fourier integral for the sensitivity of the option price with respect to perturbations of the current forward curve $\mathbf{f}(t)$, when measuring this sensitivity in terms of the Gateaux derivative. By Proposition 7.5 we can view the option price $C(t;\tau)$ as a function of $\mathbf{f}(t)$, denoted as $\overline{C}(t;\tau,\mathbf{f}(t))$. We find the following for the *delta* of the option

Lemma 7.3 *Let* $\mathbf{h} \in H$. *If,* $y \mapsto y\widehat{p}(y) \in L^1(\mathbb{R})$, *then*

$$D_{\mathbf{h}}\overline{C}(t;\tau,\mathbf{f}(t)) = (\mathrm{i}G\mathcal{S}_{\tau-t}\mathbf{h})\frac{1}{2\pi}e^{-r(\tau-t)} \int_{\mathbb{R}} \widehat{p}(y)y\exp\Big(\mathrm{i}yG\mathcal{S}_{\tau-t}\mathbf{f}(t) + \int_t^\tau \varphi_L(y\mathbf{\Sigma}(s)^*\mathcal{S}_{\tau-s}^*G^*1)ds\Big)dy$$

where $D_{\mathbf{h}}$ *is the Gateaux derivative in direction* $\mathbf{h}$ *and the further notations are as in Proposition 7.5.*

Proof The Gateaux derivative is defined as

$$D_{\mathbf{h}}\overline{C}(t;\tau,\mathbf{f}(t)) = \frac{d}{d\epsilon}\overline{C}(t;\tau,\mathbf{f}(t)+\epsilon\mathbf{h})|_{\epsilon=0}.$$

Evidently,

$$\frac{d}{d\epsilon}\exp\left(\mathrm{i}yG\mathcal{S}_{\tau-t}(\mathbf{f}(t)+\epsilon\mathbf{h})\right) = \left(\mathrm{i}yG\mathcal{S}_{\tau-t}\mathbf{h}\right)\exp\left(\mathrm{i}yG\mathcal{S}_{\tau-t}(\mathbf{f}(t)+\epsilon\mathbf{h})\right).$$

We notice that

$$\left|\exp\Big(\mathrm{i}yG\mathcal{S}_{\tau-t}(\mathbf{f}(t)+\epsilon\mathbf{h}) + \int_t^\tau \varphi_L(y\mathbf{\Sigma}(s)^*\mathcal{S}_{\tau-s}^*G^*1)ds\Big)\right| \le 1$$

and hence, by the assumption that $y \mapsto y\widehat{p}(y) \in L^1(\mathbb{R})$, we can move the derivative inside the integral by Theorem 2.27(b) in Folland [104]. The result follows. □

If $p \in L^1(\mathbb{R})$ is continuously differentiable with $p' \in L^1(\mathbb{R})$ and zero when $|y| \to \infty$, then according to Theorem 8.22(e) in Folland [104] it holds that $\mathrm{i}y\widehat{p}(y) = \widehat{p'}(y)$. In this case the assumption in the Lemma above is an integrability condition on the Fourier transform of the derivative p' of the payoff function.

7.3.1 Geometric Models and Pricing by Fourier Techniques

The case of geometric models for the forward price is somewhat more delicate. Consider $\mathbf{f}(t) = \exp(\mathbf{g}(t))$, where by (6.12),

$$\mathbf{g}(t) = \mathcal{S}_t \mathbf{g}_0 + \int_0^t \mathcal{S}_{t-s}\boldsymbol{\mu}(s)ds + \int_0^t \mathcal{S}_{t-s}\boldsymbol{\Sigma}(s)d\mathbf{L}(s) \tag{7.17}$$

for a square-integrable mean-zero Lévy process $\mathbf{L}$ on H with covariance operator Q. Furthermore, $t \mapsto \boldsymbol{\Sigma}(s)$ is a measurable function with values in $L(H)$, where we assume that $\int_0^T \|\boldsymbol{\Sigma}(s)Q^{1/2}\|_{\mathrm{HS}}^2 ds < \infty$ for all $T < \infty$. Proposition 6.3 yields the martingale drift condition relating μ to $\boldsymbol{\Sigma}$ and the covariance operator of $\mathbf{L}$.

In order for the Fourier approach to work for geometric simple HJM-models, we introduce a condition on the gathering functional G (as was also discussed for the Gaussian models above): Suppose there exists a functional $\widetilde{G} \in H^*$ such that

$$G\exp(\mathbf{g}(t)) = \exp(\widetilde{G}\mathbf{g}(t)). \tag{7.18}$$

Notice that this condition on the gathering functional is not satisfied for swaps, but it is for fixed-delivery forwards. Indeed, $\delta_x \exp(\mathbf{g}) = \exp(\mathbf{g}(x)) = \exp(\delta_x \mathbf{g})$, so $G = \widetilde{G}$ in this case. Under some natural integrability conditions on the function $x \mapsto p(\exp(x))$, we find the following option price,

Proposition 7.6 *Suppose that there exists a linear functional $\widetilde{G} \in H^*$ such that $G\exp(\mathbf{g}) = \exp(\widetilde{G}\mathbf{g})$ for every $\mathbf{g} \in H$. If $q \in L^1(\mathbb{R})$ and $\widehat{q} \in L^1(\mathbb{R})$ for $q(x) := p(\exp x)$, then*

$$C(t;\tau) = \frac{1}{2\pi}e^{-r(\tau-t)}\int_{\mathbb{R}} \widehat{q}(y)\exp\Big(\mathrm{i}y\widetilde{G}\mathcal{S}_{\tau-t}\mathbf{g}(t) \\ +\mathrm{i}y\int_t^\tau \widetilde{G}\mathcal{S}_{\tau-s}\boldsymbol{\mu}(s)ds + \int_t^\tau \varphi_L(y\boldsymbol{\Sigma}(s)^*\mathcal{S}_{\tau-s}^*\widetilde{G}^*1)ds\Big)dy$$

where φ_L is the Lévy exponent of $\mathbf{L}$ (defined in Section 2.1 of Chapter 2) and $\boldsymbol{\mu}$ is given by Proposition 6.3.

Proof By assumption on G and p, we find from Fourier inversion formula,

$$p(G\mathbf{f}(\tau)) = p(\exp(\widetilde{G}\mathbf{g}(\tau)) = q(\widetilde{G}\mathbf{g}(\tau)) = \frac{1}{2\pi}\int_{\mathbb{R}} \widehat{q}(y)\exp(\mathrm{i}y\widetilde{G}\mathbf{g}(\tau))dy.$$

It holds that

$$\mathbf{g}(\tau) = \mathcal{S}_{\tau-t}\mathbf{g}(t) + \int_t^\tau \mathcal{S}_{\tau-s}\boldsymbol{\mu}(s)ds + \int_t^\tau \mathcal{S}_{\tau-s}\boldsymbol{\Sigma}(s)d\mathbf{L}(s)$$

and we can follow the lines of the proof of Proposition 7.5 to conclude. □

The most relevant options are calls, and as we have already seen above, its payoff function $p(z) = \max(z - K, 0)$ is not integrable on $\mathbb{R}$. This non-integrability becomes even more pronounced for exponential models, as

$$p(\exp x) = \max(e^x - K, 0) \sim e^x$$

when $x \to \infty$. Again we can consider a dampened version of p, namely

$$q_\alpha(x) = e^{-\alpha x} \max(e^x - K, 0)$$

with $\alpha > 1$. Since $q_\alpha(x) = 0$ for $x < \ln K$, we find that $q_\alpha \in L^1(\mathbb{R})$. Moreover, a direct computation reveals,

$$\widehat{q}_\alpha(y) = \frac{K^{1-\alpha-\mathrm{i}y}}{((\alpha - 1) + \mathrm{i}y)(\alpha + \mathrm{i}y)}$$

from which it follows

$$|\widehat{q}_\alpha(y)| = \frac{K^{1-\alpha}}{((\alpha - 1)^2 + y^2)^{1/2}(\alpha^2 + y^2)^{1/2}}$$

which is a continuous and bounded function on $\mathbb{R}$ with $|\widehat{q}_\alpha(y)| \sim K^{1-\alpha}/y^2$ when $|y| \to \infty$. Hence, $\widehat{q}_\alpha \in L^1(\mathbb{R})$, and we can apply Fourier techniques to derive a pricing formula as in Proposition 7.6 above. Of course, we need to impose an exponential integrability condition on $\alpha \int_t^\tau \widetilde{G}\mathcal{S}_{\tau-s}\boldsymbol{\Sigma}(s)d\mathbf{L}(s)$ for $\alpha > 1$. A sufficient condition for this to hold is stated in (7.16) with $\widetilde{G}$ instead of G.

We can also in this case derive a *delta* for the option price similar to Lemma 7.3 as a sensitivity measure for perturbation in the forward curve. Recalling Proposition 7.6, the option price can be written as a function of $\mathbf{g}(t)$, $C(t;\tau) := \overline{C}(t;\tau,\mathbf{g}(t))$. Moreover, following the argument of Lemma 7.3 with the assumption that $y \mapsto y\widehat{q}(y) \in L^1(\mathbb{R})$, we derive

$$\begin{aligned} D_{\mathbf{h}}\overline{C}(t;\tau,\mathbf{g}(t)) = (\mathrm{i}\widetilde{G}\mathcal{S}_{\tau-t}\mathbf{h})\frac{1}{2\pi}e^{-r(\tau-t)} \int_{\mathbb{R}} \widehat{q}(y)y \exp\Big(\mathrm{i}y\widetilde{G}\mathcal{S}_{\tau-t}\mathbf{g}(t) \\ +\mathrm{i}y \int_t^\tau \widetilde{G}\mathcal{S}_{\tau-s}\boldsymbol{\mu}(s)ds + \int_t^\tau \varphi_L(y\boldsymbol{\Sigma}(s)^*\mathcal{S}_{\tau-s}^*\widetilde{G}^*1)ds\Big)\,dy \end{aligned}$$

for any $\mathbf{h} \in H$. The Fourier expression of the Gateaux derivative above uses no conditions on the directions $\mathbf{h}$. In Proposition 7.4 we had to impose a rather restrictive condition on $\mathbf{h}$ in order to apply the density method in infinite dimensions to compute the delta. The Fourier-based approach to option pricing also works for the Gaussian case, of course, and thus allow for rather flexible analysis of deltas.

7.4 Pricing of Options on Simple HJM-Models with Stochastic Volatility

We recall from Section 6.3 of Chapter 6 a stochastic volatility model for $\mathbf{\Sigma}$, which we consider here in the simple arithmetic HJM-dynamics. To this end, let the forward price dynamics $\mathbf{f}(t)$ follow the dynamics in (6.7) with $\mathbf{L} = \mathbf{W}$, a Wiener process with covariance operator $\mathcal{Q}$, where we assume that $t \mapsto \mathbf{\Sigma}(t)$ is given by

$$\mathbf{\Sigma}(t) = \mathcal{M}\mathcal{Y}^{1/2}(t) \tag{7.19}$$

for $\mathcal{Y}(t)$ defined in (6.20). We make the assumptions on the parameters given in Section 6.3 of Chapter 6 to make $\mathcal{Y}$ well-defined as a symmetric positive definite process in $L_{\mathrm{HS}}(H)$, the space of Hilbert-Schmidt operators on H. Moreover, $\mathcal{M} \in L(H)$, and can be thought of as the operator modelling the Samuelson effect (see discussion at the end of Section 6.3 of Chapter 6).

Using the Fourier approach, it turns out that under some additional conditions on the stochastic volatility process $\mathcal{Y}(t)$ and the covariance operator $\mathcal{Q}$ of the driving Wiener process $\mathbf{W}$ in the forward price dynamics, we can express the option price as a Fourier integral. The precise result is formulated in the following Proposition:

Proposition 7.7 *Assume $p \in L^1(\mathbb{R})$ with a Fourier transform $\widehat{p} \in L^1(\mathbb{R})$. If there exists an operator $\mathcal{B} \in L(H)$ such that $\mathcal{Y}^{1/2}(t)\mathcal{Q}\mathcal{Y}^{1/2}(t) = \mathcal{B}\mathcal{Y}(t)\mathcal{B}^*$ for all $t \geq 0$, then*

$$\begin{aligned} C(t;\tau) = \frac{1}{2\pi} e^{-r(\tau-t)} \int_{\mathbb{R}} \widehat{p}(y) \exp\left(\mathrm{i}yG\mathcal{S}_{\tau-t}\mathbf{f}(t)\right) \\ \times \exp\left(-\frac{1}{2}y^2 \langle \mathcal{Y}(t), \int_0^{\tau-t} \mathfrak{S}_s^*(\mathcal{B}^*\mathcal{M}^*\mathcal{S}_{\tau-t-s}^* G^* 1)^{\otimes 2} ds\rangle_{HS}\right) \\ \times \exp\left(\int_0^{\tau-t} \varphi_{\mathcal{L}}\left(\frac{\mathrm{i}}{2}y^2 \int_0^s \mathfrak{S}_u^*(\mathcal{B}^*\mathcal{M}^*\mathcal{S}_{s-u}^* G^* 1)^{\otimes 2} du\right) ds\right) dy, \end{aligned}$$

where $\varphi_{\mathcal{L}}$ is the Lévy exponent of $\mathcal{L}$ (defined in Section 2.1 in Chapter 2).

Proof Following the first part of the proof of Prop. 7.5 yields that

$$\begin{aligned} C(t;\tau) = \frac{1}{2\pi} e^{-r(\tau-t)} \int_{\mathbb{R}} \widehat{p}(y) \exp\left(\mathrm{i}yG\mathcal{S}_{\tau-t}\mathbf{f}(t)\right) \\ \times \mathbb{E}_{\mathbb{Q}}\left[\exp\left(\mathrm{i}yG\int_t^\tau \mathcal{S}_{\tau-s}\mathcal{M}\mathcal{Y}^{1/2}(s)d\mathbf{W}(s)\right) \mid \mathcal{F}_t\right] dy. \end{aligned}$$

The remainder of the proof consists in computing the conditional expectation in the expression above. To this end, recall that $\mathcal{L}$, and thus $\mathcal{Y}$, is independent of $\mathbf{W}$ by definition. Introduce the σ-algebra $\mathcal{F}_{t,\tau}^{\mathcal{Y}}$ generated by $\mathcal{F}_t$ and the paths of $\mathcal{Y}(s)$ for

$s \leq \tau$. Using the tower property of conditional expectation yields

$$
\begin{aligned}
&\mathbb{E}_{\mathbb{Q}}\left[\exp\left(\mathrm{i}yG\int_t^\tau \mathcal{S}_{\tau-s}\mathcal{M}\mathcal{Y}^{1/2}(s)d\mathbf{W}(s)\right)\mid\mathcal{F}_t\right] \\
&\quad=\mathbb{E}_{\mathbb{Q}}\left[\mathbb{E}_{\mathbb{Q}}\left[\exp\left(\mathrm{i}yG\int_t^\tau \mathcal{S}_{\tau-s}\mathcal{M}\mathcal{Y}^{1/2}(s)d\mathbf{W}(s)\right)\mid\mathcal{F}^{\mathcal{Y}}_{t,\tau}\right]\mid\mathcal{F}_t\right] \\
&\quad=\mathbb{E}_{\mathbb{Q}}\left[\exp\left(-\frac{1}{2}y^2\int_t^\tau G\mathcal{S}_{\tau-s}\mathcal{M}\mathcal{Y}^{1/2}(s)\mathcal{Q}\mathcal{Y}^{1/2}(s)\mathcal{M}^*\mathcal{S}^*_{\tau-s}\mathcal{G}^*1ds\right)\mid\mathcal{F}_t\right] \\
&\quad=\mathbb{E}_{\mathbb{Q}}\left[\exp\left(-\frac{1}{2}y^2\int_t^\tau G\mathcal{S}_{\tau-s}\mathcal{M}\mathcal{B}\mathcal{Y}(s)\mathcal{B}^*\mathcal{M}^*\mathcal{S}^*_{\tau-s}\mathcal{G}^*1ds\right)\mid\mathcal{F}_t\right].
\end{aligned}
$$

Here, we used in the second equality that the stochastic integral with respect to $\mathbf{W}$ conditioned on $\mathcal{F}^{\mathcal{Y}}_{t,\tau}$ becomes a Wiener integral, while the third equality follows by assumption on $\mathcal{Q}$ and $\mathcal{Y}^{1/2}(t)$. We notice that $G^*1 \in H$.

For $s \geq t$, we have from the dynamics of $\mathcal{Y}$ that

$$
\mathcal{Y}(s) = \mathfrak{S}_{s-t}\mathcal{Y}(t) + \int_t^s \mathfrak{S}_{s-u}d\mathcal{L}(u).
$$

Therefore, since $\mathcal{Y}(t)$ is $\mathcal{F}_t$-measurable and $\mathcal{L}$ is a Lévy process and thus has independent increments,

$$
\begin{aligned}
&\mathbb{E}_{\mathbb{Q}}\left[\exp\left(\mathrm{i}yG\int_t^\tau \mathcal{S}_{\tau-s}\mathcal{M}\mathcal{Y}^{1/2}(s)d\mathbf{W}(s)\right)\mid\mathcal{F}_t\right] \\
&\quad=\exp\left(-\frac{1}{2}y^2\int_t^\tau G\mathcal{S}_{\tau-s}\mathcal{M}\mathcal{B}\mathfrak{S}_{s-t}\mathcal{Y}(t)\mathcal{B}^*\mathcal{M}^*\mathcal{S}^*_{\tau-s}G^*1ds\right) \qquad (7.20) \\
&\quad\quad\times\mathbb{E}_{\mathbb{Q}}\left[\exp\left(-\frac{1}{2}y^2\int_t^\tau G\mathcal{S}_{\tau-s}\mathcal{M}\mathcal{B}(\int_t^s \mathfrak{S}_{s-u}d\mathcal{L}(u))\mathcal{B}^*\mathcal{M}^*\mathcal{S}^*_{\tau-s}G^*1ds\right)\right].
\end{aligned}
$$

Let us investigate the first exponential in (7.20). The integral can be computed as follows:

$$
\begin{aligned}
&\int_t^\tau G\mathcal{S}_{\tau-s}\mathcal{M}\mathcal{B}\mathfrak{S}_{s-t}\mathcal{Y}(t)\mathcal{B}^*\mathcal{M}^*\mathcal{S}^*_{\tau-s}G^*1ds \\
&\qquad=\int_t^\tau \langle\mathcal{S}_{\tau-s}\mathcal{M}\mathcal{B}\mathfrak{S}_{s-t}\mathcal{Y}(t)\mathcal{B}^*\mathcal{M}^*\mathcal{S}^*_{\tau-s}G^*1, G^*1\rangle_H ds \\
&\qquad=\int_t^\tau \langle\mathfrak{S}_{s-t}\mathcal{Y}(t)\mathcal{B}^*\mathcal{M}^*\mathcal{S}^*_{\tau-s}G^*1, \mathcal{B}^*\mathcal{M}^*\mathcal{S}^*_{\tau-s}G^*1\rangle_H ds
\end{aligned}
$$

$$= \int_t^\tau \langle \mathfrak{S}_{s-t}\mathcal{Y}(t), (\mathcal{B}^*\mathcal{M}^*\mathcal{S}^*_{\tau-s}G^*1)^{\otimes 2}\rangle_{\text{HS}} ds$$

$$= \langle \mathcal{Y}(t), \int_0^{\tau-t} \mathfrak{S}^*_s(\mathcal{B}^*\mathcal{M}^*\mathcal{S}^*_{\tau-t-s}G^*1)^{\otimes 2} ds\rangle_{\text{HS}}.$$

In the last equality, we did an appropriate change of variables in the Bochner integral.

The stochastic integral in the exponent inside the expectation operator in (7.20) becomes

$$\int_t^\tau G\mathcal{S}_{\tau-s}\mathcal{M}\mathcal{B}(\int_t^s \mathfrak{S}_{s-u}d\mathcal{L}(u))\mathcal{B}^*\mathcal{M}^*\mathcal{S}^*_{\tau-s}G^*1 ds$$

$$= \langle \int_t^\tau \mathcal{S}_{\tau-s}\mathcal{M}\mathcal{B}(\int_t^s \mathfrak{S}_{s-u}d\mathcal{L}(u))\mathcal{B}^*\mathcal{M}^*\mathcal{S}^*_{\tau-s}G^*1 ds, G^*1\rangle_H$$

$$= \langle \int_t^\tau \int_u^\tau \mathcal{S}_{\tau-s}\mathcal{M}\mathcal{B}(\mathfrak{S}_{s-u}d\mathcal{L}(u))\mathcal{B}^*\mathcal{M}^*\mathcal{S}^*_{\tau-s}ds G^*1, G^*1\rangle_H,$$

where we used the stochastic Fubini Theorem in the second equality (see Peszat and Zabczyk [165, Thm. 8.14] or the proof of Prop. 3.2 in Benth, Rüdiger and Süss [45]). The integration in the last expression means that we *first* integrate with respect to ds, and *then* integrate with respect to $d\mathcal{L}(u)$. It is the non-commutativity of the operators that leaves us with this interpretation of the notation of the double integral above. Notice that the $d\mathcal{L}(u)$-stochastic integral has a deterministic integrand. For a partition $\mathcal{P}$ of $[t, \tau]$ such that $|\mathcal{P}| \to 0$, we find by appealing to the independent increment property of $\mathcal{L}$,

$$\mathbb{E}_{\mathbb{Q}}\left[\exp\left(-\frac{1}{2}y^2\int_t^\tau G\mathcal{S}_{\tau-s}\mathcal{M}\mathcal{B}(\int_t^s \mathfrak{S}_{s-u}d\mathcal{L}(u))\mathcal{B}^*\mathcal{M}^*\mathcal{S}^*_{\tau-s}G^*1 ds\right)\right]$$

$$= \lim_{|\mathcal{P}|\to 0}\mathbb{E}_{\mathbb{Q}}\left[\exp\left(-\frac{1}{2}y^2\sum_{u_i\in\mathcal{P}}\right.\right.$$

$$\left.\left.\times \langle \int_{u_i}^\tau \mathcal{S}_{\tau-s}\mathcal{M}\mathcal{B}\mathfrak{S}_{s-u_i}\Delta\mathcal{L}(u_i)\mathcal{B}^*\mathcal{M}^*\mathcal{S}^*_{\tau-s}ds G^*1, G^*1\rangle_H\right)\right]$$

$$= \lim_{|\mathcal{P}|\to 0}\prod_{u_i\in\mathcal{P}}\mathbb{E}_{\mathbb{Q}}\left[\exp\left(-\frac{1}{2}y^2\right.\right.$$

$$\left.\left.\times \langle \int_{u_i}^\tau \mathcal{S}_{\tau-s}\mathcal{M}\mathcal{B}\mathfrak{S}_{s-u_i}\Delta\mathcal{L}(u_i)\mathcal{B}^*\mathcal{M}^*\mathcal{S}^*_{\tau-s}ds G^*1, G^*1\rangle_H\right)\right].$$

Let us analyse the ds-integral in the expression above. By commuting the inner product and the integral, we get

$$\langle\int_{u_i}^{\tau} \mathcal{S}_{\tau-s}\mathcal{M}\mathcal{B}\mathfrak{S}_{s-u_i}\Delta\mathcal{L}(u_i)\mathcal{B}^*\mathcal{M}^*\mathcal{S}_{\tau-s}^*dsG^*1, G^*1\rangle_H$$

$$= \int_{u_i}^{\tau} \langle\mathcal{S}_{\tau-s}\mathcal{M}\mathcal{B}\mathfrak{S}_{s-u_i}\Delta\mathcal{L}(u_i)\mathcal{B}^*\mathcal{M}^*\mathcal{S}_{\tau-s}^*G^*1, G^*1\rangle_H ds$$

$$= \int_{u_i}^{\tau} \langle\mathfrak{S}_{s-u_i}\Delta\mathcal{L}(u_i)\mathcal{B}^*\mathcal{M}^*\mathcal{S}_{\tau-s}^*G^*1, \mathcal{B}^*\mathcal{M}^*\mathcal{S}_{\tau-s}^*G^*1\rangle ds$$

$$= \int_{u_i}^{\tau} \langle\Delta\mathcal{L}(u_i), \mathfrak{S}_{s-u_i}^*(\mathcal{B}^*\mathcal{M}^*\mathcal{S}_{\tau-s}^*G^*1)^{\otimes 2}\rangle_{\mathrm{HS}} ds$$

$$= \langle\Delta\mathcal{L}(u_i), \int_{u_i}^{\tau} \mathfrak{S}_{s-u_i}^*(\mathcal{B}^*\mathcal{M}^*\mathcal{S}_{\tau-s}^*G^*1)^{\otimes 2} ds\rangle_{\mathrm{HS}}.$$

As $\Delta\mathcal{L}(u_i)$ is positive definite and $\mathfrak{S}$ is positivity preserving, we see from the above derivations that

$$\langle\Delta\mathcal{L}(u_i), \int_{u_i}^{\tau} \mathfrak{S}_{s-u_i}^*(\mathcal{B}^*\mathcal{M}^*\mathcal{S}_{\tau-s}^*G^*1)^{\otimes 2} ds\rangle_{\mathrm{HS}} \geq 0$$

from which it follows

$$\exp\left(-\frac{1}{2}y^2\langle\Delta\mathcal{L}(u_i), \int_{u_i}^{\tau} \mathfrak{S}_{s-u_i}^*(\mathcal{B}^*\mathcal{M}^*\mathcal{S}_{\tau-s}^*G^*1)^{\otimes 2} ds\rangle_{\mathrm{HS}}\right) \leq 1$$

and thus being a $\mathbb{Q}$-integrable random variable. This implies, using the definition of the Lévy exponent of $\mathcal{L}$,

$$\mathbb{E}_{\mathbb{Q}}\left[\exp\left(-\frac{1}{2}y^2\int_t^{\tau} G\mathcal{S}_{\tau-s}\mathcal{M}\mathcal{B}(\int_t^s \mathfrak{S}_{s-u}d\mathcal{L}(u))\mathcal{B}^*\mathcal{M}^*\mathcal{S}_{\tau-s}^*G^*1ds\right)\right]$$

$$= \lim_{|\mathcal{P}|\to 0}\prod_{u_i\in\mathcal{P}} \mathbb{E}_{\mathbb{Q}}\left[\exp\left(\mathrm{i}\langle\Delta\mathcal{L}(u_i), \frac{\mathrm{i}}{2}y^2\int_{u_i}^{\tau} \mathfrak{S}_{s-u_i}^*(\mathcal{B}^*\mathcal{M}^*\mathcal{S}_{\tau-s}^*G^*1)^{\otimes 2} ds\rangle_{\mathrm{HS}}\right)\right]$$

$$= \lim_{|\mathcal{P}|\to 0}\exp\left(\sum_{u_i\in\mathcal{P}} \varphi_{\mathcal{L}}\left(\frac{\mathrm{i}}{2}y^2\int_{u_i}^{\tau} \mathfrak{S}_{s-u_i}^*(\mathcal{B}^*\mathcal{M}^*\mathcal{S}_{\tau-s}^*G^*1)^{\otimes 2} ds\right)\Delta u_i\right)$$

$$= \exp\left(\int_0^{\tau-t} \varphi_{\mathcal{L}}\left(\frac{\mathrm{i}}{2}y^2\int_0^s \mathfrak{S}_u^*(\mathcal{B}^*\mathcal{M}^*\mathcal{S}_{s-u}^*G^*1)^{\otimes 2} du\right) ds\right).$$

After passing to the limit, we have performed a convenient change of variables in the last equality. The proof is complete. □

The proof of the above result is adopted from Benth and Sgarra [52, Proof of Props. 2.3 and 5.7]. We notice that the option price is dependent on both the current forward curve $\mathbf{f}(t)$ as well as the current stochastic variance operator $\mathcal{Y}(t)$. Thus, to price, we need to know the current state of the stochastic volatility and not only the price forward curve.

In the formulation of the above pricing expression of the option, we have made a specific assumption on the relationship between the covariance operator $\mathcal{Q}$ and the stochastic variance process $\mathcal{Y}$. If the two happen to commute, we have

$$\mathcal{Y}^{1/2}(t)\mathcal{Q}\mathcal{Y}^{1/2}(t) = (\mathcal{Y}^{1/2}(t)\mathcal{Q}^{1/2})(\mathcal{Q}^{1/2}\mathcal{Y}^{1/2}) = \mathcal{Q}^{1/2}\mathcal{Y}(t)\mathcal{Q}^{1/2}$$

as the square-roots commute as well. Hence, $\mathcal{B} = \mathcal{Q}^{1/2}$. The next Lemma shows a sufficient condition for when $\mathcal{Y}$ commutes with $\mathcal{Q}$:

Lemma 7.4 *Assume that $\mathcal{Q}$ commutes with $\mathcal{Y}_0$ and $\mathcal{L}(t)$ for all $t \geq 0$. If $\mathfrak{C}(\mathcal{T})\mathcal{Q} = \mathfrak{C}(\mathcal{T}\mathcal{Q})$ and $\mathcal{Q}\mathfrak{C}(\mathcal{T}) = \mathfrak{C}(\mathcal{Q}\mathcal{T})$ for every $\mathcal{T} \in L_{HS}(H)$, then $\mathcal{Y}(t)$ commutes with $\mathcal{Q}$ for every $t \geq 0$.*

Proof From the dynamics of $\mathcal{Y}$ in (6.18) and the assumption on $\mathfrak{C}$ we find

$$\mathcal{Q}\mathcal{Y}(t) = \mathcal{Q}\mathcal{Y}_0 + \int_0^t \mathfrak{C}(\mathcal{Q}\mathcal{Y}(s))ds + \mathcal{Q}\mathcal{L}(t)$$

and

$$\mathcal{Y}(t)\mathcal{Q} = \mathcal{Y}_0\mathcal{Q} + \int_0^t \mathfrak{C}(\mathcal{Y}(s)\mathcal{Q})ds + \mathcal{L}(t)\mathcal{Q}.$$

By assumptions on $\mathcal{Q}$, $\mathcal{Y}_0$ and $\mathcal{L}$, it holds that $\mathcal{Q}\mathcal{Y}_0 = \mathcal{Y}_0\mathcal{Q}$ and $\mathcal{L}(t)\mathcal{Q} = \mathcal{Q}\mathcal{L}(t)$. Define $\widehat{\mathcal{L}}(t) := \mathcal{Q}\mathcal{L}(t)$. Its conditional characteristic function becomes, after using the independent increment property of Lévy processes,

$$\begin{aligned}\mathbb{E}[e^{\mathrm{i}\langle\widehat{\mathcal{L}}(t)-\widehat{\mathcal{L}}(s),\mathcal{T}\rangle_{\mathrm{HS}}} \mid \mathcal{F}_t] &= \mathbb{E}[e^{\mathrm{i}\langle\mathcal{L}(t)-\mathcal{L}(s),\mathcal{Q}\mathcal{T}\rangle_{\mathrm{HS}}} \mid \mathcal{F}_t] \\ &= \mathbb{E}[e^{\mathrm{i}\langle\mathcal{L}(t)-\mathcal{L}(s),\mathcal{Q}\mathcal{T}\rangle_{\mathrm{HS}}}] \\ &= \exp((t-s)\Psi_{\mathcal{L}}(\mathcal{Q}\mathcal{T}))\end{aligned}$$

for $t \geq s \geq 0$ and $\mathcal{T} \in L_{\mathrm{HS}}(H)$. Here, $\Psi_{\mathcal{L}}$ is the Lévy exponent of $\mathcal{L}$, and hence, $\widehat{\mathcal{L}}$ is an $L_{\mathrm{HS}}(H)$-valued Lévy process. We find that $\mathcal{Q}\mathcal{Y}(t)$ and $\mathcal{Y}(t)\mathcal{Q}$ are mild solutions of the same equation

$$d\mathcal{Z}(t) = \mathfrak{C}(\mathcal{Z}(t))dt + d\widehat{\mathcal{L}}(t), \tag{7.21}$$

with the same initial condition. The mild solution of (7.21) is uniquely given as

$$Q\mathcal{Y}(t) = \mathcal{Y}(t)Q = \mathfrak{S}_t Q\mathcal{Y}_0 + \int_0^t \mathfrak{S}_{t-s} d\widehat{\mathfrak{L}}(s),$$

which proves the claim. □

Fix $C \in L(H)$. In the two examples $\mathfrak{C}^{(1)}\mathcal{T} = C\mathcal{T}C$ and $\mathfrak{C}^{(2)}\mathcal{T} = C\mathcal{T} + \mathcal{T}C$ given in (6.21) and (6.22), where $\mathcal{T} \in L_{\mathrm{HS}}(H)$, the required properties of $\mathfrak{C}$ in the Lemma above hold whenever C commutes with Q. A simple example of a Lévy process commuting with Q is constructed as follows: let $\mathcal{U} \in L_{\mathrm{HS}}(H)$ be self-adjoint and positive definite, and $L(t)$ a Lévy process taking values on $\mathbb{R}_+$ (i.e., a subordinator). Then $\mathcal{L} := \mathcal{U}L$ is an $L_{\mathrm{HS}}(H)$-valued Lévy process, and it commutes with Q whenever $\mathcal{U}$ commutes with Q.

In Proposition 7.7 and several other pricing formulas that we have derived, adjoint operators frequently occur. For example, in Proposition 7.7 we have the adjoint of semigroups $(\mathcal{S}_t)_{t\geq 0}$ and $(\mathfrak{S}_t)_{t\geq 0}$ as well as bounded linear operators on H like $\mathcal{M}$ and $\mathcal{B}$. If we work on the Filipović space as the state space H, the Samuelson effect operator $\mathcal{M}$ has an explicit adjoint $\mathcal{M}^*$ derived in Prop. 3.5 in the case we model it as a multiplication operator. For general bounded operators on H_w, we may resort to Prop. 3.3 for a characterization of the adjoint. Particularly, we find an explicit representation of the adjoint of the shift semigroup in (3.9). Finally, the assumptions on $\mathfrak{C}$ yields that $\mathfrak{S}_t^* = \exp(t\mathfrak{C}^*)$.

We end by remarking that the Fourier representation may be significantly generalized for infinite dimensional HJM-models with leverage, see Benth and Sgarra [52].

7.5 Options and Markovian Forward Models

Assume for $\tau \geq t \geq 0$ we have a Markovian forward price dynamics defined as the unique H-valued mild adapted RCLL solution $(\mathbf{f}(t))_{t\geq 0}$ of the SPDE (6.27), i.e., satisfying the equation in H given by

$$\mathbf{f}(\tau) = \mathcal{S}_{\tau-t}\mathbf{f}(t) + \int_t^\tau \mathcal{S}_{\tau-s}\mathbf{\Sigma}(s, \mathbf{f}(s))\, d\mathbf{W}(s)$$
$$+ \int_t^\tau \int_H \mathcal{S}_{\tau-s}\mathbf{\Gamma}(s, \mathbf{f}(s-), \mathbf{z})\widetilde{\mathbf{N}}(d\mathbf{z}, ds). \tag{7.22}$$

We assume the conditions in Chapter 6, Section 6.4, on $\mathbf{\Gamma}$ and $\mathbf{\Sigma}$ to hold, with $\mathcal{S}_t$, $\mathbf{W}$ and $\widetilde{\mathbf{N}}$ defined there. Moreover, as the drift is zero, we consider the risk-neutral dynamics of $\mathbf{f}$.

Recalling the option price $C(t;\tau)$, we find by the Markovian property that it is a function of the time-t forward curve $\mathbf{f}(t)$, $\overline{C}(t;\tau,\mathbf{f}(t))$ where

$$\overline{C}(t;\tau,\mathbf{g}) := e^{-r(\tau-t)}\mathbb{E}_{\mathbb{Q}}[p(G(\mathbf{f}(\tau)))\,|\,\mathbf{f}(t)=\mathbf{g}] \tag{7.23}$$

with $\mathbf{g}\in H$. We have the following result on the option price:

Proposition 7.8 *Assume that the payoff function p is Lipschitz continuous. Then the option price is Lipschitz continuous in the initial forward curve, i.e., for any* $\mathbf{g},\mathbf{h}\in H$

$$|\overline{C}(t;\tau,\mathbf{g})-\overline{C}(t;\tau,\mathbf{h})|\leq k|\mathbf{g}-\mathbf{h}|_H$$

for some constant $k>0$.

Proof First, denote by $\mathbf{f}^{t,\mathbf{g}}(\tau)\in H$ the random variable

$$\begin{aligned}\mathbf{f}^{t,\mathbf{g}}(\tau)&=\mathcal{S}_{\tau-t}\mathbf{g}+\int_t^\tau \mathcal{S}_{\tau-s}\boldsymbol{\Sigma}(s,\mathbf{f}^{t,\mathbf{g}}(s))d\mathbf{W}(s)\\&\quad+\int_t^\tau\int_H \mathcal{S}_{\tau-s}\boldsymbol{\Gamma}(s,\mathbf{f}(s-),\mathbf{z})\widetilde{\mathbf{N}}(d\mathbf{z},ds)\end{aligned}$$

for $\mathbf{g}\in H$. Since p is Lipschitz continuous and $G\in H^*$, there exists a constant $k>0$ (which may change in the steps below) such that

$$|p(G(\mathbf{f}^{t,\mathbf{g}}(\tau)))-p(G(\mathbf{f}^{t,\mathbf{h}}(\tau)))|\leq k|\mathbf{f}^{t,\mathbf{g}}(\tau)-\mathbf{f}^{t,\mathbf{h}}(\tau)|_H,$$

with $\mathbf{h}\in H$. Hence, by Cauchy-Schwarz' inequality,

$$|\overline{C}(t;\tau,\mathbf{g})-\overline{C}(t;\tau,\mathbf{h})|\leq k\mathbb{E}[|\mathbf{f}^{t,\mathbf{g}}(\tau)-\mathbf{f}^{t,\mathbf{h}}(\tau)|_H^2]^{1/2}.$$

The elementary inequality $(a+b+c)^2\leq 3a^2+3b^2+3c^2$, the Itô isometry in Definition 4.1 (and the discussion following it) and the Lipschitz continuity of $\boldsymbol{\Sigma}$ and $\boldsymbol{\Gamma}$ yield

$$\begin{aligned}&\mathbb{E}_{\mathbb{Q}}[|\mathbf{f}^{t,\mathbf{g}}(\tau)-\mathbf{f}^{t,\mathbf{h}}(\tau)|_H^2]\\&\qquad\leq 3|\mathcal{S}_{\tau-t}(\mathbf{g}-\mathbf{h})|_H^2\\&\qquad\quad+3\mathbb{E}_{\mathbb{Q}}[|\int_t^\tau \mathcal{S}_{\tau-s}(\boldsymbol{\Sigma}(s,\mathbf{f}^{t,\mathbf{g}}(s))-\boldsymbol{\Sigma}(s,\mathbf{f}^{t,\mathbf{h}}(s))d\mathbf{W}(s)|_H^2]\\&\qquad\quad+3\mathbb{E}_{\mathbb{Q}}[|\int_t^\tau\int_H \mathcal{S}_{\tau-s}(\boldsymbol{\Gamma}(s,\mathbf{f}^{t,\mathbf{g}}(s-),\mathbf{z})\\&\qquad\quad-\boldsymbol{\Gamma}(s,\mathbf{f}^{t,\mathbf{h}}(s-),\mathbf{z})\widetilde{\mathbf{N}}(d\mathbf{z},ds)|_H^2]\end{aligned}$$

$$\leq 3\|\mathcal{S}_{\tau-t}\|_{\text{op}}^2|\mathbf{g}-\mathbf{h}|_H^2$$
$$+3\mathbb{E}_{\mathbb{Q}}[\int_t^\tau \|\mathcal{S}_{\tau-s}(\boldsymbol{\Sigma}(s,\mathbf{f}^{t,\mathbf{g}}(s))-\boldsymbol{\Sigma}(s,\mathbf{f}^{t,\mathbf{h}}(s))Q^{1/2}\|_{\text{HS}}^2 ds]$$
$$+3\mathbb{E}_{\mathbb{Q}}[\int_t^\tau\int_H |\mathcal{S}_{\tau-s}(\boldsymbol{\Gamma}(s,\mathbf{f}^{t,\mathbf{g}}(s),\mathbf{z})-\boldsymbol{\Gamma}(s,\mathbf{f}^{t,\mathbf{h}}(s),\mathbf{z})|_H^2\nu(d\mathbf{z})ds]$$
$$\leq 3\|\mathcal{S}_{\tau-t}\|_{\text{op}}^2|\mathbf{g}-\mathbf{h}|_H^2$$
$$+3\int_t^\tau \|\mathcal{S}_{\tau-s}\|_{\text{op}}^2K^2(s)\mathbb{E}_{\mathbb{Q}}[|\mathbf{f}^{t,\mathbf{g}}(s)-\mathbf{f}^{t,\mathbf{h}}(s)|_H^2](1+\|Q^{1/2}\|_{\text{HS}}^2)\,ds.$$

Now, the operator norm of the quasi-contractive semigroup $\mathcal{S}_t$ is bounded by $\|\mathcal{S}_{\tau-t}\|_{\text{op}} \leq \exp(\eta(\tau-t)) \leq \exp(\eta\tau)$ for an $\eta \in \mathbb{R}$ and any $\tau \geq t \geq 0$. Moreover, $\|Q^{1/2}\|_{\text{HS}}^2 = \text{Tr}(Q) < \infty$. Since the Lipschitz constant of $\boldsymbol{\Sigma}$ and $\boldsymbol{\Gamma}$ satisfies $K \in L^2_{\text{loc}}(\mathbb{R}_+)$, we have by the Gronwall inequality

$$\mathbb{E}_{\mathbb{Q}}[|\mathbf{f}^{t,\mathbf{g}}(\tau)-\mathbf{f}^{t,\mathbf{h}}(\tau)|_H^2] \leq k|\mathbf{g}-\mathbf{h}|_H^2\exp(k\int_t^\tau K^2(s)ds)$$

for a constant $k > 0$. This proves the assertion. □

In practice, when modelling the forward curve dynamics, one may apply various smoothing techniques to obtain the current forward price curve $x \mapsto f(t,x)$ (recall the discussion in Section 6.8 of Chapter 6). The Lipschitz continuity of the option price with respect to the current forward curve suggests a robustness in the price with respect to small variations in the initial curve, and therefore the price will not be very sensitive to small perturbations due to, say, different smoothing techniques. We recall that call and put options have Lipschitz continuous payoff functions p, and that options on delivery period forwards of geometric HJM-type models indeed are depending on the current forward curve and not on a particular forward price.

Let us discuss another application of the Lipschitz continuity of the option price. In practice, we only have available a finite number of forward prices at the current time t. In typical commodity markets like metals or oil, there will be prices available for a finite set of given delivery times, which can be next week, month, or year, say. In gas and power, we will have a finite set of prices for forwards with given delivery periods. Hence, for an ONB $(\mathbf{e}_i)_{i\in\mathbb{N}}$ of H, we can view the situation as having knowledge of $\mathcal{P}_n\mathbf{g}$ rather than $\mathbf{g} \in H$, where,

$$\mathcal{P}_n\mathbf{g} = \sum_{i=1}^n \langle \mathbf{g}, \mathbf{e}_i\rangle_H \mathbf{e}_i \tag{7.24}$$

is the projection operator from H into H^n, with H^n being the Hilbert space spanned by $\{\mathbf{e}_1,\ldots,\mathbf{e}_n\}$. But this means that we will compute a price $\overline{C}_n(t;\tau,\mathbf{g}) := \overline{C}(t;\tau,\mathcal{P}_n\mathbf{g})$ rather than the true price $\overline{C}(t;\tau,\mathbf{g})$. However, from Proposition 7.8 it follows that

$$|\overline{C}_n(t;\tau,\mathbf{g}) - \overline{C}(t;\tau,\mathbf{g})| \leq k|\mathbf{g} - \mathcal{P}_n\mathbf{g}|_H$$

for some constant $k > 0$. Since

$$|\mathbf{g} - \mathcal{P}_n\mathbf{g}|_H^2 = \sum_{i=n+1}^{\infty} \langle \mathbf{g}, \mathbf{e}_i \rangle_H^2 \to 0$$

as $n \to \infty$, we have that $\overline{C}_n(t;\tau,\mathbf{g})$ is close to $\overline{C}(t;\tau,\mathbf{g})$ if there are sufficiently many traded forwards in the market (i.e., sufficiently high n).

We observe that with

$$\widehat{C}_n(t;\tau,x_1,\ldots,x_n) := \overline{C}(t;\tau,\sum_{i=1}^{n} x_i\mathbf{e}_i)$$

we have

$$\overline{C}_n(t;\tau,\mathbf{g}) = \widehat{C}_n(t;\tau,\langle \mathbf{g},\mathbf{e}_1\rangle_H,\ldots,\langle \mathbf{g},\mathbf{e}_n\rangle_H).$$

Indeed, $\widehat{C}_n$ can be seen as the option price on $\mathbf{f}$ started in the finite dimensional subspace H^n. Although $\mathbf{f}^{t,\mathcal{P}_n\mathbf{g}}(t) = \mathcal{P}_n\mathbf{g} \in H^n$, we have no guarantee that $\mathbf{f}^{t,\mathcal{P}_n\mathbf{g}}(\tau) \in H^n$. On the contrary, we expect in general that $\mathbf{f}^{t,\mathcal{P}_n\mathbf{g}}(\tau) \notin H^m$ for any $m \in \mathbb{N}$, that is, a truly infinite dimensional object. As $\mathcal{P}_n$ does not commute with the semigroup $\mathcal{S}_t$, we will not have that $s \mapsto \mathbf{F}(s,T) := \delta_{T-s}\mathcal{P}_n\mathbf{f}^{t,\mathcal{P}_n g}(s)$, $t \leq s \leq T$ defines a $\mathbb{Q}$-martingale dynamics, and therefore $\mathcal{P}_n\mathbf{f}^{t,\mathcal{P}_n g}(s)$ is not an arbitrage-free model. Therefore, $\widehat{C}_n(t;\tau,\langle \mathbf{g},\mathbf{e}_1\rangle_H,\ldots,\langle \mathbf{g},\mathbf{e}_n\rangle_H)$ will possibly yield arbitrage opportunities. For arbitrage-free approximations, we refer to Benth and Krühner [36].

7.6 Bibliographical Remarks

Carr and Madan [71] is the seminal paper on Fourier methods applied to pricing financial derivatives, in particular options on non-Gaussian models. Their methods have been put into an energy derivatives context in Benth, Šaltytė Benth and Koekebakker [49]. The Fourier approach separates the payoff function from the random price in a Fourier integral, were the two components are represented by their characteristic functional and the Fourier transform of the payoff. Using the numerical technique of fast Fourier transform, these integrals can efficiently be computed in practical situations. We refer to Eberlein [89] for a survey of these methods in mathematical finance, with a particular view towards interest rate theory and term structure models. Kwok, Leung and Wong [144] discuss in a recent survey computational issues on the fast Fourier transform in financial derivatives pricing. Spread options, which are very common in commodities and energy, have been priced by Fourier methods in Dempster and Hong [84], Hurd and Zhou [126] and Benth and Zdanowicz [55].

The infinite dimensional models that we work with require knowledge of inner products between infinite dimensional objects and Lévy exponents for operators (recall Prop. 7.7, for example) in order to be made operative in a fast Fourier implementation. A natural practical approach could be to resort to finite dimensional approximations of these terms, which raises new and interesting questions around the quantification of the approximation error and speed of convergence. We have touched upon these questions already at the end of Section 7.5 of this chapter. Referring to Benth and Krühner [36], arbitrage-free representations are developed for infinite dimensional HJM-models which provides a reasonable starting point for computations.

The background for much of the material in this chapter on option pricing for Gaussian models is found in Benth and Krühner [37]. Our results on infinite dimensional Gaussian forward curve models extend the classical Black-76 formula for call and put options on forward and futures contracts (see Black [60]) and Margrabe's formula for spread options (see Margrabe [155]). These formulas have been developed and extended in various ways for both Gaussian and non-Gaussian models, in particular for Ornstein–Uhlenbeck processes and for HJM-type dynamics (see, e.g., Benth, Šaltytė Benth and Koekebakker [25]).

The Greeks of options play an important role in hedging and risk assessment, and as we have seen, their calculation involves differentiation of the price functional with respect to the parameter in question. The parameter resides in the payoff function of the option, which typically is not smooth. Numerical differentiation using Monte Carlo methods require a high number of samples to reduce the uncertainty in the estimate, and analytical formulas are highly attractive (possibly combined with numerical methods). In Prop. 7.4 we generalised the so-called *density method* to derive an expression for the delta in a Gaussian model context. The density method has been presented and applied in Broadie and Glasserman [65] and Glasserman [116], and is a method that moves differentiation from the payoff function, which may even be non-differentiable, to the density function, which is usually smooth. As long as one has available the density of the underlying asset at exercise time, one may apply the density method also for non-Gaussian models.

Since the seminal papers of Fournié *et al.* [105] and [106], the *Malliavin approach* has gained popularity due to its applicability to handle the computation of Greeks when the underlying asset dynamics follows a stochastic differential equation. The approach is based on the close relationship between the derivative, the stochastic derivative (i.e., the Malliavin derivative) and an integration-by-parts formula. The result is that many Greeks can be represented as the expected value of the randomly weighted payoff function, expression which are suitable for Monte Carlo simulation. We refer to Cufaro Petroni and Sabino [80] for a recent application of the Malliavin approach to computing the delta for Asian and basket options using quasi-Monte Carlo simulation. Benth, Di Nunno and Simonsen [28] derive theoretically the Greeks in an infinite dimensional forward dynamics model with stochastic volatility. Their approach relies on Malliavin Calculus in infinite dimensions.

Appendix A
Collection of Some Fundamental Properties of the Filipović Space

Abstract Some fundamental properties of the Filipović space with proofs are provided. It is shown that this state space of forward term structures is a separable Hilbert space, and that the shift operator is a C_0-semigroup with the derivative operator as densely defined generator.

Here we have collected some further basic properties with proofs of the Filipović space that are used and mentioned in Chapter 3 for the convenience of the interested reader. The results and their proofs are collected from Filipović [99].

Lemma A.1 *$\langle \cdot, \cdot \rangle_w$ defines an inner product on H_w.*

Proof Remark that by Cauchy-Schwarz' inequality, $|\langle f, g\rangle_w| \leq 2|f|_w|g|_w < \infty$, so $\langle f, g\rangle_w$ is well-defined for all $f, g \in H_w$. Obviously, for any $f, g \in H_w$, $\langle f, g\rangle_w = \langle g, f\rangle_w$ and $\langle cf, g\rangle_w = c\langle f, g\rangle_w$. Moreover, if $h \in H_w$, then it straighforwardly holds that $\langle f + g, h\rangle_w = \langle f, h\rangle_w + \langle g, h\rangle_w$. Also, $\langle f, f\rangle_w = |f|_w^2 \geq 0$ and $\langle 0, 0\rangle_w = 0$. It remains to show that if $\langle f, f\rangle_w = 0$, then $f(x) = 0$ for all $x \in \mathbb{R}_+$. But if $\langle f, f\rangle_w = |f|_w^2 = 0$ for $f \in H_w$, we have $f(0) = 0$ and from Lemma 3.1 $|f'|_{L^1([0,x])} \leq \sqrt{x}|f|_w = 0$. Thus, by the fundamental theorem of calculus in Lemma 3.1

$$|f(x)| = |f(0) + \int_0^x f'(y)\,dy| \leq \int_0^x |f'(y)|\,dy = 0$$

for all $x \geq 0$. Hence, $f(x) = 0$ for all $x \in \mathbb{R}_+$ and we conclude that $\langle \cdot, \cdot\rangle_w$ is an inner product. □

We find that $|\cdot|_w$ is a norm on H_w, induced from the inner product $\langle \cdot, \cdot\rangle_w$. We have (Filipović [99, Thm. 5.1.1]),

Proposition A.1 *$(H_w, |\cdot|_w)$ is a separable Hilbert space*

Proof Introduce the space $H := \mathbb{R} \times L^2(\mathbb{R}_+)$ equipped with the norm $\|(c, f)\|^2 := c^2 + |f|^2_{L^2(\mathbb{R}_+)}$ for $(c, f) \in H$. This forms a separable Hilbert space, being the

F. E. Benth, P. Krühner, *Stochastic Models for Prices Dynamics in Energy and Commodity Markets*, Springer Finance,
https://doi.org/10.1007/978-3-031-40367-5

product space of $L^2(\mathbb{R}_+)$ with the real line. Introduce the linear operator $\mathcal{T}: H_w \to H$ by

$$\mathcal{T}g = (g(0), g'w^{1/2}), \qquad g \in H_w .$$

It holds that

$$\|\mathcal{T}g\|^2 = g^2(0) + \int_0^\infty (g'(y)w^{1/2}(y))^2\, dy = |g|_w^2 ,$$

so $\mathcal{T}$ is an isometric operator between H and H_w. Its inverse is

$$\mathcal{T}^{-1}(c, f)(x) = c + \int_0^x f(y)w^{-1/2}(y)\, dy .$$

The result follows. □

Lemma A.2 *Assume w is a continuously differentiable function on $\mathbb{R}_+$. Then $(\mathcal{S}_t)_{t\geq 0}$ is a C_0-semigroup on H_w.*

Proof Fix $t \geq 0$. If $f \in H_w$, we know from the fundamental theorem of calculus in Lemma 3.1 that

$$\begin{aligned}
\mathcal{S}_t f(x) &= f(x+t) \\
&= f(0) + \int_0^{x+t} f'(y)\, dy \\
&= f(t) + \int_t^{x+t} f'(y)\, dy \\
&= f(t) + \int_0^x f'(z+t)\, dz .
\end{aligned}$$

Hence, $x \mapsto \mathcal{S}_t f(x)$ is absolutely continuous, with weak derivative $f'(\cdot + t)$. Trivially, $\mathcal{S}_t(f + g) = \mathcal{S}_t f + \mathcal{S}_t g$ for $f, g \in H_w$, thus a linear operator. Since w is a non-decreasing function with $w(0) \geq 1$, we derive using Lemma 3.1 and the Cauchy-Schwarz inequality,

$$\begin{aligned}
|\mathcal{S}_t f|_w^2 &= f^2(t) + \int_0^\infty w(x) f'(x+t)^2\, dx \\
&= (f(0) + \int_0^t f'(y)\, dy)^2 + \int_t^\infty w(y-t) f'(y)^2\, dy \\
&\leq 2f^2(0) + 2(\int_0^t w^{-1/2}(y)w^{1/2}(y) f'(y)\, dy)^2 + \int_t^\infty w(y) f'(y)^2\, dy
\end{aligned}$$

$$
\begin{aligned}
&\leq 2f^2(0) + 2\int_0^t w^{-1}(y)\,dy \int_0^t w(y)f'(y)^2\,dy + \int_t^\infty w(y)f'(y)^2\,dy \\
&\leq 2f^2(0) + 2t\int_0^t w(y)f'(y)^2\,dy + \int_t^\infty w(y)f'(y)^2\,dy \\
&\leq 2\max(t,1)|f|_w^2 < \infty\,.
\end{aligned}
$$

Thus, $\mathcal{S}_t f \in H_w$ and $\mathcal{S}_t$ bounded operator, and we conclude that $\mathcal{S}_t \in L(H_w)$.

For $s, t \geq 0$, we see that $\mathcal{S}_s\mathcal{S}_t f(x) = f(x+s+t) = \mathcal{S}_{t+s} f(x)$, so $(\mathcal{S}_t)_{t\geq 0}$ satisfies the semigroup property. Moverover, $\mathcal{S}_0 f = f$, and thus $\mathcal{S}_0$ is the identity operator. We show strong continuity of the shift operators:

First, introduce the set $\mathcal{D} \subset H_w$

$$
\mathcal{D} = \{f \in C^2(\mathbb{R}_+) \mid f' \in C_c^1(\mathbb{R}_+)\}
$$

where $C^2(R_+)$ denotes functions which are twice continuously differentiable on $\mathbb{R}_+$ and $C_c^1(\mathbb{R}_+)$ functions which are once continuously differentiable on $\mathbb{R}_+$ with compact support. For $g \in H_w$, we recall from the proof of Prop. A.1 that $g'\sqrt{w} \in L^2(\mathbb{R}_+)$. From [104, Prop. 7.9], $C_c^1(\mathbb{R}_+)$ is dense in $L^2(\mathbb{R}_+)$, so we choose a sequence $(f_n)_{n=1}^\infty \subset C_c^1(\mathbb{R}_+)$ such that $f_n \to g'\sqrt{w}$ in $L^2(\mathbb{R}_+)$. Define $g_n(x) := g(0) + \int_0^x f_n(y)w^{-1/2}(y)\,dy$, which (again referring back to the proof of Prop. A.1), will be a sequence in H_w converging to g. Moreover, since $f_n \in C_c^1(\mathbb{R}_+)$ and $w \in C^1(\mathbb{R}_+)$, being non-decreasing and $w(0) = 1$, it follows that $g_n \in \mathcal{D}$. Thus, $\mathcal{D}$ is dense in H_w.

Notice that for any $g \in \mathcal{D}$, we have $g' \in H_w$ and the fundamental theorem of calculus yields (after a change of variables)

$$
g'(x+t) - g'(x) = t\int_0^1 g''(x+st)\,ds\,.
$$

It follows from the Cauchy-Schwarz inequality that

$$
\begin{aligned}
|\mathcal{S}_t g - g|_w^2 &= |g(t) - g(0)|^2 + \int_0^\infty w(x)|g'(x+t) - g'(x)|^2\,dx \\
&= |\int_0^t g'(x)\,dx|^2 + \int_0^\infty w(x)t^2(\int_0^1 g''(x+st)\,ds)^2\,dx \\
&\leq \int_0^t w^{-1}(x)\,dx \int_0^t w(x)g'(x)^2\,dx \\
&\quad + t^2\int_0^1\int_0^\infty w(x)g''(x+st)^2\,dx\,ds \\
&\leq t|g|_w^2 + t^2\int_0^1 |\mathcal{S}_{st} g'|_w^2\,ds\,.
\end{aligned}
$$

Using the estimate from the boundedness of $\mathcal{S}_t$ above, we find $|\mathcal{S}_{st}g|_w^2 \leq 2\max(st, 1)|g'|_w^2$. Hence,

$$|\mathcal{S}_t g - g|_w^2 \leq t|g|_w^2 + 2t^2 \max(t, 1)|g'|_w^2 ,$$

and we find strong continuity on $\mathcal{D}$. Let $f \in H_w$ and $g \in \mathcal{D}$. Then the triangle inequality yields

$$\begin{aligned}|\mathcal{S}_t f - f|_w &\leq |\mathcal{S}_t(f - g)|_w + |\mathcal{S}_t g - g|_w + |f - g|_w \\ &\leq \sqrt{2\max(t, 1)}|f - g|_w + \sqrt{t}\sqrt{|g|_w^2 + 2\max(t, 1)|g'|_w^2} + |f - g|_w .\end{aligned}$$

Consider now $t < 1$. For given $\epsilon > 0$, we can by density of $\mathcal{D}$ choose g such that $|f - g| < \epsilon/(2(1 + \sqrt{2})$ and $t < \epsilon^2/(4(|g|_w^2 + 2|g'|_w^2))$, which show that $|\mathcal{S}_t f - f| < \epsilon$ for $t < \delta$, with the given $\delta > 0$. Strong continuity follows. □

Lemma A.3 *The C_0-semigroup $(\mathcal{S}_t)_{t\geq 0}$ has generator $(d/dx, Dom(d/dx))$, where the dense domain is*

$$Dom(d/dx) = \left\{ f \in H_w \mid f' \in H_w \right\} .$$

Proof Assume $f \in \mathrm{Dom}(d/dx) \subset H_w$. Then, $\mathcal{S}_t f \in H_w$, and by Engel and Nagel [94, Lemma 1.3, Ch. I] we have that

$$\frac{d}{dt}\mathcal{S}_t f = \mathcal{S}_t \mathcal{A} f \in H_w$$

where $\mathcal{A}$ is the generator of $(\mathcal{S}_t)_{t\geq 0}$. Then

$$\frac{d}{dt}\mathcal{S}_t f = \frac{d}{dt} f(\cdot + t) = f'(\cdot + t).$$

But the evaluation map δ_0 is a continuous linear functional on H_w, so

$$\delta_0(\frac{d}{dt}\mathcal{S}_t f) = f'(t) = \delta_0(\mathcal{S}_t \mathcal{A} f) = \mathcal{A} f(t).$$

Thus, for $f \in \mathrm{Dom}(d/dx)$ we find that $\mathcal{A} = d/dx$. Moreover, since $\mathcal{A} f \in H_w$, $f' \in C(\mathbb{R}_+)$. Thus, $\mathrm{Dom}(d/dx) \subset \{f \in H_w \cap C^1(\mathbb{R}_+) \mid f' \in H_w\}$. Recall that the domain of the generator d/dx is dense in H_w (see, e.g., Engel and Nagel [94, Thm. 1.4, Ch. II]).

From above we have that $\mathrm{Dom}(d/dx) \subset \{f \in H_w \mid f' \in H_w\}$, where f' in general is the weak derivative of f (which we recall always exists on H_w). Consider now $f \in \{f \in H_w \mid f' \in H_w\}$. By Lemma 3.1, we find that

$$f(x + t) - f(x) = t \int_0^1 f'(x + st)\, ds$$

and

$$f'(x+t) - f'(x) = t\int_0^1 f''(x+st)\,ds$$

after a change of variables. We recall that as $f' \in H_w$, f'' exists as the weak derivative. Thus, by Cauchy-Schwarz' inequality,

$$\begin{aligned}
&\left|\frac{\mathcal{S}_t f - f}{t} - f'\right|_w^2 \\
&\quad = \left|\frac{f(t)-f(0)}{t} - f'(0)\right|^2 + \int_0^\infty w(y)\left|\frac{f'(y+t)-f'(y)}{t} - f''(y)\right|^2 dy \\
&\quad \leq \int_0^1 \left|f'(st) - f'(0)\right|^2 ds + \int_0^1\int_0^\infty w(y)\left|f''(y+st) - f''(y)\right|^2 dy\,ds \\
&\quad \leq 2\int_0^1 |\mathcal{S}_{st} f' - f'|_w^2\,ds
\end{aligned}$$

As $f' \in H_w$ and $\mathcal{S}_t$ is strongly continuous, we find by the dominated convergence theorem that

$$\left|\frac{\mathcal{S}_t f - f}{t} - f'\right|_w^2 \to 0$$

as $t \downarrow 0$. Hence, $f \in \mathrm{Dom}(d/dx)$ and $\mathcal{A}f = f'$. The Lemma follows. □

References

1. Aïd, R. (2015). *Electricity Derivatives*, SpringerBriefs in Quantitative Finance. Springer-Verlag, Cham, Heidelberg, New York, Dordrech, London.
2. Aïd, R., Campi, L., Nguyen Huu, A., and Touzi, N. (2009). A structural risk–neutral model of electricity prices. *International Journal of Theoretical and Applied Finance*, **12**, pp. 925–947.
3. Alasseur, C., and Féron, O. (2018). Structural price model for coupled electricity markets. *Energy Economics*, **75**, pp. 104–119.
4. Albert, A. (1972). *Regression and the Moore-Penrose Pseudoinverse*. Academic Press, New York.
5. Andersen, A., Koekebakker, S., and Westgaard, S. (2010). Modeling electricity forward prices using the multivariate normal inverse Gaussian distribution. *Journal of Energy Markets*, **3**(3), pp. 3–25.
6. Ané, T., and Geman, H. (2000). Order flow, transaction clock, and normality of asset returns. *Journal of Finance*, **55**(5), pp. 2259–2284.
7. Applebaum, D. (2009). *Lévy Processes and Stochastic Calculus*. Cambridge University Press, Cambridge.
8. Applebaum, D., and Riedle, M. (2010). Cylindrical Lévy processes in Banach spaces. *Proceeding of the London Mathematical Society*, **101**(3), pp. 697–726.
9. Audet, N., Heiskanen, P., Keppo, J., and Vehvilainen, I. (2002). Modelling of electricity forward curve dynamics in the Nordic electricity market. In: *Modelling Prices in Competitive Electricity Markets*, D. W. Bunn (editor), pp. 251–265, John Wiley & Sons, Chichester.
10. Baeumer, B., Kovács, M., and Meerschaert, M. M. (2008). Subordinated multiparameter groups of linear operators: Properties via the transference principle. In *Functional Analysis and Evolution Equations. The Günter Lumer Volume*, Amann, H., Arendt, W., Hieber, M., Neubrander, F., Nicaise, S. and J. von Below (eds.), Birkhäuser Verlag, Basel Boston Berlin, pp. 35–50.
11. Barndorff-Nielsen, O. E. (1977). Exponentially decreasing distributions for the logarithm of particle size. *Proceedings of the Royal Society of London*, Series A, **353**(1674), pp. 401–409.
12. Barndorff-Nielsen, O. E. (1998). Processes of normal inverse Gaussian type. *Finance & Stochastics*, **2**(1), pp. 41–68.
13. Barndorff-Nielsen, O. E., Benth, F. E., and Veraart, A. (2013). Modelling energy spot prices by volatility modulated Lévy-driven Volterra processes. *Bernoulli*, **19**, pp. 803–845.
14. Barndorff-Nielsen, O. E., Benth, F. E., and Veraart, A. (2014). Modelling electricity futures by ambit fields. *Advances in Applied Probability*, **46**(3), pp. 719–745.

F. E. Benth, P. Krühner, *Stochastic Models for Prices Dynamics in Energy and Commodity Markets*, Springer Finance,
https://doi.org/10.1007/978-3-031-40367-5

15. Barndorff-Nielsen, O. E., Benth, F. E., and Veraart, A. (2018). *Ambit Stochastics*, Springer-Verlag, Cham.
16. Barndorff-Nielsen, O. E., Pedersen, J., and Sato, K. (2001). Multivariate subordination, self-decomposability and stability. *Advances in Applied Probability*, **33**(1), pp. 160–187.
17. Barndorff-Nielsen, O. E., and Schmiegel, J. (2004), Lévy-based tempo-spatial modelling with applications to turbulence. *Uspekhi Mat. NAUK*, **59**, pp. 65–91.
18. Barndorff-Nielsen, O. E., and Shephard, N. (2001). Non-Gaussian Ornstein–Uhlenbeck-based models and some of their uses in economics. *Journal of the Royal Statistical Society*, Series B, **63**(2), pp. 167–241 (with discussion).
19. Barndorff-Nielsen, O. E., and Stelzer, R. (2011). Multivariate supOU processes. *Annals of Applied Probability*, **21**(1), pp. 140–182.
20. Barth, A., and Benth, F. E. (2014). The forward dynamics in energy markets – infinite-dimensional modelling and simulation. *Stochastics*, **86**(6), pp. 932–966.
21. Benth, F. E. (2011). The stochastic volatility model of Barndorff-Nielsen and Shephard in commodity markets. *Mathematical Finance*, **21**(4), pp. 595–625.
22. Benth, F. E. (2013). Stochastic volatility and dependency in energy markets: Multi-factor modelling. In *Paris-Princeton Lectures on Mathematical Finance*, Springer-Verlag, Berlin Heidelberg, pp. 109–167.
23. Benth, F. E. (2015). Kriging smooth futures curves. *Energy Risk*, February, pp. 64–69.
24. Benth, F. E. (2021). Pricing of commodity and energy derivatives for polynomial processes. *Mathematics*, **9**, p. 124.
25. Benth, F. E., Šaltytė Benth, J., and Koekebakker, S. (2008). *Stochastic Modelling of Electricity and Related Markets.* World Scientific, Singapore.
26. Benth, F. E., Cartea, A., and Kiesel, R. (2008). Pricing forward contracts in power markets by the certainty equivalence principle: explaining the sign of the market risk premium. *Journal of Banking & Finance*, **32**(10), pp. 2006–2021.
27. Benth, F. E., Christensen, T. S., and Rohde, V. (2021). Multivariate continuous-time modeling of wind indexes and hedging of wind risk. *Quantitative Finance*, **21**(1), pp. 165–183.
28. Benth, F. E., Di Nunno, G., and Simonsen, I. (2021). Sensitivity analysis in the infinite dimensional Heston model. *Infinite Dimensional Analysis, Quantum Probability and Related Topics*, **24**(2), paper 2150014.
29. Benth, F. E., and Harang, F. A. (2021). Infinite dimensional pathwise Volterra processes driven by Gaussian noise – probabilistic properties and applications. *Electronic Journal of Probability*, **26**, article no. 114.
30. Benth, F. E., Kallsen, J., and Meyer-Brandis, T. (2007). A non-Gaussian Ornstein–Uhlenbeck process for electricity spot price modelling and derivatives pricing. *Applied Mathematical Finance*, **14**(2), pp. 153–169.
31. Benth, F. E., Klüppelberg, C., Müller, G., and Vos, L. (2014). Futures pricing in electricity markets based on stable CARMA spot models. *Energy Economics*, **44**, pp. 392–406.
32. Benth, F. E., and Koekebakker, S. (2008). Stochastic modeling of financial electricity contracts. *Energy Economics*, **30**(3), pp. 1116–1157.
33. Benth, F. E., Koekebakker, S., and Ollmar, F. (2007). Extracting and applying smooth forward curves from average-based commodity contracts with seasonal variation. *Journal of Derivatives*, **15**, pp. 52–66.
34. Benth, F. E., and Krühner, P. (2014). Representation of infinite dimensional forward price models in commodity markets. *Communications in Mathematics and Statistics*, **2**(1), pp. 47–106.
35. Benth, F. E., and Krühner, P. (2015) Subordination of Hilbert space valued Lévy processes. *Stochastics*, **87**(3), pp. 458–476.
36. Benth, F. E., and Krühner, P. (2018) Approximation of forward curve models in commodity markets with arbitrage-free finite-dimensional models. *Finance & Stochastics*, **22**(2), pp. 327–366.
37. Benth, F. E., and Krühner, P. (2015). Derivatives pricing in energy markets: an infinite dimensional approach. *SIAM Journal of Financial Mathematics*, **6**(1), pp. 825–869.

38. Benth, F. E., and Lempa, J. (2014). Optimal portfolios in commodity markets. *Finance & Stochastics*, **18**(2), pp. 407–430.
39. Benth, F. E., and Meyer-Brandis, T. (2009). The information premium for non-storable commodities. *Journal of Energy Markets*, **2**(3), pp. 111–140.
40. Benth, F. E., and Ortiz-Latorre, S. (2014). A pricing measure to explain the risk premium in power markets. *SIAM Journal of Financial Mathematics*, **5**, pp. 685–728.
41. Benth, F. E., and Ortiz-Latorre, S. (2015). A change of measure preserving the affine structure in the BNS model for commodity markets. *International Journal of Theoretical and Applied Finance*, **18**(6), 1550038.
42. Benth, F. E., and Paraschiv, F. (2018). A structural model for electricity forward prices. *Journal of Banking & Finance*, **95**, pp. 203–216.
43. Benth, F. E., and Pircalabu, A. (2018). A non-Gaussian Ornstein–Uhlenbeck model for pricing wind power futures. *Applied Mathematical Finance*, **25**(1), pp. 36–65.
44. Benth, F. E., and Rohde, V. (2019). On non-negative modeling with CARMA processes. *Journal of Mathematical Analysis and Applications*, **476**(1), pp 196–214.
45. Benth, F. E., Rüdiger, B., and Süss, A. (2018). Ornstein–Uhlenbeck processes in Hilbert space with non-Gaussian stochastic volatility. *Stochastic Processes and their Applications*, **128**, pp. 461–486.
46. Benth, F. E., and Šaltytė Benth (2004). The normal inverse Gaussian distribution and spot price modelling in energy markets. *International Journal of Theoretical and Applied Finance*, **7**(2), pp. 1–16.
47. Benth, F. E., and Šaltytė Benth (2009). Dynamic pricing of wind futures. *Energy Economics*, **31**(1), pp. 16–24.
48. Benth, F. E., and Šaltytė Benth (2012). Stochastic modelling of temperature variations with a view towards weather derivatives. *Applied Mathematical Finance*, **12**(1), pp. 53–85.
49. Benth, F. E., and Šaltytė Benth (2012). *Modelling and Pricing in Financial Markets for Weather Derivatives*. World Scientific, Singapore.
50. Benth, F. E., Schroers, D., and Veraart, A. (2020). A weak law of large numbers for realized covariation in a Hilbert space setting. *Stochastic Processes and their Applications*, **145**, pp. 241–268.
51. Benth, F. E., and Sgarra, C. (2012). The risk premium and the Esscher transform in power markets. *Stochastic Analysis and Applications*, **30**, pp. 20–43.
52. Benth, F. E., and Sgarra, C. (2021). A Barndorff-Nielsen and Shephard model with leverage in Hilbert space for commodity forward markets. *E-print*, available at ssrn:abstract=3835053.
53. Benth, F. E., and Simonsen, I. C. (2018). The Heston stochastic volatility model in Hilbert space. *Stochastic Analysis and Applications*, **36**(4), pp. 733–750.
54. Benth, F. E., and Vos, L. (2013). Pricing of forwards and options in a multivariate non-Gaussian stochastic volatility model for energy markets. *Advances in Applied Probability*, **45**, pp. 572–594.
55. F. E. Benth and H. Zdanowicz (2016). Pricing and hedging of energy spread options and volatility modulated Volterra processes. *International Journal of Theoretical and Applied Finance*, **19**(1), paper 1650002.
56. Bertoin, J. (1996). *Lévy Processes*. Cambridge University Press, Cambridge.
57. Bessembinder, H., and Lemon, M. L. (2002). Equilibrium pricing and optimal hedging in electricity forward markets. *Journal of Finance*, **57**(3), pp. 1347–1382.
58. Björk, T. (2009). *Arbitrage Theory in Continuous Time*. Oxford University Press, Oxford.
59. Björk T., and Landen C. (2002). On the term structure of futures and forward prices. In: *Mathematical Finance – Bachelier Congress 2000*, Geman H., Madan D., Pliska S.R., Vorst T. (eds.), Springer Finance. Springer-Verlag, Berlin Heidelberg, pp. 111–149.
60. Black, F. (1976). The pricing of commodity contracts. *Journal of Financial Economics*, **3**, pp. 167–179.
61. Bochner, S. (1949). Diffusion equation and stochastic processes. *Proceedings of the National Academy of Sciences of the United States of America*, **35**, (7), pp. 368–370.

62. Borovkova, S., Permana, F. J., and Pavlyukevich, I. (2009). Modeling electricity prices by potential Lévy diffusions. *Journal of Energy Markets*, **2**(3), pp. 83–110.
63. Borovkova, S., and Schmeck, M. D. (2017). Electricity price modeling with stochastic time change. *Energy Economics*, **63**, pp. 51–65.
64. Brigo, D., and Mercurio, F. (2001). *Interest Rate Models. Theory and Practice.* Second Edition. Springer-Verlag, Berlin Heidelberg New York.
65. Broadie, M. and Glasserman, P. (1996). Estimating security price derivatives using simulation, *Management Science* **42**, pp. 269–285.
66. Callegaro, G., Mazzoran, A., and Sgarra, C. (2022). A self-exciting modeling framework for forward prices in power markets. *Applied Stochastic Models in Business and Industry*, **38**(1), pp. 27–48.
67. Carmona, R. (2015). Financialization of the commodities markets: a non-technical introduction. In *Commodities, Energy and Environmental Finance*, R. Aid, M. Ludkovski and R. Sircar (eds.), Springer-Verlag, New York, pp. 3–37.
68. Carmona, R., Coulon, M., and Schwarz, D. (2013). Electricity price modeling and asset valuation: a multi–fuel structural approach. *Mathematics and Financial Economics*, **7**, pp. 167–202.
69. Carmona, R., and Teranchi, M. (2006). *Interest Rate Models: an Infinite Dimensional Stochastic Analysis Perspective.* Springer-Verlag, Berlin Heidelberg New York.
70. Carr, P., Geman, H., Madan, D. B., and Yor, M. (2002). The fine structure of asset returns: an empirical investigation. *Journal of Business*, **75**(2), pp. 305–332.
71. Carr, P., and Madan, D. B. (1998). Option valuation using the fast Fourier transform. *Journal of Computational Finance*, **2**, pp. 61–73.
72. Clewlow, L., and Strickland, C. (2000). *Energy Derivatives. Pricing and Risk Management.* Lacima Publications, London.
73. Conway, J. (1990). *A Course in Functional Analysis*, 2nd Ed., Springer-Verlag, New York.
74. Cox, J. C., Ingersoll, J. E., and Ross, S. A. (1985). A theory of the term structure of interest rates. *Econometrica*, bf 53(2), pp. 385–408.
75. Cox, S., Karbach, S., and Khedher, A. (2022). Affine pure-jump processes on positive Hilbert-Schmidt operators. *Stochastic Processes and their Applications*, **151**, pp. 191–229.
76. Cox, S., Karbach, S., and Khedher, A. (2022). An infinite-dimensional affine stochastic volatility model. *Mathematical Finance*, **32**(3), pp. 878–906.
77. Cressie, N. A. C. (1993). *Statistics for Spatial Data.* John Wiley & Sons, New York.
78. Cressie, N. A. C., and Wikle, C. K. (2011). *Statistics for Spatio-Temporal Data.* John Wiley & Sons, New York.
79. Cuchiero, C. (2011). *Affine and Polynomial Processes*. PhD thesis dissertation. ETH Zürich, Switzerland.
80. Cufaro Petroni, N., and Sabino, P. (2013). Multidimensional quasi-Monte Carlo Malliavin greeks. *Decisions in Economics and Finance*, **36**, pp. 199–224.
81. Da Prato, G., and Zabczyk, J. (1992). *Stochastic Equations in Infinite Dimensions.* Cambridge University Press, Cambridge.
82. Deschatre, T., Féron, O., and Gruet, P. (2021). A survey of electricity spot and futures price models for risk management applications. *Energy Economics*, **102**, article 105504.
83. Deschatre, T., and Veraart, A. E. D. (2017). A joint model for electricity spot prices and wind penetration with dependence in the extremes. In *Forecasting and Risk Management for Renewable Energy*, Drobinski, Ph., Mougeot, M., Picard, D., Plougonven, R., and Tankov, P. (eds.). Springer-Verlag, Cham, pp. 185–207.
84. Dempster M. A. H. and Hong S. S. G. (2002). Spread option valuation and the fast Fourier transform. In: *Mathematical Finance – Bachelier Congress 2000*, Geman H., Madan D., Pliska S.R., Vorst T. (eds.). Springer Finance. Springer-Verlag, Berlin Heidelberg. pp. 203–220.
85. Di Persio, L., and Perin, I. (2015). An ambit stochastic approach to pricing electricity forward contracts: The case of the German energy market. *Journal of Probability and Statistics*, 2015 (2015).

86. Diebold, F. X., and Rudebusch, G. D. (2013). *Yield Curve Modeling and Forecasting: The Dynamic Nelson-Siegel Approach*. Princeton University Press, Princeton.
87. Dinculeanu, N. (2000). *Vector Integration and Stochastic Integration in Banach Spaces*. Wiley-Interscience. John Wiley & Sons, New York.
88. Duffie, D. (1992). *Dynamic Asset Pricing Theory*. Princeton University Press, Princeton.
89. Eberlein, E. (2014). Fourier-based valuation methods in mathematical finance. In *Quantitative Energy Finance – Modelling, Pricing and Hedging in Energy and Commodity Markets*, F. E. Benth, V. A. Kholodnyi and P. Laurence (eds.). Springer-Verlag, New York. pp. 85–114.
90. Eberlein, E., Jacod, J., and Raible, S. (2005). Lévy term structure models: No-arbitrage and completeness. *Finance & Stochastics*, **9**, pp. 67–88.
91. Eberlein, E., and Keller, U. (1995). Hyperbolic distributions in finance. *Bernoulli*, **1**, pp. 281–299.
92. Eberlein, E., and Stahl, G. (2003). Both sides of the fence: A statistical and regulatory view of electricity risk. *Energy Risk*, **8**, pp. 34–38.
93. Ekeland, I., and Taflin, E. (2005). A theory of bond portfolios. *Annals of Applied Probability*, **15**(2), pp. 1260–1305.
94. Engel, K.-J., and Nagel, R. (2000). *One-Parameter Semigroups for Linear Evolution Equations*, Springer-Verlag, New York.
95. Esscher, F. (1932). On the probability function in collective theory of risk. *Skandinavisk Aktuarietidsskrift*, **15**, pp. 175–195.
96. Ethier, S. and Kurtz, T. (1986). *Markov Processes. Characterization and Convergence*. John Wiley & Sons, New York.
97. Eydeland, A., and Wolyniec, K. (2003). *Energy Risk and Power Risk Management – New Developments in Modeling, Pricing, and Hedging*. Wiley-Finance. John Wiley & Sons, Hoboken, New Jersey.
98. Eyjolfsson, H., and Tjøstheim, D, (2018). Self-exciting jump processes with applications to energy markets. *Annals of the Institute of Statistics and Mathematics*, **70**, pp. 373–393.
99. Filipović, D. (2001). *Consistency Problems for Heath-Jarrow-Morton Interest Rate Models*. Lecture Notes in Mathematics, Vol 1760. Springer-Verlag, Berlin Heidelberg.
100. Filipović, D. (2009). *Term Structure Models – A Graduate Course*. Springer-Verlag, Berlin Heidelberg.
101. Filipović, D., Pelger, M., and Ye, Ye (2022). Shrinking the term structure. *Swiss Finance Institute Research Paper* No. 61. Available at SSRN: https://ssrn.com/abstract=4182649 or http://dx.doi.org/10.2139/ssrn.4182649
102. Filipović, D., Tappe, S., and Teichmann, J. (2010). Jump-diffusions in Hilbert spaces: existence, stability and numerics. *Stochastics*, **82**(5), pp. 475—520.
103. Filipović, D., and Willems, S. (2018). Exact smooth term structure estimation. *SIAM Journal on Financial Mathematics*, **9**(3), pp. 907–929.
104. Folland, G. B. (1984). *Real Analysis – Modern Techniques and Their Applications*. John Wiley & Sons, New York.
105. Fournié, E., Lasry, J.-M., Lebuchoux, J., Lions, P.-L., Touzi, N. (1999). Applications of Malliavin calculus to Monte-Carlo methods in finance. *Finance & Stochastics*, **3**(4), pp. 391–412.
106. Fournié, E., Lasry, J.-M., Lebuchoux, J., and Lions, P.-L. (2001). Applications of Malliavin calculus to Monte-Carlo methods in finance. II. *Finance & Stochastics*, **5**(2), pp. 201–236.
107. Frestad, D., Benth, F. E., and Koekebakker, S. (2010). Modeling term structure dynamics in the Nordic electricity swap market. *Energy Journal*, **31**(2), pp. 53–86.
108. Friz, P. K., and Hairer, M. (2014). *A Course on Rough Paths*, Second Edition, Springer-Verlag, Cham.
109. Garcia, I., Klüppelberg, C., and Müller, G. (2011). Estimation of stable carma models with an application to electricity spot prices. *Statistical Modelling*, **11**, pp. 447– 470.
110. Gawarecki, L., and Mandrekar, V. (2011). *Stochastic Differential Equations in Infinite Dimensions*, Springer-Verlag, Berlin Heidelberg.

111. Geman, H. (2005). *Commodities and Commodity Derivatives*. Wiley-Finance. John Wiley & Sons, Chichester.
112. Geman, H., Madan, D. B., and Yor, M. (2001). Time changes for Lévy processes. *Mathematical Finance*, **11**(1), pp. 79–96.
113. Geman, H., and Roncoroni, A. (2006). Understanding the fine structure of electricity prices. *Journal of Business*, **79**, pp. 1225–1261.
114. Geman, H., and Vašíček, O. (2001). Forwards and futures on non storable commodities: the case of electricity. *RISK*, August issue.
115. Gerber, H. U., and Shiu, E. S. W. (1994). Option pricing by Esscher transform. *Transaction of the Society of Actuaries*, **46**, pp. 91–191 (with discussion).
116. Glasserman, P. (2003). *Monte Carlo Methods in Financial Engineering*. Springer-Verlag, New York.
117. Goutte, S., Oudjane, N., and Russo, F. (2013). Variance optimal hedging for discrete time processes with independent increments. Application to electricity markets. *Journal of Computational Finance*, **17**(2), pp. 71–111.
118. Goutte, S., Oudjane, N., and Russo, F. (2014). Variance optimal hedging for continuous time additive processes and applications. *Stochastics*, **86**, pp. 147–185.
119. Grønborg, N. S., and Lunde, A. (2016). Analyzing oil futures with a dynamic Nelson-Siegel model. *Journal of Futures Markets*, **36**(2), pp. 153–173.
120. Féron, O., and Gruet, P. (2020). Estimation of the number of factors in a multi-factorial Heath-Jarrow-Morton model in electricity markets. Preprint, available at hal-02880824.
121. Heath, D., Jarrow, R., and Morton, A. (1992). Bond pricing and the term structure of interest rates: a new methodology for contingent claims valuation. *Econometrica*, **60**(1), pp. 77–105.
122. Heston, S. L. (1993). A closed-form solution for options with stochastic volatility with applications to bond and currency options. *Review of Financial Studies*, **6**(2), pp. 327–343.
123. Hikspoor, S, and Jaimungal, S. (2008). Asymptotic pricing of commodity derivatives using stochastic volatility spot models. *Applied Mathematical Finance*, **15**(5–6), pp. 449–477.
124. Hinderks, W. J., and Wagner, A. (2020). Factor models in the German electricity market: Stylized facts, seasonality, and calibration. *Energy Economics*, **85**, article 104351.
125. Holden, H., Øksendal, B., Ubøe, J., and Zhang, T. (2010). *Stochastic Partial Differential Equations*, Second Edition, Springer-Verlag, New York.
126. Hurd, T. R. and and Zhou, Z. (2010). A Fourier transform method for spread option pricing. *SIAM Journal of Financial Mathematics*, **1**(1), pp. 142–157.
127. Jacod, J. (1979). *Calcul Stochastique et Problèmes de Martingales*. Lecture Notes in Mathematics, Vol. 714. Springer-Verlag, Berlin Heidelberg.
128. Jacod, J., and Shiryaev, A. (2003). *Limit Theorems for Stochastic Processes*, 2nd Ed. Springer-Verlag, Berlin Heidelberg.
129. Jakubowski, A., and Riedle, M. (2017). Stochastic integration with respect to cylindrical Lévy processes. *Annals of Probability*, **45**(6B), pp. 4273–4306.
130. Jamshidian, F. (1992). Commodity option valuation in the Gaussian futures term structure model. *Review of Futures Markets*, **10**, pp. 61–86.
131. Janczura, J., and Weron, R. (2010). An empirical comparison of alternate regime-switching models for electricity spot prices. *Energy Economics*, **32**, pp. 1059–1073.
132. Jiao, Y., Ma, C., Scotti, S., and Sgarra, C. (2019). A branching process approach to power markets. *Energy Economics*, **79**, pp. 144–156.
133. Karatzas, I., and Shreve, S. E. (1991). *Brownian Motion and Stochastic Calculus*, Second Edition. Springer-Verlag, New York.
134. Kholodnyi, V. A. (2004). Valuation and hedging of European contingent claims on power with spikes: a non-Markovian approach. *Journal of Engineering Mathematics*, **49**, pp. 233–252.
135. Kholodnyi, V. A. (2011). Modeling power forward prices for power spot prices with upward and downward spikes in the framework of the non-Markovian approach. *Journal of Mathematics in Engineering, Science and Aerospace*, **2**(2), pp. 217–232.

136. Kholodnyi, V. A., and Lukic, M. N. (1997). Random field formulation for the term structure of interest rates. In J.F. Price (Ed.): *Derivatives and Financial Mathematics*, Nova Science Publishers, Inc., Commack, New York, pp. 139–143.
137. Kiesel, R., Börger, R., and Schindlmayr, G. (2009). A two-factor model for the electricity forward market. *Quantitative Finance*, **9**(3), pp. 279–287.
138. Kiesel, R., and Kustermann, M: (2016). Structural models for coupled electricity markets. *Journal of Commodity Markets*, **3**, pp. 16–38.
139. Kiesel, R., Paraschiv, F., and Sætherø, A. (2019). On the construction of hourly price forward curves for electricity prices. *Computational Management Finance*, **16**(1-2), pp. 345–369.
140. Kleisinger-Yu, X., Komaric, V., Larsson, M., and Regez, M. (2020). A multi-factor polynomial framework for long-term electricity forwards with delivery period. *SIAM Journal of Financial Mathematics*, **11**(3), pp. 928–957.
141. Koekebakker, S., and Ollmar, F. (2005). Forward curve dynamics in the Nordic electricity market. *Managerial Finance*, **31**(6), pp. 74–95.
142. Kremer, M., Benth, F. E., Felten, B., and Kiesel, R. (2020). Volatility and liqudity on high-frequency electricity futures markets: empirical analysis and stochastic modeling. *International Journal of Theoretical and Applied Finance*, **23**(4), pp. 1–38.
143. Kristoufek, L. (2014). Leverage effect in energy futures. *Energy Economics*, **45**, pp. 1–9.
144. Kwok, Y. K., Leung, K. S., and Wong, H. Y. (2012). Efficient options pricing using the fast Fourier transform. In *Handbook of Computational Finance*, J.-C. Duan, W. K. Härdle and J. E. Gentle (eds.) Springer-Verlag, Berlin Heidelberg. pp. 579–604.
145. Ladokhin, S., Schmeck, M. D., and Borovkova, S. (2021). Commodity forward curves with stochastic time change. *E-print* July, available on ssrn: 3871680.
146. Larsson, K., Green, R., and Benth, F. E. (2023). A stochastic time-series model for solar irradiation. *Energy Economics*, **117**, Article 106421.
147. Latini, L., Piccirilli, M., and Vargiolu, T. (2019). Mean-reverting no-arbitrage additive models for forward curves in energy markets. *Energy Economics*, **79**, pp. 157–170.
148. Linde, W. (1983). *Infinitely Divisible and Stable Measures on Banach Spaces.*. B. G. Teubner Verlagsgesellschaft, Leipzig.
149. Lucia, J., and Schwartz, E. S. (2002). Electricity prices and power derivatives: evidence from the Nordic power exchange. *Review of Derivatives Research*, **5**(1), pp. 5–50.
150. Lucia, J. J., and Schwartz, E. S. (2002). Electricity prices and power derivatives: Evidence from the Nordic power exchange. *Review of Derivatives Research*, **5**, pp. 5–50.
151. Madan, D., Carr, P., and Chang, E. (1998). The variance gamma process and option pricing. *European Finance Review*, **2**, pp. 79–105.
152. Madan, D., and Seneta, E. (1990). The VG model for share market returns. *Journal of Business*, **63**(4), pp. 511–524.
153. Madan, D., and Yor, M. (2008). Representing the CGMY and Meixner Lévy processes as time changed Brownian motions. *Journal of Computational Finance*, **12**(1), pp. 27–47.
154. Mandrekar, V., and Rüdiger, B. (2015). *Stochastic Integration in Banach Spaces*, Springer-Verlag, Cham.
155. Margrabe, W. (1978). The value of an option to exchange one asset for another. *Journal of Finance*, **33**, pp. 177–187.
156. Mendoza-Arriaga, R., and Linetsky, V. (2016). Multivariate subordination of Markov processes with financial applications. *Mathematical Finance*, **26**(4), pp. 699–747.
157. Musiela, M. (1993). Stochastic PDEs and term structure models. *Journées Internationales de Finance,* IGR-AFFI, La Baule.
158. Nelson, C., and Siegel, A. F. (1987). Parsimonious modeling of yield curves. *Journal of Business*, **60**, pp. 473–489.
159. Nomikos, N. K., and Andriosopoulos, K. (2012). Modelling energy spot prices: empirical evidence from NYMEX. *Energy Economics*, **34**(4), pp. 1153–1169.
160. Øigård, T. A., and Hanssen, A. (2002). The multivariate normal inverse Gaussian heavy-tailed distribution: simulation and estimation. In *2002 IEEE International Conference on Acoustics, Speech, and Signal Processing (ICASSP)*, Vol. 2, pp. II–1489 - II–1492.

161. Paraschiv, F., Fleten, S.-E., and Schürle, M. (2015). A spot-forward model for electricity prices with regime shifts. *Energy Economics*, **47**, pp. 142–153.
162. Parthasarathy, K. R. (1967). *Probability Measures on Metric Spaces*. Academic Press, New York-London.
163. Pedersen, G. K. (1989). *Analysis Now*, Springer-Verlag, New York,
164. Pérez-Abreu, V., and Rocha-Arteaga, A. (2005). Covariance-parameter Lévy processes in the space of trace-class operators. *Infinite Dimensional Analysis, Quantum Probability and Related Topics*, **8**(1), pp. 33–54.
165. Peszat, S., and Zabczyk, J. (2007). *Stochastic Partial Differential Equations with Lévy Noise*. Cambridge University Press, Cambridge.
166. Piccirilli, M., Schmeck, M. D:, and Vargiolu, T. (2021). Capturing the power options smile by an additive two-factor model for overlapping futures prices. *Energy Economics*, **95**, article 105006.
167. Pirrong, C. (2011). *Commodity Price Dynamics: A Structural Approach*. Cambridge University Press, Cambridge.
168. Prévôt, C., and Röckner, M. (2007). *A Concise Course on Stochastic Partial Differential Equations*, Springer-Verlag, Berlin Heidelberg.
169. Protter, Ph. (1990). *Stochastic Integration and Differential Equations*. Springer-Verlag, Berlin Heidelberg.
170. Quiao, H., and Wu, J.-L. (2019). On the path-independence of the Girsanov transformation for stochastic evolution equations with jumps in Hilbert spaces. *Discrete and Continuous Dynamical Systems – B*, **24**(4), pp. 1449–1467.
171. Rowinska, P. A., Veraart, A. E. D. , and Gruet, P. (2021). A multifactor approach to modelling the impact of wind energy on electricity spot prices. *Energy Economics*, **104**, article 105640.
172. Rydberg, T. (1997). The normal inverse Gaussian Lévy process: simulation and approximation. *Communications in Statistics. Stochastic Models*, **13**, pp. 887–910.
173. Samuelson, P. (1965). Proof that properly anticipated prices fluctuate randomly. *Industrial Management Review*, **6**, pp. 41–44.
174. Samuelson, P. (1976). Is real-world price a tale told by the idiot of chance? *The Review of Economics and Statistics*, **58**(1), pp. 120–123.
175. Sato, K. (1999). *Lévy Processes and Infinitely Divisible Distributions*. Cambridge University Press, Cambridge.
176. Schmidt, T., Tappe, S., and Yu, W. (2020). Infinite dimensional affine processes. *Stochastic Processes and their Applications*, **130**(12), pp. 7131–7169.
177. Schwartz, E. S. (1997). The stochastic behaviour of commodity prices: implications for valuation and hedging. *Journal of Finance*, **LII**(3), pp. 923–973.
178. Schwartz, E. S., and Smith, J. E. (2000). Short-term variations and loing-term dynamics in commodity prices. *Management Science*, **46**(7), pp. 893–911.
179. Skorohod, A. V. (1991). *Random Processes with Independent Increments*. Kluwer Academic Press, Mathematics and its Applications Vol. 47.
180. Skov, J. B., and Skovmand, D. (2021). Dynamic term structure models for SOFR futures. *Journal of Futures Markets*, **41**(10), pp. 1520–1544.
181. Svensson, L. E. O. (1994). Estimating and interpreting forward interest rates: Sweden 1992–4. *NBER Working Paper Series*, no 4871, September.
182. Tappe, S. (2012). Some Refinements of Existence Results for SPDEs Driven by Wiener Processes and Poisson Random Measures. *International Journal of Stochastic Analysis* Volume 2012, Article ID 236327, 24 pages.
183. van Neerven, J., Veraar, M., and Weis, L. (2015). Stochastic integration in Banach spaces – a survey. In *Stochastic Analysis: A Series of Lectures*, R. C. Dalang, M. Dozzi, F. Flandoli, and F. Russo (eds.). Progress in Probability 68, Birkhäuser, pp. 297–332. Springer-Verlag, Basel.
184. Vargiolu, T. (1999). Invariant measures for the Musiela equation with deterministic diffusion term. *Finance & Stochastics*, **3**(4), pp. 483–492.
185. Veraart, A. E. D. (2011). Likelihood estimation of Lévy-driven stochastic volatility models through realized variance measures. *Econometrics Journal*, **14**,(2), pp. 204–240.

186. Veraart, A. E. D., and Veraart, L. A. M. (2014). Risk premiums in energy markets, *Journal of Energy Markets*, **6**(4), pp. 91–132.
187. Ware, T. (2019). Polynomial processes for power prices. *Applied Mathematical Finance*, **26**(5), pp. 453–474.
188. Weron, R. (2006). *Modeling and Forecasting Electricity Loads and Prices – A Statistical Approach*. John Wiley & Sons, Chichester.

Index

F. E. Benth, P. Krühner, *Stochastic Models for Prices Dynamics in Energy and Commodity Markets*, Springer Finance,
https://doi.org/10.1007/978-3-031-40367-5

GPSR Compliance

The European Union's (EU) General Product Safety Regulation (GPSR) is a set of rules that requires consumer products to be safe and our obligations to ensure this.

If you have any concerns about our products, you can contact us on ProductSafety@springernature.com

In case Publisher is established outside the EU, the EU authorized representative is:

Springer Nature Customer Service Center GmbH
Europaplatz 3
69115 Heidelberg, Germany

Batch number: 10732090

Printed by Printforce, the Netherlands